TEXTBOOK OF BIODIVERSITY

K.V. KRISHNAMURTHY

CRC Press
Taylor & Francis Group
Boca Raton London New York

CRC Press is an imprint of the
Taylor & Francis Group, an informa business

Cover photograph: *Strobilanthes kunthiana* (Nees) T. Anderson ex Benth. A plant which flowers once in twelve years in the Western ghats of India and was of great cultural interest to the Ancient Tamils (*'Kurunji'*).

CRC Press
Taylor & Francis Group
6000 Broken Sound Parkway NW, Suite 300
Boca Raton, FL 33487-2742

First issued in hardback 2017

ISBN 13: 978-1-138-40771-8 (hbk)
ISBN 13: 978-1-57808-325-1 (pbk)

Visit the Taylor & Francis Web site at
http://www.taylorandfrancis.com

and the CRC Press Web site at
http://www.crcpress.com

Library of Congress Cataloging-in-Publication Data

Krishnamurthy, K.V.
Textbook of biodiversity/K.V. Krishnamurthy.
p. cm.
Includes bibliographical references and index.
ISBN 1-57808-325-7
1. Biological diversity. 2. Biological diversity conservation. I. Title.

QH541.15.B56K75 2003
333.92--dc22

2003055631

Dedicated to
Late Prof. B.G.L. Swamy (1918 - 1980)
who introduced me to the
World of Biodiversity

PREFACE

"The most wonderful mystery of life may well be the means by which it created so much diversity from so little physical matter."
— E.O. Wilson 1992

Exactly thirty-two years ago I had to enter an undergraduate classroom to begin teaching courses on Plant Morphology, Taxonomy and Diversity. I had just then completed my doctoral degree and was recruited as a teacher in a government college. All my senior colleagues in the Department of Botany had opted to teach 'more rewarding and modern courses', relegating the job of teaching the aforesaid subjects, despised as 'classical and old-fashioned', to me. The task of teaching these subjects to the undergraduate and subsequently to postgraduate students became an intimate part of my academic career all those thirty-two years, but with the tag 'classical and old-fashioned teacher' increasingly glued to my name. During this long experience of teaching, not only did I realise but my students came to realise that these subjects are not 'classical and old-fashioned' as thought by others out of ignorance. On the contrary, they are fundamental to the very existence and growth of other disciplines. In fact, Biodiversity science has grown more modern than most other disciplines with inputs from several other subjects. Biodiversity has also become an essential condition for the very continuance of life on Earth. Its relevance extends far into the future. It forms the important link between the evolutionary past, through the 'current period of attrition and depletion', to future survival. Biodiversity, in fact, has grown into a key component in the constellation of subjects that vie for everyone's attention today.

In the first few years of teaching biodiversity science, I was more than generally satisfied by picking up whatever teaching material I could lay my hands on and discussing them in the class. However, in subsequent years, I slowly became aware of the increasing outpouring of information that vastly transgressed different disciplines, from Morphology to Molecular Biology, from Systematics to Sociopolitical Economics and from Taxonomy to very advanced technologies. The subject matter had become more and more weighty and, in hindsight, I was witnessing the exponential growth and development of a subject area in which concepts, ideas and methodologies were building up at a very great speed.

For several years my students have been asking me whether I could recommend some single book dealing comprehensively with all aspects of biodiversity science. They received only an apologetic headshake in response. Books have been published in the last ten years on several aspects of biodiversity but to the best of my knowledge a comprehensive text/ reference book on biodiversity catering to the needs of students and young researchers has yet to appear. Slowly I was forced, not only out of compulsion from my students, but my own innate urge, to start preparations for this book. Once decided, I was determined to write a single author book, although I knew this would be a daunting task as the subject matter is so broad and quick-paced in advancement. But, in spite of possible drawbacks in subject coverage, it seemed to me that the themes and ideas could be more homogeneously presented and glaring overlaps avoided in a single-author book than in a multiauthor work. However, in reality this book is a multiauthor work since its contents and sequence greatly reflect the input from several batches of my students. The reader will also find that the book is a little biased in the sense that plants and microbes have been given greater attention. This could not be avoided for two reasons: one, I am basically a botanist, and secondly the manuscript is essentially based on lecture notes that were constantly updated for a course on Biodiversity for students specializing in Plant Science and Biotechnology. However, the contents of the book can equally satisfy the requirements of students of Zoology, the shortage of animal examples notwithstanding.

I started working on this book seven years back and for the last two years have been reluctant to send the manuscript for publication for the simple reason that it was not yet 'complete'. Now I force myself to release the manuscript for publication remembering the words of the 13th century Chinese scholar Tai T'ung: 'Were I to await perfection, my book would never be finished'.

K.V. KRISHNAMURTHY

ACKNOWLEDGEMENTS

Very few authors can produce the finished manuscript for a book entirely unaided and this author is no exception. Many people have helped me, too many to mention individually, with the preparation of this manuscript. They all made useful suggestions and I actually absorbed a few, but definitely not enough. I am grateful to my numerous students and fellow scientists in this regard. I am particularly thankful to Dr. G. Subramanian and Dr. M.A. Akbarsha for their invaluable suggestions made during our daily 'lunch-table academic discussions'. I am also extremely grateful to Dr. Muthiah Mariappan, an eminent Environmental Engineer and the Vice-chancellor of our university for his generous support, advice and encouragement. Dr. M. Selvaraj, Registrar of our University was a source of great inspiration to me during the final days of great tension when the book was being completed. The longhand writing of mine was skillfully and correctly deciphered by my doctoral students Dr. A.V.P. Karthikeyan and Dr. G.V.Kumar who patiently typeset the whole manuscript and I am grateful to both. My gratitude is beyond bounds for the first and the only lady in my life, my wife Brindha. She has gracefully and patiently tolerated my negligence of the family during the writing of this book and been my strongest pillar of support. I am further indebted to my son Arvind, daughter-in-law Anusha and grandson Sundar not only for their forbearance and support, but for foregoing the time they could otherwise have spent with me.

CONTENTS

1

BIODIVERSITY SCIENCE: DEFINITION SCOPE AND CONSTRAINTS

Introduction

Biodiversity is the abbreviated word for **Biological Diversity**. The latter usage appears to have come into prominence around 1980, when Norse and McManus (1980) first defined it. Its abbreviation into 'biodiversity' was apparently made by Walter G. Rosen in 1985 during the first planning meeting of the 'National Forum on Biodiversity' held at Washington DC in September 1986 (UNEP 1995). The published proceedings of this meeting in a book entitled *Biodiversity* (Wilson and Peters 1988) introduced the notion of biodiversity and popularised this word among the scientific community as well as the public. Since then, not only the number of publications on biodiversity, but also of people interested in the subject for one reason or the other has steadily increased (Harper and Hawksworth 1994). The United Nations Conference on Environment and Development (UNCED) held in 1992 at Rio de Janeiro (**Rio Summit** or **Earth Summit**) has also substantially elevated the status of Biodiversity.

Biodiversity—Concept and Definition

The word Biodiversity is now very widely used not only by the scientific community, but also the general public, environmental groups, conservationists, industrialists and economists. It has also gained a very high profile in the national and international political arena. In fact, the term has become very fashionable with no clear understanding of what it means. Such loose usage has given the word so many different meanings, connotations and intentions that the actual concept of biodiversity has been lost in obfuscation and confusion. Hence there is a real need to unequivocally define the concept of biodiversity, which is today a recognised separate science with its own principles and facts, and to define the scope of this new science as well.

Biodiversity is generally considered an 'umbrella term' referring to organisms found within the living world, i.e., the number, variety and variability of living organisms. It may thus be assumed to be a synonym for 'Life on Earth', 'variety of life and its processes' (Keystone Center 1991), 'condition of being different' (Gove *et al.* 1996), or what Darwin (1859) exclaimed as 'Life's endless forms'. Taken in this general sense, biodiversity is indeed 'the essence of life' (Frankel 1970). In reality, however, biodiversity is a very vast and complex concept and its ramifications extend deep into all spheres of human life and activity.

Biodiversity is normally treated in terms of genes, species and ecosystems in correspondence with the three fundamental hierarchical levels of biological organisation; these three diversities are respectively referred to as Genetic, Species and Ecosystem diversities. According to Harper and Hawksworth (1994), it was Norse *et al.* (1986) who first expanded the traditional use of the term biological diversity to the three levels of biological organisation (see also OTA 1987). Diversity within species is **Genetic Diversity**, diversity between species is **Species Diversity** (also often referred to as **Taxonomic** or **Organismal Diversity**), and diversity at the ecological or habitat level is **Ecosystem Diversity** (also known as **Ecological Diversity**). Noss (1992, 1996), Szaro and Shapiro (1990), Szaro and Salwasser (1991) and Wilson (1988 a,b), among many others, have included a fourth form of biodiversity called **Landscape Diversity**. Landscape is 'a heterogeneous land area composed of a cluster of interacting ecosystems that is repeated in similar form throughout' (Forman and Godron 1986); it is also defined as 'a mosaic of heterogeneous land forms, vegetation types and land uses' (Urban *et al.* 1987). Landscapes therefore have a pattern and this pattern consists of repeated habitat components. For example, a landscape may be interspersed with grasslands, meadows, ponds, streams, shrubby areas and forests. Thus, landscape diversity is **Pattern Diversity** (Scheiner 1992). The inclusion of landscape diversity as a fourth form of diversity was emphasised by Odum (1992) when he listed the following as one of his 20 great ideas in Ecology: 'An expanded approach to biodiversity should include genetic and landscape diversity, not just species diversity'. Ray (1996) is also very much in favour of including landscape diversity as the fourth category, based on his studies on coastal marine regions.

The complexity of the biodiversity concept is reflected in the existence of numerous definitions for this word, of which Jutro (1993) identified at least 14. Two among these 14 definitions are largely used, quoted and even officialised, since they have been approved by several countries based on worldwide negotiations, agreements and strategies. The first most-used definition is sponsored by the United Nations (UN) and was included in the Convention on Biological Diversity (CBD) (UNEP 1992). According to this definition, Biodiversity refers to: 'The variability among living, *inter alia*, terrestrial, marine and other aquatic systems and the ecological complexes of which they are part; this includes diversity within species, between species and of ecosystems'. The second most-used definition of Biodiversity is sponsored by the Global Biodiversity Strategy (WRI, IUCN, and UNEP 1992) and is as follows: 'The totality of genes, species and ecosystems in a region'.

These two definitions, according to di Castri and Younès (1996), pay very little attention to the interactions within, between and among the various levels of biodiversity recognised. According to these authors, interaction is the principal intrinsic mechanism that shapes the characteristics and functions of biodiversity. Another problem they find in the various definitions, is their ignorance of the **notion of scale**; di Castri and Younès argue that structural and functional attributes of biodiversity can only be determined by employing appropriate scales of space and time (see also Lugo 1996). Consequently, according to di Castri and Younès (1996), biodiversity should not be construed as a 'simple umbrella covering a mosaic of heterogeneous activities', but should represent a composite entity 'shaped by the continuum of all its elements and their interactions'. These interactions, according to them, are of a hierarchical nature, and by interlocking the genetic, species and ecosystem diversities one can achieve the 'classical zooming effect of hierarchical theory' (Fig. 1.1). The important outcome of such an approach is that the properties of biodiversity that do not 'occur' at a lower scale of integration (say gene levels) will 'appear' at a higher scale (say species

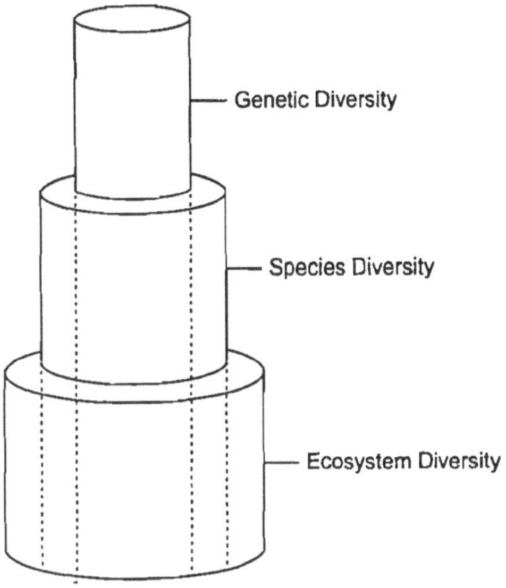

Fig. 1.1 The three hierarchical scales of biodiversity and their interrelationships (adapted from di Castri and Younès 1996)

or ecosystem level). The properties that are 'evident' at a higher scale 'disappear' at a lower scale. The hierarchical concept, so obtained, has been expanded and made more accurate by di Castri and Younès (1996) (Box 1.1); here the hierarchical patterns of biodiversity are shown as the interactions of the three different scales of organisation: Genetic, Species and Ecosystem. Each of these scales has different levels of organisation. For example, the Gene scale has the following levels of organisation/integration: Community, Population, Individual organism, Cell, and Molecule. Based on this, di Castri and Younès (1996) provided the following definition: Biodiversity is 'the ensemble and the hierarchical interactions of the genetic, taxonomic and ecological scales of organisation, at different levels of integration'. Indeed, populations (with their gene pools), species and ecosystems are respectively the cornerstones at

Box 1.1 The Scales of Organisation and Levels of Interaction of Biodiversity (adapted from di Castri and Younès, 1996)

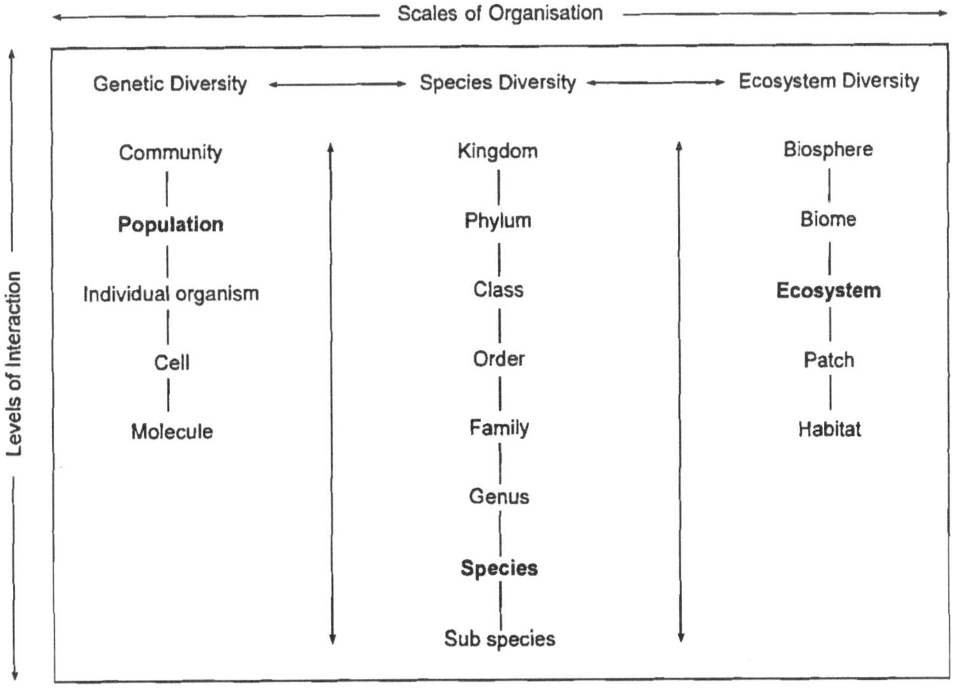

the interaction points (i.e. integration points) of the three scales (Solbrig 1991).

A slightly different composition of the various scales of biodiversity (Genetic, Organismal and Ecological) was suggested by UNEP (1995) (Box 1.2). It is interesting to find Population as a common component/unit at all three scales of biodiversity. However, Heywood (1997a) stated that population is one of the most difficult to assess, whatever may be the scale. In addition, **cultural diversity** and human interactions are noticed at all these levels. Various cultures and societies place different values, exert different driving forces and influences, and practice different measures for conserving and sustaining biodiversity. All these represent Cultural Diversity, which therefore recognises the pivotal role of sociological, ethical, religious and ethnic values in human efforts concerning biodiversity.

Attention of readers has already been drawn to the fact that many biologists are inclined to add Landscape as a fourth level of biodiversity. Consequently, a revised and comprehensive approach to the biodiversity concept, incorporating the multiple levels of organisation and many different spatial and temporal scales, was successfully attempted (Noss 1990, 1994). Noss (1994) suggested that each of these scales (genetic, species, ecosystem and landscape) should be further subdivided into compositional, structural and functional components (see details in Fig. 1.2).

Scope of Biodiversity Science

In a detailed discussion on the perspectives of Biodiversity Science, di Castri and Younès (1996) have listed at least six major reasons for considering Biodiversity as a new science. These six reasons incidentally explain the scope of this branch of science.

(i) Biodiversity is the unifying driving force (*leitmotiv*), all along much needed, to provide a continuum within the broad field of Biology. The readers are quite aware of the fact that Biology is now fragmented into a large number of disciplines. Although originally done to facilitate understanding the various facets of Biology, this fragmentation has diminished, even eradicated the connections among the different disciplines as well as engendered an unhealthy competition among them, leading consequently to underestimation of each other. In most countries, Molecular Biology and Biotechnology have become the most sought-after disciplines of Biology, relegating Morphology and Taxonomy to the lowest level of preference; the latter also hold little or no career incentives. The fact has been obscured that these two groups of disciplines of Biology are equally necessary and that a close co-operation and interaction among the various disciplines are absolutely important for the welfare of human kind. Such interaction and co-operation are also needed to address all the problems of Biodiversity with greater ease and

Box 1.2. Composition and Scales of Biodiversity (adapted from UNEP, 1995)

Genetic diversity	**Organismal diversity**	**Ecological diversity**
Populations		
Individuals	Kingdoms	Biomes
Chromosomes	Phyla	Bioregions
Genes	Orders	Landscape
Nucleotides	Families	**Ecosystems**
	Genera	Habitats
	Species	Nichés
	Subspecies	Populations
	Populations	
Cultural diversity and human interactions at all levels.		

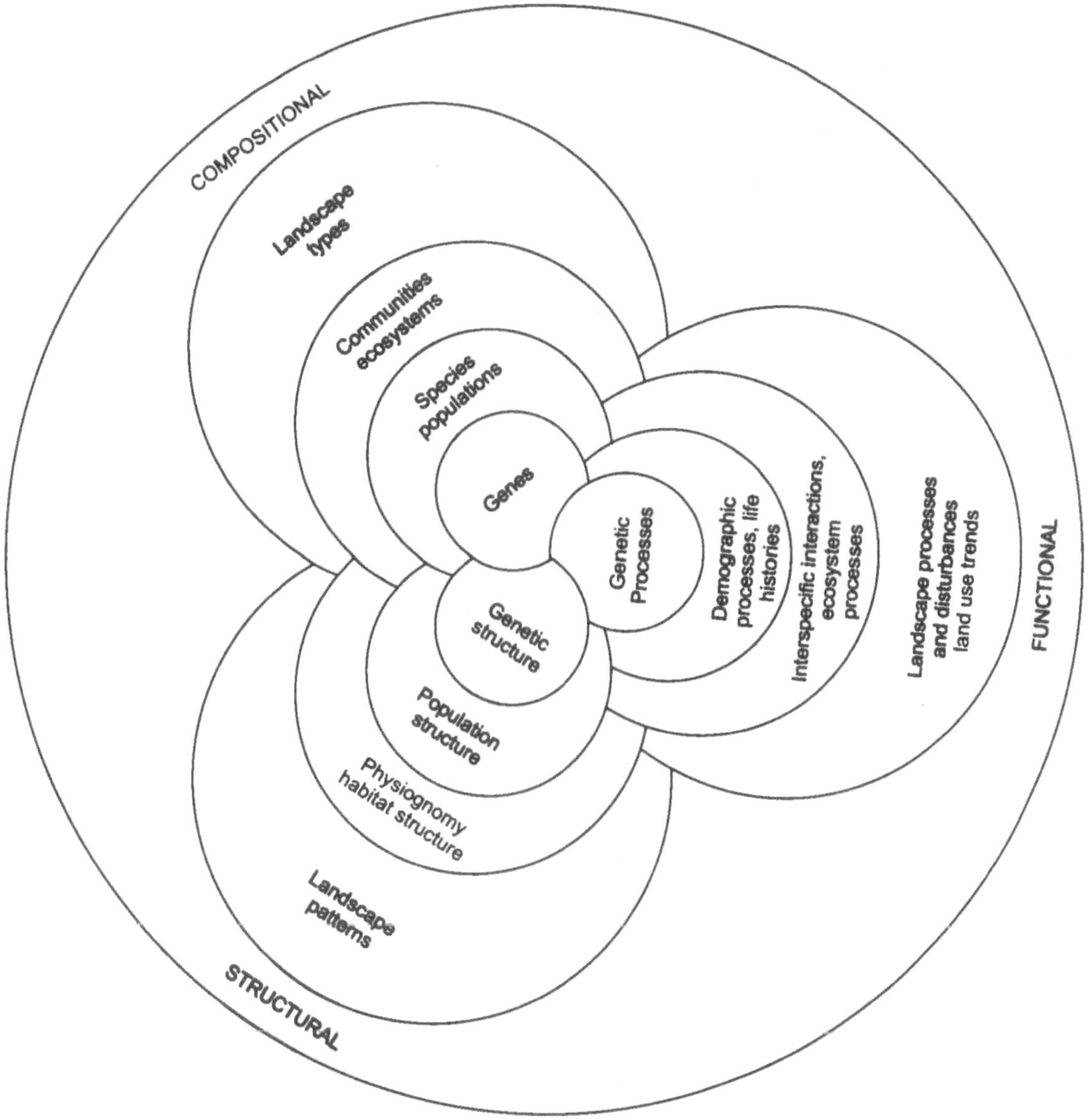

Fig. 1.2 The biodiversity concept incorporating the multiple levels of organisation and many different spatial and temporal scales, as per Noss (1990).

perfection. Thus, the science of Biodiversity has the potential to unify all fragmented disciplines of Biology and bring together the activities of all scientists professing these disciplines. This is, as a matter of fact, one of the important goals of the International Union of Biological Sciences (IUBS).

(ii) Biodiversity is the backbone for Agriculture, Aquaculture, Animal Husbandry, Forestry and a host of other applied branches of

Biology. Hence it stands at the very foundation of development, especially in a rapidly changing world. Population growth has not declined to a desired level. By the year 2005, the UN predicts that today's population of about 7 billion will likely have increased to 8 billion. We are not going to simply expand the load on the land presently available, but actually multiply the load. Evidence suggests that the Green Revolution is likely to peter out. Grain and pulse yields have stopped rising as fast as in earlier years. Supplies of fresh water are growing scarce. Human beings currently appropriate 54% of accessible freshwater run-off, but in the next 25 years, the projected increment in demand is more than 70% (Ayensu *et al* 1999). Soil quality has already started deteriorating the world over. There is very little unplanted arable land left for further agricultural exploitation. Hence, new varieties of useful plants and new breeds of domesticated animals have to be constantly evolved for increased yield/productivity, desired lifetime, disease resistance etc. For example, by 2020, world demand for rice, wheat and maize is projected to increase by ~ 40% and livestock production by more than 60%. Such an effort to increase the quality and quantity of bioresources is possible only if we have adequate information and knowledge about their wild relatives, which form the genetic source for their further improvement through conventional or biotechnological methods. Since biodiversity deals with biological resources, it is particularly important that various ecosystems be critically assessed for their useful plant/animal germplasms, especially for the presence of wild relatives. However, for this to materialise the world should engage itself in a 'gigantic' and 'multibillion dollar scientific effort'.

(iii) It is well known to all readers that intense globalisation of trade and markets has occurred during the last decade, resulting in very rapid and dramatic changes in land-use patterns and regional developmental activities. Consequently, pronounced deforestation (at the rate of some 14 million hectares each year), alarming desertification and substantial global climatic changes have taken place. About 40 to 50% of the land on Earth has been irreversibly transformed through change in land cover or degradation by human beings. Models based on UN's intermediate population projection mentioned in the paragraph above suggest that an additional one-third of global land cover will be transformed over the next one hundred years (Ayensu *et al.* 1999). It has now become highly obligatory for mankind to not only check these alarming changes, but also to reconstruct and restore the changed ecosystems to their original state. Any new change in the existing landscape must be properly planned in order to make it environment-friendly. Such undertakings require a deep understanding of biodiversity in all its aspects.

(iv) The first half of the present century, as in the last half of the previous century, is definitely going to be dominated by Biology. Biodiversity is fast becoming the fundamental requirement on which the new industrial developments and innovations are going to be based. Biodiversity will offer in the coming years, new sources of food, medicine and other human requirements. Therefore, industrial development will become possible only by exploring the great potential of the still unknown biological resources. For this, in-depth knowledge of biodiversity is imperative.

(v) Globalisation of information and communications has markedly increased. Furthermore, a substantial human migration to various parts of the world is anticipated in the next one or two decades. There is also anticipation of substantial movement of plants and animals to different parts of the globe. All these processes will definitely lead to profound changes not only in the existing society and culture, but also in the landscape of different parts of the world. Under these circumstances, the study of biodiversity cannot be treated in isolation from the anticipated human dimension. Thus, biodiversity will become the only purposeful scientific tool with which one can bridge the social and cultural world.

(vi) Biodiversity is the resource on which all human existence depends, i.e., it is the pillar of human development. Consequently, a sustainable exploitation of bioresources should be practised. Sustainable development has been compared to a chair with four legs of similar length and strength (di Castri 1995). These four legs respectively denote the economic, environmental, social and cultural facets of biodiversity. Unless all the four dimensions of biodiversity are equally strong, sustainable development cannot result. Therefore, biodiversity is vital for sustainable development.

To these six reasons why biodiversity should be considered a science of utmost importance, the following may be added:

(vii) Biodiversity has already proved itself to be a fundamental concept for a world just entering the twenty-first century. The range of influence of Biodiversity is much more general and wider than any other field of either Biology or other scientific disciplines. Its numerous manifestations are already found in inconspicuous places, for example in the **technosphere** and in the **noosphere** (the sphere of minds uniting everyone on the planet through different types of communication networks). Biodiversity has come to include not only living species, but also the multitude of human inventions based on bioresources. In summary, Biodiversity has become the basis of a general law of dynamic stabilisation of complex systems, a law fundamental to homeostasis and applicable to the functioning of the planet as a whole (de Rosnay 1996). This is also the essence of the **Gaia Hypothesis** (Lovelock 1988 a,b), which emphasises that the physical structures and processes of this planet are regulated by Biodiversity (Myers 1985 b).

Constraints of Biodiversity Science

To meet and implement the scope and action plans mentioned in the previous section, the science of biodiversity has to remove, overcome or at least address a number of constraints (di Castri and Younès 1996):

(i) The foremost and most difficult constraint to overcome is the current status of Taxonomy. It is well known that Taxonomy is the most essential infrastructure for biodiversity development (Janzen 1993a,b) and that the recognition and characterisation of biodiversity depends critically on Taxonomy as it provides the reference system for depicting the pattern of biodiversity. Only a very limited number of all species believed to exist on this earth are known to us. Many species are yet to be discovered and described (for more details see Chapter 3). Even for the known species, the information available, especially on functional attributes, is extremely meagre. In spite of this, the number of new taxonomists the world over is very small, due primarily to lack of career incentives[1]. Further, the geographic distribution of even these few taxonomists is lop-sided. While it is estimated that in the developed countries there is one taxonomist for every 10 species occurring there, in many developing countries there is only one taxonomist for every 1000 species, even if we include taxonomists of below average competency (Manilal 1997). There is also a maldistribution of taxonomists, due to which 'the amount of taxonomic effort' made so far 'varies widely from group to group' among the Biota of the world (May 2002). This maldistribution 'reflects the vagaries of intellectual fashion, and most certainly does not

[1]Janzen and Gaméz (1997) have, however, cautioned that we should not openly admit to an insufficient number of taxonomists. They feel that such statements might generate the wrong notion that the world will immediately rush to buy more taxonomists, when *de facto* the world will interpret such statements as *prima facie* evidence that taxonomists are not needed. It is a general principle that society stops paying for what it does not need.

reflect the relative importance' of the different groups 'in maintaining the structure and functions of the ecosystems' (May 2002). Good taxonomists have indeed become a highly endangered category among biologists (Khoshoo 1995).

The status of Herbaria and Museums world over is fast deteriorating academically and financially. Their number is also slowly shrinking for want of support. Given these two conditions, it is going to be an uphill or even utopian task to recognise all the species supposedly contained in the world (for more details on this issue see Chapter 3).

(ii) Measuring biodiversity is the second major constraint (Hawksworth 1994), closely related to the first. There is considerable disagreement over how to measure biodiversity (Hurlbert 1971; Norton 1986). The currently practised measures select very different components of the ecosystem for emphasis. Potential indicators include the total number of species or 'richness' (Magurran 1988; Scott *et al.* 1987), abundance and distribution of populations (Krebs 1972; Westman 1990), number of endangered species, centres of species-richness with high endemism (Myers 1988a), and degree of genetic variability (Allen 1963; Ruffie 1982). Other approaches treat ecosystem functions (Ray 1988), interactions (Janzen 1988), natural communities (TNC 1975; Western *et al.* 1989), successional stages (Franklin 1988) or ecological redundancy (Walker 1992) as key measures of diversity.

Genetic diversity is almost fully known only for a few taxa such as *Arabidopsis*, maize and *E coli*. The problems involved in measuring species diversity are alluded to above. Ecosystem diversity is measured to some extent by direct field surveys, remote-sensing techniques or by fractal analysis. But it is really difficult to link these three diversities.

Related to the above is the assessment of biodiversity loss. There are many problems in measuring loss of biodiversity, which is an important requirement in Conservation Biology. There is widespread agreement that global biodiversity is reducing at an accelerated rate (Myers 1980a; Wilson 1988b). But there is less agreement about the actual quantum of such loss (Harwood 1982; Lovejoy 1986), compounded by the wide range of operational measures, variation between biomes, and lack of baseline knowledge about the number of species already available (Freedman 1989). Even when a common measure is employed (i.e., species richness), differences in estimated rates of loss as computed by different people are large. Lugo (1988a,b), for example, compared the several estimates of species loss in the tropics and found they ranged from 15% to 50% for all species by the year 2000.

(iii) A dichotomy exists between biodiversity agenda and priorities of developed and developing countries. It is very difficult to resolve this dichotomy at present, in spite of the Rio Summit and other efforts. Khoshoo (1996) has critically summarised this dichotomy between two sets of countries, commonly referred to as North and South, its reasons and effects, as well as strengths and weaknesses (Table 1.1). Biodiversity is an important biological resource and strength in the developing countries, while the developed countries are technically developed but wanting in biodiversity. As di Castri and Younès (1996) have emphasised, 'a reconciliation of national prerogatives with a global interest, based on principles of overall equity' is immediately needed. Immediacy of achievement is a very tall order, however.

(iv) The lack of adequate knowledge about biodiversity among the people is another important constraint. As already indicated, the concept and definition of biodiversity have largely been misunderstood or poorly understood by many people, including some biologists. It is our primary duty to educate people and clear all myths that pervade the biodiversity concept. It is also important to negate the assumption that biodiversity science

Table 1.1 Comparison of biodiversity and its potential between developing and developed countries (adapted from Khoshoo, 1996)

Developing Countries	*Developed countries*
• Biodiversity rich	• Relatively poor Biodiversity
• Vavilovian Centres of Diversity	• Nil to almost nil
• Backed by indigenous people, local technical knowledge and indigenous systems	• Largely non-existent
• Biodiversity supported by cultural diversity	• Largely non-existent
• Genetics, breeding and biotechnology base poor	• Rich base in Technology
• Largely *in situ* conservation	• Largely *ex situ*, but *in situ* for their own non-agricultural biodiversity
• Conservation indigenous-science-based	• Largely modern science-based
• Largely subsistence or intensive agriculture	• Largely modern science-based
• Sustainable utilisation of biodiversity: not possible without capacity building	• Capacity exists
• Research and development, education and training, and demonstration and extension need enhancement	• Rich base
• Poverty	• Rich base
• Largely bioindustrial development	• Largely industrial development

is 'intrinsically opposite to economy and development' (di Castri and Younès 1996). To attain this task, not only should children be taught about biodiversity at an early age, but also adults sufficiently educated through media and other programmes.

2

GENETIC DIVERSITY

Introduction

In the last chapter it was recorded that Biodiversity can be studied at four levels: Gene, Species, Ecosystem and Landscape. In this chapter attention will be focused on details of **genetic diversity**, which is also referred to as **within-species diversity**, or **intra- or infra-specific diversity**. A number of infra-specific categories have often been recognised and most of them also enjoy taxonomic implications without necessarily being defined in genetic terms (UNEP 1995): subspecies, varieties, land races, clines, cultivars, ecotypes, chemotypes, cytotypes, hybrids, polytypes, polyploid complexes, aggregated species, etc. The recognition of these 'taxonomic' categories often poses problems in defining and conceptualizing genetic diversity. It should thus be emphasised that 'there is no single definition of genetic diversity that can be used for all purposes' (UNEP 1995, p. 213).

Nature and Origin of Genetic Variations

It is a well-known fact that the blueprints for all living beings are genes and that they consist of discrete segments of deoxyribonucleic acid (DNA). Meadows (1990) was correct in making the following remark: 'Nature's knowledge is contained in the DNA within living cells'. DNA is a linear molecule composed of sequences of four different nucleotide bases: adenine, guanine, thymine and cytosine. These four bases form the four base pairs: adenine-thymine, guanine-cytosine, thymine-adenine and cytosine-guanine. Genes are 'linearly arranged' along the length of the DNA molecule. All observed variations are invariably due to variations in the sequences of the four base pairs of the DNA molecule. The number of possible combinations of these base pairs exceeds the number of atoms in the universe. From this, one can imagine the magnitude of variations that can be produced. The combinations of these four base pairs in various permutations result in the **Genetic code**. The genetic code distributes 64 triplet codons to 20 amino acids, including initiation signals and three termination signals for the construction of protein molecules with specific sequences of amino acids.

The pool of genetic diversity of a species can exist at three different levels: genetic diversity within individuals, often referred to as **heterozygosity**, genetic diversity among individuals within a population, and genetic variations among populations. Thus, genetic diversity can be defined in more general terms as the heritable variations observed within and between populations of organisms. Consequently the building blocks of DNA include the following: diversity encoded by specific genes that some organisms/individuals possess but others lack, the differences in sequences that regulate gene expression, and the variations arising from differing copies of homologous related DNA sequences (for example, allelic differences), as well as diversity due to translocation of a sequence from one chromosomal site to another (for example, position effect).

New genetic variations can arise in individuals of a species by gene and chromosomal mutations. In organisms that exhibit a sexual mode of reproduction, the resultant genetic variations can spread through the population by recombination. In inbreeding members of a population, the pool of genetic variation already present is acted upon by selection. Differential survival results in changes in the frequency of genes within this pool, in turn resulting in population evolution. Therefore, the main significance of genetic variation in a population is that it enables natural evolutionary changes to take place since the rate of these changes is proportional to the amount of genetic diversity available. Genetic Diversity is a resource for the species' own survival and future evolution; it also promotes selective breeding (see Groombridge 1992). In addition, genetic variation also confers fitness advantages upon the members of a population. Another reason why it should be considered a resource is that many genes are potential sources for improving the productivity of other members of a population or of a species. For example, genes from wild plants have become very important today for improving domesticated taxa (for more details, see Chapters 4 and 10).

The global pool of genetic diversity represents all the information pertinent to all biological structures, functions and processes on this planet. In other words, every process and every pattern is encoded in a genetic 'library' of unimaginable global extent. But only a negligible fraction (often less than 1%) of the genetic material of higher organisms is expressed in the form and functions of organisms (see Groombridge 1992). This means that much of the DNA in each cell is not coded (Thomas 1992). The role of such 'silent' DNA in the expression of variations still remains unclear. Again, of the estimated 10^9 different genes in the world biota (of which about 10^5 are distributed in higher organisms) not one has been found to duplicate the contribution of another to the overall genetic diversity. Genes that control fundamental life processes such as photosynthesis and respiration are strongly conserved across different taxa and generally exhibit little variation (see Groombridge 1992).

Measurement of Genetic Diversity

Introduction

No one will dispute the fact that all other aspects of biodiversity are a consequence of genetic diversity. But despite this self-evident fact, genetic diversity studies were rarely attempted in the past and still remain few at present. Obviously the lack in earlier years of rapid and effective techniques for measuring genetic diversity accounts for the paucity of such studies then (Bachmann 1994) but is no excuse now. Assessments of genetic variation are useful for two reasons: (i) they test the mathematical and statistical theories proposed so far to explain the nature of the forces acting on genetic variants; and (ii) they serve as tools for understanding the relationships that exist within the individuals of a species, their similarities and divergences (Thomas 1992).

Before assessing the measures of genetic diversity, one must understand two basic notions in genetic diversity: **richness** and **evenness** (Frankel *et al.* 1995). Richness refers to the total number of different genotypes present in a population or a sample from it. Evenness refers to the frequency of the different genotypes present in the same population or a sample from it. Under the first notion, a population of a plant species with three different flower colours is considered genetically more diverse than a population with only two different flower colours. Under the second notion, a population of a species with two equally frequent flower colours, say 0.50

and 0.50, is considered genetically more diverse than a population with three flower colours in uneven frequencies, say 0.80, 0.15 and 0.05.

Methods Based on DNA and Chromosomes

There are several methods, including molecular, for assessing the levels of genetic variations. The most common method considers genetic diversity in terms of the diversity of genes that an individual possesses. The number of genes is believed to range from around 1000 in a bacterium, 10,000 in some fungi, to 400,000 or more in many flowering plants (UNEP 1995).

Genetic diversity is also measured in terms of the amount of DNA per cell. The actual quantity of DNA per cell of different species of Eukaryotes varies over three orders of magnitude (Fig. 2.1) (Thomas 1992). It has been shown that within the angiosperms the genomes can range in size from the relatively small 1C value of 0.15 to 0.2 pg in *Arabidopsis* to over 31.8 pg in *Tulipa* species depending on measurement technique (Arumuganathan and Earl 1991; Bennet and Smith 1991; see also Table 2.1).

Genetic diversity is also measured in terms of chromosome structure, size, shape and number (often, referred to as karyotype variation) in the case of Eukaryotes. The parameters in such assessments include changes in whole chromosome complement number (polyploidy), the number of some individual chromosomes (polysomy, monosomy, nullisomy, addition and substitution) or in the number of supernumerary or B-chromosomes. Chromosome banding techniques of various types such as Q-, C- and G-, are often followed in such assessments.

Levels of genetic diversity can also be assessed from DNA elements (= genetic markers) of mitochondria (mt DNA) or chloroplasts (cp DNA). These organellar DNA molecules are very useful because the organelles occur in high copy numbers and each copy is identical, as they are uniparentally inherited. The cp DNA is highly conserved, while mt DNA is highly variable. The size variations of selected plastid DNAs are shown in Table 2.2.

Table 2.1 Size of the nuclear genomes of some representative angiosperm taxa (after Arumuganathan and Earl 1991; (cited from Westhoff *et al.* 1998). The DNA content of nuclei was estimated in pg after staining with propidium iodide in flow cytometry using fluorescence spectroscopy and then converted to number of base pairs as well. 1 pg equals 965 million base pairs (mbp). 1 C indicates the DNA content of a haploid genome.

S.No.	*Species*	*DNA Content in picograms (pg)*	*Genomic size (mbp/1C)*
1.	*Arabidopsis thaliana*	0.15	145
2.	*Oryza sativa* sp. *indica*	0.43-0.48	419-463
3.	*Sorghum bicolor*	0.78-0.80	748-772
4.	*Lycopersicon esculentum*	0.94-1.03	907-1000
5.	*Solanum tuberosum* (2n = 4x)	1.7-1.9	1597-1862
6.	*Nicotiana plumbaginifolia*	2.4	2287
7.	*Zea mays*	2.4-2.8	2292-2716
8.	*Pisum sativum*	4.1-4.6	3947-4397
9.	*Nicotiana tabacum* (2n = 4x)	4.4-4.8	4221-4646
10.	*Hordeum vulgare*	5.0	4873
11.	*Triticum aestivum* (2n = 6x)	16.5	15,966
12.	*Tulipa sp.*	25.6-31.8	24,704-30,687

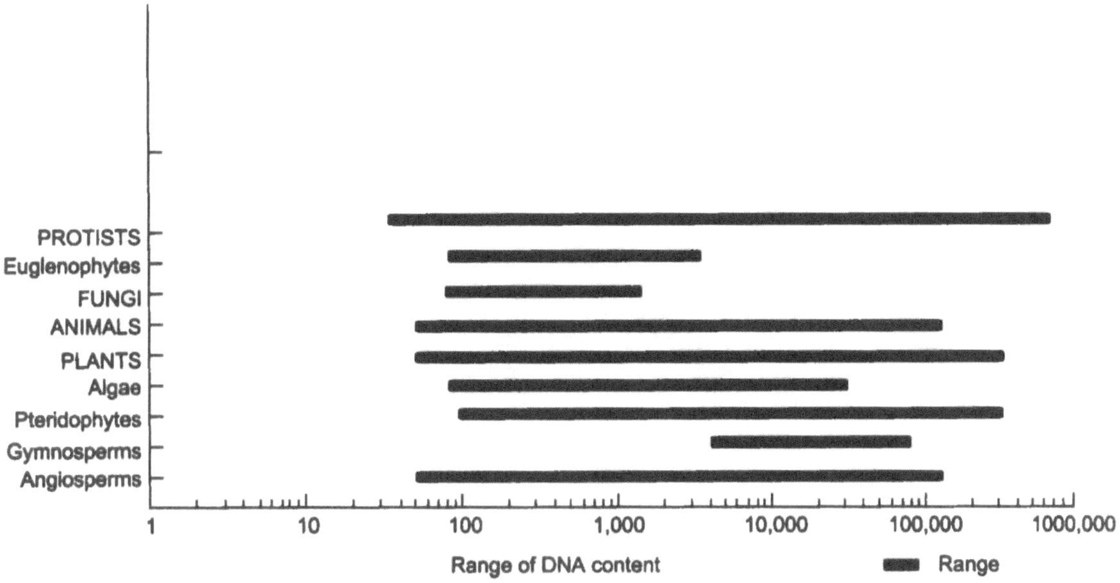

Fig. 2.1 DNA content from selected microbe groups, fungi, entire animal and plant groups and individual plant groups (after data tabulated in Li and Graur 1991).

Table 2.2 Size variation of plastid DNAs of selected plants (from Palmer 1991)

S.No.	*Species*	*Genomic size (in kb)*
1.	*Nicotiana tabacum*	156
2.	*Spinacia oleracea*	150
3.	*Pelargonium hortorum*	217
4.	*Pisum sativum*	120
5.	*Epifragus virginiana*	70
6.	*Oryza sativa*	134
7.	*Pinus* sp.	120
8.	*Ginkgo biloba*	158
9.	*Osmunda cinnamomea*	144
10.	*Marchantia polymorpha*	121
11.	*Codium fragile*	85
12.	*Chlamydomonas reinhardtii*	195
13.	*Chlamydomonas moewusii*	292
14.	*Dictyota dichotoma*	123

Molecular Marker Techniques

Allozyme Method

First let us discuss methods to measure genetic variation contributed by a single genetic locus in a plant. Although not easy to define, a genetic locus can be considered, in classical terms, as possessing alleles inherited according to Mendel's laws and acting to produce either visible differences among individuals (= visible polymorphism) or different protein profiles during electrophoretic separation.

Allozymes are a subset of isozymes, which are variants of polypeptides representing different allelic alternatives of the same gene locus. The allozyme method involves protein electrophoresis and is based on the differences in net electrical charges between variants of specific enzymes (allozymes) such as peroxidase and esterase coded for by a single gene. These charge variants, depending on their net charge and molecular weight, migrate at different rates in the protein separation gel subjected to an electrical field and hence can be distinguished from one another. Variants in enzymic proteins result when an amino acid substitution occurs in them, thereby affecting their net charge, or result due to conformation changes in these proteins, thus altering the rate of migration in the electrophoretic gel. This method revealed for the first time that, on average, 20 to 30% of the enzymic proteins of most organisms exist in more than one allelic form. The frequency of different variants in individuals collected from different regions/ areas can be measured, then the degree of genetic variation within and between individuals quantified. This in turn provides an idea of the geographic structure of the species in genetic terms.

The major advantages of the allozyme method are as follows. (i) It is cost-effective and technically simple. (ii) It provides a fairly good approximation of DNA-level changes, at least infra-specifically. Consequently this method is very useful for the study of within-species variation. (iii) Allozymes are almost invariably codominant; as a result, the phenotypes of heterozygotes differ from those of homozygotes. This enables easy calculation of gene frequencies, which can then be used for comparing populations of a species in genetic terms. (iv) This method can handle large samples with relatively greater efficiency.

The disadvantages of the allozyme method are as follows: (i) Allozymes are a 'limited and biased sample' of the plant genetic structure (Frankel *et al.* 1995). (ii) The method can measure variations only 'in the structural gene of a restricted set of enzyme loci' (Frankel *et al.* 1995). (iii) The number of enzymes is limited for which highly reliable and specific staining protocols resulting in the formation of coloured products in an electrophoretic gel can be formulated (Newbury and Ford-Lloyd 1997). Consequently, the overall numbers of markers that can be obtained and the degree of diversity assessments achieved are very limited. (iv) Allozyme data cannot show that two morphs are conspecific, but only that no significant differences could be found between them; it is always possible that differences existed, but could not be detected by the allozyme procedure (UNEP 1995).

A very good illustration of the use of this method is the work of Hamrick and Godt (1989) (see also Brown and Schoen 1992). This work summarizes the results obtained from more than 450 taxa of plants based on allozyme data. The authors classified each of the 450 taxa according to the following attributes: taxonomic status, life form, geographic range, distributional latitude, breeding type, seed dispersal and successional status. They found that of these attributes, the geographic range had the greatest effect on genetic diversity of a species. Plant taxa with small geographic ranges showed less genetic diversity (H_s) within populations, on average (H_s = 0.06), than species with a wider geographic range (H_s = 0.16). They also found that wind-pollinated species had higher levels of heterozygosity (H_s = 0.15 to 0.20) within populations, on average, than animal-pollinated taxa (H_s = 0.09 to 0.12); both showed greater levels of heterozygosity than self-pollinated taxa (H_s = 0.07). Self-pollinated plants exhibited a greater degree of genetic differences among populations (F_{ST} =

0.51) than species with mixed breeding systems (F_{ST} = 0.10 to 0.22) or outcrossing taxa (F_{ST} = 0.10 to 0.20). (F_{ST}= 0.51) is a measure of genetic diversity among populations).

The allozyme method has helped substantially in the classification of rice germplasms into six major groups based on eight enzyme activities encoded at 15 loci (Glaszmann 1987, 1988). The *indica* and *japonica* types form groups I and IV respectively. This grouping has subsequently proved to be very useful in rice breeding programmes.

Attention is drawn again to the two important notions in genetic diversity discussed earlier, namely, richness and evenness. In the case of alleles at a single genetic locus, richness refers to allelic richness. The measures for this are the 'total number of distinct alleles at that locus' in a population or its sample (Frankel *et al.* 1995). Evenness refers to the 'evenness or lack of variation among the frequency of alleles' (Frankel *et al.* 1995). It is measured by the coefficient of gene diversity (Nei 1973), which indicates the 'probability that two gametes randomly chosen from the population or sample will differ at a locus' (Frankel *et al.* 1995). Both these notions of genetic diversity at the allelic level can be demonstrated by the allozyme method in suitably selected taxa.

DNA-based Marker Techniques

It is needless to mention here that plant populations are characterized by different kinds and levels of molecular diversity, especially in DNA molecule. Several major classes of DNA sequences have been recognised, such as single-copy genes, multigene families, hypervariable mini- and microsatellite sequences and cpDNA and mtDNA sequences. Variation within any of these major classes of DNA sequences can be surveyed at the level of restriction site or fragment length variations, or comparative maps of the various restriction sites or at the level of even complete DNA sequences. The major advantage of using DNA molecular diversity is that it precludes the 'potential difficulties with differential gene expressions due to developmental state or environmental influences' (Newbury and Ford-Llyod 1997). For nuclear DNA sequences of low copy number, diversity can be measured by RFLP or RAPD methods. Empirical data gathered to date bears out the theoretical prediction that levels of intraspecific genetic variability in cpDNA (and likely in mtDNA also) are typically lower than those detected in the nuclear genome.

Restriction Fragment Length Polymorphism (RFLP) method

RFLP is a DNA-based technique to study genetic variation within and between populations of species. It is a well-known fact that various species of bacteria specifically produce enzymes to protect themselves from infecting viruses by cutting (= restricting) the viral DNA; these enzymes are consequently known as **restriction enzymes**. The restriction enzymes are highly specific in the DNA sequence they recognize and cut. Now, dozens of restriction enzymes are commercially available and these can be independently used to analyse variants at specific loci. This property of the restriction enzymes can be advantageously used to detect genetic variability in a population. RFLP procedure provides data in two ways (Swofford and Olsen 1990). The first pertains to **restriction site data** and the second **restriction fragment length data**. Since utilisation of the first type of data can be difficult and time-consuming (Murphy *et al.* 1990), people prefer to use the second type. The presence or absence of restriction fragments of a given length is often exploited. If DNA from one individual of a population of a plant species is extracted and cut with a restriction enzyme and the resultant fragments separated in an electrophoretic gel by length, a particular pattern of separation will be obtained. Another individual of the same population may reveal, using the same enzyme, a different pattern, which may slightly or greatly deviate from the pattern obtained for the first individual. This method can be repeated in several individuals of this population and, based on the results obtained, an estimation of the amount of variation in DNA sequence among the individuals can be obtained through computation of genetic distances from the proportion of fragments shared. If the DNA from one organism is digested with one or more restriction enzymes, the total number of fragments obtained are too many to be analysed; therefore small regions of the taxon's genome are generally used. Analysis based on the banding patterns is done only if comparison is to be made between just a small number of individuals.

As already indicated, RFLP analysis can be carried out in both nuclear and organellar DNA. If the latter is used, it is better to employ fresh tissues in large amounts. The problems in employing organellar DNA for RFLP analysis are that (i) the results obtained are not applicable to nuclear DNA, and (ii) since organelles are maternally transmitted, the organelle haplotypes will not behave in a Mendelian way. Therefore, organellar DNA data may provide only false information about population structuring (Avise 1994).

To study diversity realistically, however, as mentioned earlier, mere comparison of bands does not suffice. Large numbers of individuals are to be studied and consequently analysis will involve numbering or labelling the total number of bands produced employing one or more probes/primers which occur in different positions on the detection gels, scoring whether one is present or absent for each individual plant analysed. So RFLP fragments must be visualised with molecular probes (**DNA fingerprinting**) such as cloned maize alcohol dehydrogenase (*Adh*) genes as used by Gepts and Clegg (1989) in *Pennisetum glaucum* or anonymous random genomic or cDNA clones (Miller and Tanksley 1990 in species of *Lycopersicon*). A small proportion of these fragments is then selected. Next homologous regions of DNA are visualised differentially using the probe technique. The DNA fragments are subjected to electrophoresis, then transferred to a membrane (Southern blotting), after which they are hybridized with pieces of cloned DNA (probes) for one given region of the genome. These DNA probes can be labelled isotopically or chemically for subsequent identification using autoradiography or specific dyes. Subsequently, the probes are stripped of the membrane, which can then be re-probed for another segment of the genome. This is a very sensitive and simple method, but the results are dependent on suitable probes. RFLP protocols can also be applied to DNA fragments generated by polymerase chain reaction (PCR) (see RAPD technique).

The RFLP method has been very useful in determining the geographic structure of a population in genetic terms; it can also be used to estimate the degree of gene flow in a population. This method has been used to estimate genetic diversity in *Pennisetum glaucum* (Gepts and Clegg 1989) and species of *Lycopersicon* (Miller and Tanksley 1990) and to study intraspecific variation in cpDNA in over 50 species of plants, including barley, lupine, maize, pine etc. (Soltis *et al.* 1992), where both changes in restriction sites and fragment lengths due to insertions or inversions have been detected. Hollingsworth *et al.* (1999) investigated the genetic diversity in British samples of *Fallopia japonica* var. *japonica, F. japonica* var. *compacta* and *F. saccharinensis* through the use of the PCR RFLP analysis of the *trn*K intron to identify the markers that distinguish between chloroplast genomes. Breiman *et al.*(1991) estimated mtDNA variation in four wild diploid relatives of wheat from populations in the Middle East.

Polymerase Chain Reaction (PCR) and Random Amplified Polymorphic DNA (RAPD) techniques

The production of large amounts of a small segment of DNA molecule is called

amplification. This can be done by **recombinant cloning** or by replication *in vitro* using DNA polymerase (PCR technique). In the former method, the chosen DNA fragment is inserted into a plasmid or viral genome (a process referred to as **ligation**) and this plasmid or viral genome is then inserted into actively reproducing bacterial cells that cause the reproduction (**transformation**) of the DNA fragment thousands of times. Subsequently the plasmid or viral DNA is extracted from the bacterial cells and subjected to treatments with restriction enzymes. In the PCR technique, several cycles of heat are used to denature the DNA molecule in order to separate its two strands, followed by cooling for annealment with specific primers. DNA polymerase treatment then results in the synthesis of a new DNA fragment in such a way that in one cycle of about 3-5 minutes the number of copies of the region of DNA between the two primers is doubled. The cycle is repeated several times to produce large numbers of the copies of DNA. This process is called PCR and it is an extremely sensitive and simple technique compared to recombinant cloning. Further, the PCR technique has the advantage of permitting amplification of a DNA sequence even from very minute amounts of starting material and, therefore, has great importance in studying the genetic diversity of rare and endangered taxa. The products of the PCR are subsequently cut with restriction enzymes for RFLP analysis. The products of the PCR may be employed *per se* if they are of variable length or can be sequenced (Avise 1994).

Because RFLP analysis requires a relatively large amount of unsheared DNA and also because RFLP is a slow and expensive technique, the use of **random primers** or RAPD technique in combination with PCR is now-a-days preferred for estimation of genetic diversity (Clegg and Durbin 1990). The advent of arbitrarily primed PCR (AP-PCR) has made genetic polymorphism studies easy and cost-effective. In this procedure, arbitrary oligonucleotide primers, usually 9 to 10 base pairs long, containing no palindromic sequences and having a G+C content of 50 to 80%, are individually added to the sample of nuclear DNA. During PCR, wherever a primer can hybridize to both strands of target DNA in the proper orientation and the two sites are about 100 to 3000 base pairs from each other, the intervening DNA region will be amplified. A characteristically sized fragment can then be visualized in agarose gel electrophoresis. Even a single substitution in a primer will result in a different RAPD pattern.

Chalmers *et al.* (1992) employed the RAPD technique to measure genetic variation among and within populations of two tropical leguminous species, *Gliricidia sepium* and *G. maculata*. They found between populations of these species almost 60% variation in the presence of RAPD fragments. Huff *et al.* (1993) demonstrated a similar proportion of variation in RAPD fragment markers in outbreeding dioecious populations (of different geographic regions) of the grass *Buchloe dactyloides*. Mosseler *et al.* (1992) estimated the levels of genetic diversity in red pine using RAPD markers and found them to be very low.

More recently, RAPD analyses have been upgraded by employing specific primers, such as micro- and minisatellite complementary oligonucleotides (also called 'simple sequence repeats or SSRs', 'simple tandem repeats or STRs', or 'simple sequences') either alone or in combination with arbitrary primers (for details, see Weising *et al.* 1998). These microsatellites have become the molecular marker target sequences of choice for measuring genetic diversity. The microsatellites consist of 'head-to-tail tandem arrays of short DNA motifs (usually 1-5 bases)' (Weising *et al.* 1998). The following are the most important microsatellite-based techniques: (i) oligonucleotide fingerprinting, (ii) microsatellite-primed PCR with unanchored primers (MP-PCR; SPAR),

(iii) microsatellite-primed PCR with anchored primers (AMP-PCR; INTER-SSR-PCR), (iv) random amplified polymorphic microsatellites (RAMP), (v) hybridization of microsatellite probes to RAPD or MP-PCR fragments (RAMPO) and (vi) locus-specific microsatelitte analysis (STMS). These satellite-based molecular marker techniques have several advantages over the earlier simple RAPD technique, viz: (i) they can detect high levels of polymorphism, (ii) have high reproducibility, (iii) are less time-consuming, simple and effective, (iv) do not require radioactivity, (v) micro- and minisatellites are distributed evenly throughout the genome, and (vi) they have no pleiotropic effect.

Wolff *et al.* (1994) studied the genetic variation in the populations of three species of *Plantago* using minisatellites detected by the human M 13 probe. They found that breeding systems have a great influence on the genetic diversity of this taxon. The RAMP technique was used by Wu *et al.* (1994) to obtain data on screenable microsatellite polymorphism from different *Arabidopsis thaliana* strains and ecotypes; it was likewise used by Becker and Heun (1995) to study polymorphism in barley. Hüttel (1996) used this technique to study genetic variability in chickpea. For more recent examples, see Amaral (2001), Lakshmikumaran (2001) and Ratnam (2001).

Comparisons have been made on the measurement of genetic diversity obtained through molecular variation on the one hand and allozyme variation on the other (Dong and Wagner 1993; Frankel *et al.* 1995; Liu and Furnier 1993; Strauss *et al.* 1993; Wolff *et al.* 1994) (Table 2.3). The results of such comparisons show that molecular variations indicate a greater proportion of divergence between populations than allozyme variations do. It is therefore advisable to estimate genetic diversity of populations using the more modern molecular techniques detailed above.

Determinants of Genetic Diversity

Genetic variation within populations of any species is the outcome of the sum of three interacting factors: abiotic, biotic and intrinsic. Abiotic factors include climatic variables and edaphic features; biotic factors include interactions between organisms such as competition, symbiosis, antagonism etc; and intrinsic factors comprise population size, mating system, mutations, migrations, ecesis, etc. All three factors individually and in combination affect the genetic diversity of a

Table 2.3 Comparison between molecular marker techniques (based on Scott and Williams, 1994 and Westneat and Webster, 1994)

Technical Criterion	*Isozymes*	*Multi-locus Microsatellite*	*Single locus probes*	*PCR- amplified minisatelllites*	*RAPD*
Development time	1 month	2-4 weeks	1-3 months	2-6 months	1-3 weeks
Amount of material needed for analysis	About 100 mg of plant tissue	ng - µg DNA	pg-ng DNA	pg-ng DNA	pg-ng DNA
Processing time	2 days	2-4 weeks	2-4 weeks	< 1 weeks	< 1 week
Ease of use	+++	+	++	++	+++
Identification of allelic states	Yes	No	Yes	Yes	No
Rate of detection of genetic variation	Low	High	Moderate	High	High
Sensitivity of Techniques to slight changes in protocol	Low	Low	Low	Moderate	High

population stochastically as well as deterministically (for more details, see Chapter 7).

The length of a species' existence on this earth also determines its genetic diversity. The extent of genetic diversity exhibited by microorganisms is vast compared to that of macroorganisms, because the former have existed on this earth for a longer time. The earliest bacteria probably arose about 3.5 billion years ago while the earliest land plants did not originate until about 0.4 billion years ago. This underscores the fact that microorganisms have had nearly ninefold more time to diverge genetically than land plants. The greater diversity of microbes is illustrated not only by the number of their phyla recognised to date, but also by the molecular diversity they exhibit. Of the 95 phyla of living organisms proposed by Margulis and Schwartz (1988), 52 belong to microorganisms (the authors excluded viruses). Their study of 16S- like rRNAs in prokaryotes suggests that these organisms should be separated into two groups, Archaebacteria and Eubacteria, which together are roughly equivalent in size to the Eukaryotes, emphasising once again the great diversity of microbes. Woese *et al.* (1990) assigned these three groups respectively to three domains—Archaea, Bacteria and Eucarya. The great genetic diversity of microbes is also evidenced in, for example, *Chlamydomonas reinhardtii* in which at least 159 mutant lines have already been identified, in *Neurospora crassa* with more than 3000 mutants and in *Salmonella* with more than 3500 serotypes.

Genetic Diversity vs Transgenic Organisms

A transgenic organism can be defined as one that has received a very specific gene/groups of genes from another, usually distant organism through biotechnological breeding with a view to improving its performance/productivity. Concerns have been expressed from a few quarters about the possibility of negative effects of transgenic organisms on the genetic diversity of agricultural crops and their wild relatives. Most transgene researchers argue that none will occur. They also argue that transgene technology amounts to 'artificial selection' and that it definitely adds to the already existing pool of genetic diversity available in such crops. Added to this is the fact that the modified crops/microbes will transfer their genetic elements (through pollen in the former) to their wild relatives, which will result in the introduction of a new gene. The most important requirement now is to assess the impact of a transgenic organism on the already existing genetic diversity of the unmodified organisms of the same species (for further discussion, see Chapter 10).

3

SPECIES DIVERSITY: WILD TAXA

Introduction

Systematics is that branch of Biology, which is involved in the recognition, comparison, classification and naming of the millions of organisms that existed and exist at present on the Earth. Thus the basic framework for the whole of Biology is provided by Systematics (Vane-Wright 1992). Consequently, Systematics is also a very fundamental aspect of Biodiversity science. It becomes imperative for Systematics to document and understand the extent and significance of biodiversity by carrying out the following functions: recognition of taxa (Differentiation), universal diagnosis of taxa (Identification), providing universally accepted names to taxa (Nomenclature), analysing relationships of taxa (Comparison) and assembling and grouping taxa on the basis of relationships (Classification) (Vane-Wright 1992). The initially recognised groups are then assembled into more and more inclusive higher groups. Thus different levels of groups are produced as a series of hierarchical categories. The resultant structure is often called **taxonomic structure**. In this structure, the **species** is now almost universally accepted as the basic unit of the hierarchy. As detailed below, the species is also considered one of the leading players in biodiversity, conceptually, biologically and legally; it has also been almost universally used as the unit in which biodiversity is measured for all practical purposes.

The basic question of what is a species, however, has teased both biologists and non-biologists for more than two centuries and continues to do so even today. Most practising taxonomists often have an intuitive feeling for species of plants and are able to identify and name many species 'apparently with ease and considerable confidence'. But, the species is far from easy to define. In fact, many conflicting definitions of species are in vogue and universal agreement not yet in sight. Currently several **species concepts** are in use, among which the Morphological, Biological, Phylogenetic, Evolutionary, Ecological and Cohesive concepts are the most common [see UNEP, 1995 and *Systematics Botany* 20(4), 1995 for more details on definitions, comments and criticisms of these concepts]. The most important outcome of discussions on this topic is that species and species concepts are highly heterogeneous both in theory and practice and that this is essentially due to the problem of variation, so ubiquitous in living organisms. Such being the case, we have no option but to accept the reality that the species currently recognised in different groups of plants and microbes are actually not comparable entities. In other words, the basis of species distinctions in Algae is not the same as in Angiosperms. This further means that the named species in different groups of plants are comparable only by designation but not really in terms of their degree of evolutionary or phenetic differentiation (Heywood 1997b). The genetic variability within a single microbial species and that between different species of a microbial genus can also be vast compared to any macrobial taxon. This is especially evident at the DNA homology level, where only 20-50% similarities are regularly seen between any two microbial species, whereas in some higher plant groups two 'species' may still be considered distinct from one another, although sharing more than 90% DNA homology. Thus we are forced to not only accept at present the species as proposed for the different groups of plants and microbes, but also to consider these as units on which biodiversity of these groups is to be assessed and measured. As May (1995) suggested, the general public, on whom we are ultimately dependent for financial support for biodiversity maintenance and conservation, also find it easy to understand biodiversity as immanent in species, especially those of the **charismatic** or **flagship** type.

Species Inventory

How Good Should an Inventory be?

An **inventory** is a formal surveying, sorting, cataloguing, quantifying and mapping of the occurrence of defined elements of biodiversity such as genes, individuals, populations, species, habitats, ecosystems and landscapes at a particular point of time in a defined geographical unit (spatial scales range from nanometres to countries or even continents). Here we are concerned with the inventory of species. This inventory must be done for specific purposes and according to standard and well-established field procedures. It must also be done in accordance with statistically valid sampling designs and 'using rigorous quality control and data administration practices' (Dennis and Ruggiero 1996). Due to lack of funds, time and trained personnel, this inventory almost invariably constitutes a sampling rather than a complete listing of the species of an area (di Castri *et al.* 1992). Thus no inventory is ever complete, as there will always be additions and disappearances, as well as changes in abundance.

The uses of biodiversity inventories and monitoring are recorded in Box 3.1. Based on species inventory, one can study biodiversity at the global or national/regional levels for the whole plant kingdom or specific groups of plants. Dennis and Ruggiero (1996) suggested four possible approaches for orienting an inventory: (i) survey of major elements; (ii) identification of **keystone species** and **indicator elements**; (iii) identification of targeted elements, such as threatened species; and (iv) comprehensive assessment of all other important elements, such as **Exotic** or **Alien invasives**, **Flagship species**, and economically useful taxa. The purpose and orientation of an inventory will determine the choice of methods, which in turn will influence the completeness of the inventory in terms of taxonomy, community/ecosystem representation, geographical space, and seasonal/temporal representation (Solbrig 1991; Stohlgren and Quinn 1991).

An inventory which records all occurrences of chosen taxonomic groups of plants can be claimed to be taxonomically complete but not necessarily either ecologically or spatially. An inventory that covers inventorying each type of ecological element in a study area can be considered as ecologically complete but not necessarily taxonomically or spatially. Similarly, an inventory that covers every grid of a geographic area will be complete geographically but not necessarily taxonomically or ecologically (Dennis and Ruggiero 1996).

Dennis and Ruggiero (1996) mention three general levels of intensity in an inventory. A qualitative inventory will merely provide information about the presence or absence of a biodiversity element. A quantitative inventory will detail population sizes, frequency

Box 3.1 Use of inventories and monitoring data in basic and applied sciences (based on NRC, 1993)

Inventorying and monitoring of biological diversity provide information that may be used:

- To provide a basis for the scientific research necessary for understanding the world in which we live (both inside and outside protected areas);
- To define the current and future options available for meeting human needs; and
- To guide immediate and long-term management, policy and decision-making.

Areas in which inventories and monitoring activities are important include:

- Providing information for determining and conserving biological diversity;
- Providing information necessary for the sustainable management of natural resources;
- Identifying economically valuable products from wild species (bioprospecting);
- Maintaining or increasing the productivity of agricultural systems through the identification of (i) new varieties or new species of use to humans and (ii) beneficial and harmful organisms;
- Improving human health through the identification of pest organisms and beneficial organisms;
- Understanding ecosystem processes so that ecological services essential for human survival can be maintained;
- Defining the impact of human activities on biodiversity so as to help avoid undesirable effects on the environment;
- Understanding the potential effects and impact of climate change and other forms of natural environmental change;
- Determining the aesthetic benefits of diversity so as to preserve the quality of human life.

distribution, or coverage of an element of biodiversity. A relational inventory will combine either or both the above two inventories with an inventory of other biotic or abiotic elements with a view to studying the factors affecting the distribution and abundance of a desired element of biodiversity.

Several considerations influence a good inventory: (i) The existing knowledge base on which the proposed inventory is to be commenced; the greater the existing knowledge, the better the inventory. (ii) The level of expertise of personnel and technical capabilities available; the sounder these are, the better the inventory. For example, a team having advanced facilities in remote sensing, field knowledge, optics, GIS, computing, statistics and laboratory would definitely compile a better inventory than a team without these facilities. (iii) The level of funding; the larger the funding, the better the inventory. (iv) The purpose and intensity of inventory. (v) The presence of multiple performers 'contributing to a common network of data administration and analysis' will promote greater success in the inventory. (vi) Lastly, the level of enthusiasm, dedication and commitment of the personnel and institutions involved in the inventory significantly determine its coverage (Dennis and Ruggiero 1996).

Problems in Inventorying Species

Inventories can be undertaken at any geographical scale: local, regional, national and international. A critical analysis of all the inventories so far undertaken in different parts of the world in various groups of organisms has indicated the following problems:

(i) The total number of species currently recognised in various groups of microbes and plants is imperfectly known.

(ii) Many areas in the world and many ecosystems have either not been inventoried at all or not been fully explored.

(iii) Many groups of microbes and plants, both at the global and regional levels, have not been properly inventoried.

(iv) The number of species described and the number currently recorded as effective and valid are not precisely known for many groups of microbes and plants.

The existing variations in published figures of total number of species are largely due to differences in whether certain taxa are recorded as good species or not (Hammond 1992). In some groups, the apparent discrepancies in number are due to differences in the year up to which counts were done, or to miscalculations or to oversight. Most parts of existing inventories contain substantial amounts of unrecognised or at least unreported synonymies (Altaba 1996; May and Nee 1995; Patterson 1996). According to Solow *et al.* (1995), the true synonomy rate may be around 40% in the named species. In certain groups, the difference in species counts is due to the lack of agreement among workers as to what constitutes a species (Heywood 1997a).

Monitoring

Monitoring should always go hand in hand with any inventory. Monitoring consists of periodic surveillance to ascertain the extent of 'compliance with a predetermined standard or degree of deviation from an expected norm' (UNEP 1995; Hellawell 1991). Monitoring biodiversity 'aims to develop a strategic framework for predicting the behaviour of key variables in order to improve management, increase management options and provide an early warning of system change' (UNEP 1995). It is important to mention here that many recent international agreements and strategies have recognised the absolute need for inventorying and monitoring biodiversity, and have requested many countries to initiate these two activities immediately and wholeheartedly.

Total number of Species of Microbes and Plants

Largely due to the reasons mentioned above, the figures published for the Biota as a whole in recent years vary markedly— from around 1.4 million to at least 5 million (Given 1996; May 1992 b,c, 1995, 2002). The roughly 1.5 to 5 million species of living organisms known to date are considered to be probably fewer than 15% of the actual number (the 'Grail number', according to Tinker 1996). According to some extreme estimates, the number of species presently known to exist on the Earth may be even fewer than 2%, i.e., approximately 30 million (Erwin 1982; Given 1996; Tinker 1996; Wilson 1988b). A conservative estimate is around 12.5 million. The maximum number contemplated is 100 million species (Ehrlich and Wilson 1991; May 2002).

Table 3.1 provides details on the total number of species recorded thus far and the number believed to exist on the Earth among Viruses, Bacteria, Fungi (including lichens), Algae, Bryophytes, Pteridophytes, Gymnosperms and Angiosperms.

Viruses

A comprehensive catalogue of the known viruses is not yet available. It is evident from Table.3.1 that about 4000-5000 different viruses are known to date, although a conservative estimate of the possibly existing viruses in the world places the figure at around 400,000-500,000. This means we know only 1% of the viruses. The remaining are likely to be discovered as more plant, animal and microbial taxa are analysed for their viral population. Especially important in this respect are the marine organisms.

Bacteria

To date, approximately 4000 species of bacteria (including Cyanobacteria) have been discovered and recorded. The conservative estimate of the total number of bacterial species on the Earth is around 400,000. Like viruses, new bacterial species will be brought to light once all organisms are thoroughly analysed for their bacterial population in all environments, especially in marine atmosphere, hot springs and fumaroles, high salinity salt lakes, low pH acidic crater lakes, acidic mine waters, and low

Table. 3.1. The extent of described (i.e. recorded) and estimated number of species in different microbial and plant groups, as per three estimates

	Viruses	*Bacteria*	*Fungi*	*Algae*	*Bryophytes*	*Pteridophytes*	*Gymnosperms*	*Angiosperms*
Hammond (1992)	5000*	4000*	70,000	40,000		250,000* (b-a)		
	500,000+[1]	3,000,000+[1]	1,500,000+[1]	10,000,000+[1]		500,000+[1] (b-a)		
	500,000[3]	400,000[3]	1,000,000[3]	200,000[3]		300,000[3] (b-a)		
UNEP (1995)	ca 4000*	ca 4000*	72,000*	40,000*	16,000*	10,000*	240,000* (g-a)	
	1,000,000[1]	3,000,000[1]	2,700,000[1]	1,000,000[1]	500,000[1] (b-a)			
	50,000[2]	50,000[2]	200,000[2]	150,000[2]	300,000[2] (b-a)			
	400,000[3]	1,000,000[3]	1,500,000[3]	400,000[3]	320,000[3] (b-a)			
Hammond (1995)	4000*	4000*	72,000*	40,000*		270,000* p-a		
	1,000,000[1]	1,000,000[1]	2,000,000[1]	1,000,000[1]		500,000[1] p-a		
	50,000[2]	50,000[2]	200,000[2]	150,000[2]		300,000[2] p-a		
	400,000[3]	1,000,000[3]	1,500,000[3]	400,000[3]		320,000[3] p-a		

*Described species
[1]Highest Estimate
[2]Lowest Estimate
[3]Working Figure

b-a = Bryophytes to Angiosperms combined
g-a = Gymnosperms and Angiosperms combined
p-a = Pteridophytes to Angiosperms combined

temperature Arctic, Antarctic and Alpine regions. More archaebacterial and cyanobacterial species will be discovered through these attempts (Edwards 1990; Postgate 1994). New bacterial species are bound to be discovered through the use of newer techniques of analysis (Fliermans and Balkwill 1989; Wilson 1992).

Fungi and Lichens

To date, more than 70,000 species of fungi, including about 13,500-17,000 species of lichens, are known. The conservative figure for fungi existing on the Earth is about 1.5 million while the extreme estimate is 2.7 million; the number of lichen species likely to exist is around 25,000 (Hawksworth 1991). In a small area such as the British Isles, the currently recorded species of fungi is around 12,000. The yet to be described fungal species are likely to come from the tropical countries and the marine environment; both free-living fungi as well as those associated with living organisms (as parasites and symbionts) are likely to be discovered in large numbers from these environments.

Lichens are composite organisms consisting of a **mycobiont** and one or more **photobionts**, both leading a symbiotic life. Lichen photobionts come from a small group of algal or cyanobacterial genera, most of which occur widely in nature. The lichen mycobionts, on the other hand, are exclusively lichen-forming (i.e., not free-living) and are taxonomically diverse. The involved fungus is invariably an Ascomycete, and only rarely a Basidiomycete. Of the 46 orders of Ascomycetes, 16 have lichenised taxa; of the 238 families of Ascomycetes, 81 consist entirely of lichenised fungi or at least have some lichenised taxa (Krishnamurthy and Upreti 2001). The USA and Canada together account for nearly 400 genera. The other lichen-rich countries include Australia, the United Kingdom, New Zealand, India, China and Sweden.

Algae

So far 40,000 algal species have been described and another 360,000 species are believed to exist on the Earth, as per conservative estimates. Algae yet to be described are likely to come from barks and rocks as well as from the marine environment; a good contribution is also expected from the Polar regions. The green algae are cosmopolitan and occur in marine, brackish, fresh waters and soil environments; there are about 1040 species and 170 genera in about 8 orders (Silva 1982). Cladophoraceae is the largest family among green algae with about 300 species. The brown algae are global in distribution and essentially marine. There are 265 genera and more than 1500 species accommodated in about 14 orders (Wynne 1982). The brown algae reach their greatest diversity in the Japanese region of the Pacific and the North Atlantic. South Australia shows the highest number of endemic taxa of brown algae. The red algae form the largest group and, again, are predominantly marine. There are approximately 555 genera (Dixon 1982). Like brown algae, the red algae are also the richest in the Japanese region of the Pacific Ocean and the Western Atlantic region; again, South Australia is full of endemic red algal taxa. The Charophyta are predominantly freshwater, although some taxa occur in brackish water; they comprise six genera, in which about 125 species are endemic

It is very difficult to comment on the distribution and number of microalgae since information on them is very scant. Desmids and Diatoms have been surveyed extensively on regional scales in certain parts of the world, but these data are hardly suffice for drawing any generalisation.

Bryophytes

Bryophytes are a diverse group of plants containing several classes. So far 14,000-16,000 species of bryophytes are known, of which about 8000-9000 species (under 425 genera) are mosses and 6000-7000 species liverworts. The number of bryophyte species is likely to increase to about 30,000, if more regions of the world are subjected to serious inventory. Although, as a group, bryophytes occur throughout the world, the majority of the taxa are reported from cooler regions, especially in moist tropical forests and temperate woodlands. The maximum diversity is noticed in regions where cool climate has persisted over prolonged geological time. The Indo-Australian archipelago and South America are the richest areas for bryophytes, closely followed by South Australia, North America, North-east Asia, the Himalayas, East Africa and Europe (including the British Isles). *Plagiochila* with about 500 species and *Frullania* with about 400 species are the largest genera among bryophytes.

Pteridophytes

The Pteridophytes are vascular land plants and together with Gymnosperms and Angiosperms dominate the terrestrial environment of the Earth. There are about 15,000 species of Pteridophytes, of which many are native to moist tropical forests. The fern-allies of Pteridophyta consist of the Psilophytales (the earliest known vascular land plants) with two living genera, *Psilotum* (tropical in distribution) and *Tmesipterus* (confined to Australia and New Zealand), the Lycophytales with five living genera (*Lycopodium, Phylloglossum, Selaginella, Isoetes* and *Stilites*) and the Sphenophytales represented by a single living genus, *Equisetum*. The ferns are a very diverse group and exhibit a great range of form—from very tiny and delicate filmy ferns (Hymenophyllaceae) to very tall tree ferns (Cyatheaceae and Dicksoniaceae) that may attain several metres in height. Ferns are cosmopolitan in distribution, although very dominant in moist tropics, particularly as epiphytes. It has been estimated that 12.5% of the world's fern species are found in Papua New Guinea (Johns and Bellamy 1979).

Gymnosperms

Most species of Gymnosperms are trees, although a few are shrubs. *Welwitschia mirabilis* is described as 'neither a tree, nor a shrub, nor a herb' because of its very peculiar habit. This species is acaulescent and has a very thick rootstock, which produces at the ground level only a pair of leaves that persist and grow throughout the life of the plant. There are about 500 species of Conifers, 100 species of Cycads and 71 species of Gnetales. The number of species yet to be discovered in Gymnosperms is likely to be very few. *Ginkgo biloba*, a species native to and endemic in China, is a **living fossil**. Conifers occur worldwide but predominantly in the temperate and alpine regions of the tropics. The genus *Podocarpus* is predominantly tropical and subtropical. *Podocarpus wallichi* is endemic to the southern Western Ghats of India. The Gnetales consist of three living genera, *Gentum, Ephedra* and *Welwitschia*. *Gnetum* is restricted to moist and semimoist forests of the tropics, *Ephedra* is mostly arid in distribution while *Welwistchia* is endemic to some southern areas in the African continent.

Angiosperms

The angiosperms or flowering plants constitute an extremely diverse group of vascular plants. There are about 235,000 to 300,000 species (Gentry 1996; Heywood 1997a). Another 500,000 species are estimated to be present on the Earth, awaiting discovery. Angiosperms are the most recent group of plants to evolve in geological history, having made their probable first appearance around 35 million years ago. But within a very short span of geological time, they have become the most dominant of all plant elements on the globe, probably because of their great evolutionary capabilities. Most of the food and other requirements for humans come from angiosperms. The plants range in size from 1 mm across (*Wolffia* species) to over 100 metres tall (species of *Eucalyptus*). The flowers can reach more than a 1-metre span in *Rafflesia arnoldii* growing in Sumatra and Borneo.

The 235,000 to 300,000 species of flowering plants are grouped in about 17,000 genera under about 200-600 families depending on the classification system. Orchidaceae with 25,000 to 35,000 species and Leguminosae (*sensu lato*) with about 15,000 species are the largest families among angiosperms. In fact, approximately 30 families account for almost 62% of the known angiosperms; 36 families are unispecific (e.g. Adoxaceae).

Species Diversity

History and Origin of Species Diversity

Fossil records not only provide us the necessary background for analysing the history of diversity through geological time, but also clues to the origin of this diversity. Although incomplete and biased, fossil records provide us enough information to understand ancient biodiversity in terms of taxonomic richness, particularly at higher taxonomic levels. They also permit historical reconstruction of the origins of modern biodiversity. The records further reveal biotic events that could never be deducted from living biota alone, especially with reference to vast evolutionary radiations and mass extinctions. Fossil diversity is basically measured from family or generic diversity rather than from actual species diversity, since the concept of a fossil species is not equivalent to the concept of a living one. However, fossil generic and family diversity trends are considered useful proxies for species diversity trends (Gaston *et al.* 1995; Raup 1979a; Sepkoski 1992) from the Cambrian to the Miocene epochs (Erwin 1996). Although detailed patterns of taxonomic richness through the Earth's history remain debatable, the overall outline is generally accepted. Presumably, during the Palaeozoic and early Mesozoic eras relatively few species existed on the Earth. Since then, that is, for the past hundred million years, diversity has increased substantially, with a particularly sharp increase since the Jurassic and Cretaceous periods. Large numbers of new evolutionary species lineages emerged three times during the history of life on this Earth (Meffe and Carroll 1994): (i) Cambrian explosion, which happened about half a billion years ago; (ii) a second explosion *circa* 60 million years later; and (iii) Triassic explosion that led to the evolution of modern plants. There were also extinction events, some major and others minor (Signor 1990). In spite of these extinctions, diversification appears to have continued unabated, with the world apparently reaching its highest ever level of species richness during the Pliocene and Pleistocene epochs, at which time dramatic changes in climate followed by the advent of human activity, finally halted the diversification process.

Two factors are believed to be primarily responsible for the continual rise in number of plant species during the evolution of the planet Earth: (i) The initial break-up of **Pangaea** and subsequently of **Gondwanaland**, resulting in increased provinciality in the earth's landmass. Following this major separation of landmasses, evolution of plant life on each of the new landmasses produced several new species that subsequently became continental endemics. This increase in species number due to provincialisation was particularly striking during the Coenozoic era; and (ii) The evolution of different plant communities within each of these provincialised landmasses favoured speciation at a very rapid rate. For example, the late Coenozoic plant communities comprised almost twice as many species than did earlier communities.

It is generally accepted that the first vascular terrestrial plants arose in the Silurian period (Niklas *et al.* 1985), although some argue for their late Ordovician origin (Gray 1985). Diversity in vascular plant groups slowly increased in the Silurian and then very rapidly in the Devonian period, at which time the first seed-bearing plants made their appearance. Species of at least 40 or more genera of vascular plants have been recorded for the late Devonian (Knoll 1986). Species diversity then declined slightly, only to markedly increase during the Carboniferous period; at least 200 species were recorded by the mid-Carboniferous. Following this, diversity increased quite slowly until the end of the Permian, when a minor decrease in diversity occurred that coincided with or preceded the mass extinction recorded for that period. Biodiversity increased to previous levels immediately after the Permian and slowly rose again up to the early Cretaceous when about 250 species were recorded. But from the mid-Cretaceous, diversity again increased very rapidly (Fig. 3.1).

Diversity Indices Based on Species

The literature on diversity measurement based on species is immense. Numerous indices and methods are available (Grassle *et al.* 1979; Hurlbert 1971; Magurran 1988; Pielou 1974, 1975, 1977). Ecologists have estimated biodiversity by three measures: (i) **species richness**, which is indicated by the total number of species in an area (Ashton 1992); (ii) **species abundance**, which is indicated by the total number of individuals of a species in an area; and (iii) **species evenness**, which represents equitability of species as given by their relative abundance. Sometimes one or more of these indices are combined, especially species

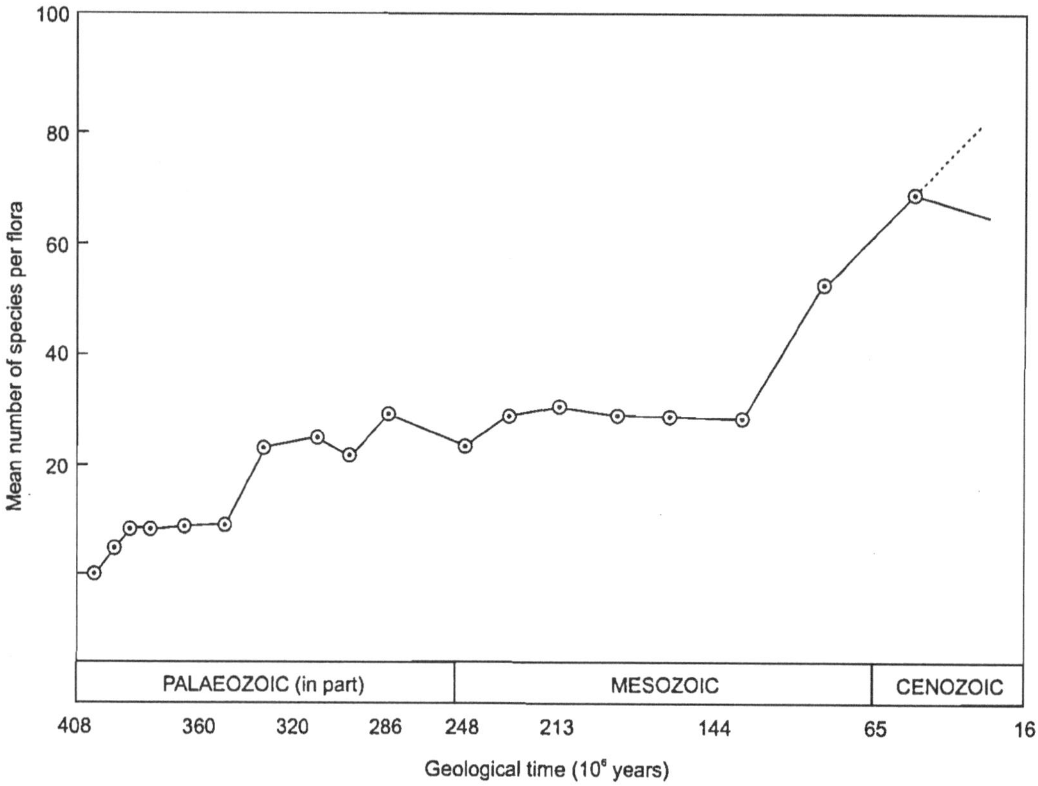

Fig. 3.1 Changes in land plant diversity over geological time (adapted from Knoll 1986).

richness and abundance. Some attempts have also been made to estimate biodiversity by measuring changes in species richness between areas, the resultant product being referred to as **species turnover**. Each index has its own relative advantages and disadvantages. Their value is also dependent on the purpose at hand. For example, species turnover has an important say in estimations of species extinctions, especially under fragmented habitats. Since all these indices are based on species, biodiversity is now very commonly considered a synonym for species diversity. Ecologists, in particular those interested in functional aspects of ecosystems, more often use the **Shannon-Wiener index, Simpson index, Fisher's alpha-log series** or modifications of the same, which again are based on species abundance. For a more detailed treatment of these indices see Magurran (1988) and Gove *et al.* (1996).

A new measure of biodiversity was recently proposed (Ganeshaiah *et al.* 1997) called the 'Avalanche Index'. This index uses, in addition to species numbers and frequencies, the biological and ecological differences among species comprising a community. It attempts to integrate, over all possible species combinations, the biological differences among species comprising a community. For more details on these indices see Ganeshaiah *et al.* (1997).

Species Richness

Species richness, as already indicated, has become an important component of biodiversity assessment. Now it is very commonly used as a synonym of species diversity. Similarly, global biodiversity is very often considered in terms of global number of species in each of the different taxonomic groups. In other words, measures of biodiversity for particular areas, habitats or ecosystems are often largely reduced to a straightforward measure of species richness (Jenkins 1992). Ideally, species richness should be estimated by cataloguing all species occurring in the area under consideration. Practically speaking, this is unrealistic since even with small sites a complete cataloguing of all species available is not possible. Hence species richness is measured on samples carefully chosen in a particular area.

Although species richness data may provide relatively little ecologically significant information, in practice such data are the most easily derived. Thus they are perhaps the most useful indices for comparisons of diversity on a larger geographical scale. Moreover, at present, species richness data are the only type of information available for most areas of the world (Jenkins 1992). Such data are also important for prioritising conservation strategies since they allow identification of geographic regions of the world with exceptional or with very poor diversity.

Species Abundance

As just mentioned, simple species-richness data may have very limited ecological value. More meaningful are measures of species abundance, especially relative abundance of different species in an area. In general, as remarked earlier, , the more equally abundant the various species in an area or ecosystem, the more diverse it is considered.

Taxic Diversity

One of the major limitations with species diversity measures is that they treat all species (even within a specific group of organisms) equally, i.e., they take no account of differences between species in relation to their place in a natural hierarchical system. The taxic diversity approach, therefore, is based on the view that 'individual species vary enormously in the contribution they make to diversity because of their taxonomic position' (UNEP 1995). Taxonomically isolated species or species of taxonomically isolated genera are of very great value (e.g. *Ginkgo biloba*) in diversity assessment in an area. An area containing taxonomically diverse species is considered to have greater diversity than an area with closely related species in equal numbers.

How to measure taxic diversity in an ecosystem is a question that has not been seriously considered by many biologists. Williams and Humphries (1994), and Williams *et al.* (1991) suggested a **cladistic** approach to solving this problem. Even in this approach, five measures or diversity indices can be applied: root weight, higher-taxon richness, unrooted spanning-subtree length, rooted spanning-subtree length, and dispersion (Fig. 3.2). Root weight places the highest individual value on taxa which separate closest to the root of the **cladogram**; this measure gives high weightage to relict taxa. Higher-taxon richness selects taxa according to their rank and number of species included. Unrooted spanning-subtree length measure counts the number of nodes on the taxonomic (cladistic) tree needed to link the taxa; each node is counted only once. Dispersion is the most complex of the measures suggested so far and endeavours to select an even spread of taxa across the hierarchy.

Comparisons of Species Diversity of Different Sites

Species/Area Relationships

Once the measure of diversity based on species has been decided, it is possible to compare the diversity of different regions. As already indicated, the most useful and practical measure at present is species richness. The relative diversity of different regions will depend on the scale at which species richness is measured. Thus, one m^2 area of a tropical rain forest will have less diversity than one km^2 area of the same type of forest. This is due to the fact that the number of species increases with the size of the area sampled and studied. The rate of increase in species richness, however, varies from area to area. For example, the rate of increase in species richness in a tropical rain forest will be much greater as the area studied increases, while the rate of increase in diversity in a temperate forest will invariably be less as its area of study increases. This species-area relationship has been shown to apply generally both to enumeration data, when each and every

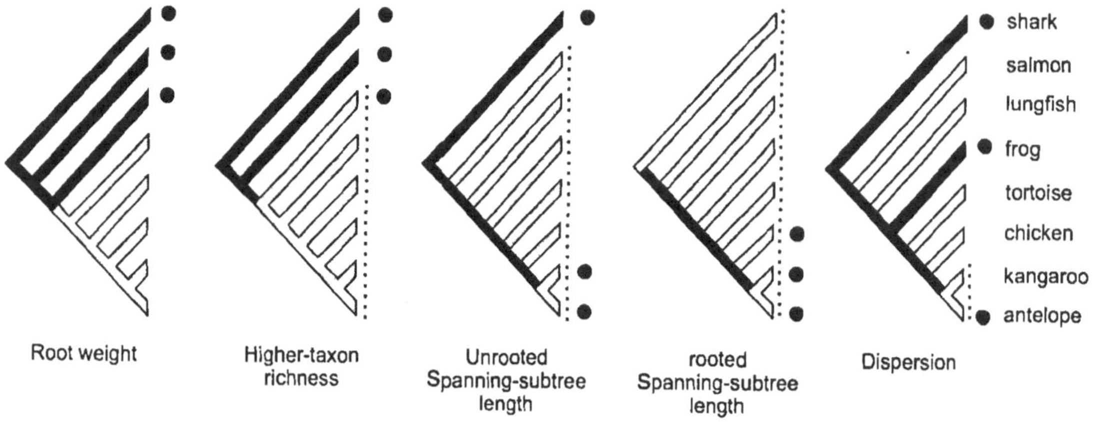

Fig. 3.2 Five different measures of taxic diversity using a cladistic approach (adapted from Williams and Humphries 1994).

species present is counted, and to sampling data, derived by counts of the number of species occurring in sample plots of various sizes.

The species-area relationship has been shown to follow the Arrhenius equation (Arrhenius 1921). This equation can be written as follows:

$$\log S = c + z \log A$$ [1]

where S = number of species (species richness); A = area of study; and c, z are constants.

The equation is graphically represented in Fig. 3.3. On a logarithmic scale, the relationship can be plotted as a straight line, where c is the Y-intercept and z is the slope of the line. The slope of the line (i.e., z) varies considerably between surveys, although it generally falls between 0.15 and 0.35 in natural ecosystems (Connor and McCoy 1979; Wilson 1992). It is closer to 0.25 for several theoretical models and for much field data (Crawley and Harral 2001). An important, but commonly overlooked corollary of this is that species density (i.e., the number of species per unit area) declines with increasing area. The relationship between size of an area and the number of species it supports, which is most readily observed using data from Islands, is a key empirical generalisation in the development of the **Theory of Island Biogeography** (MacArthur 1985; MacArthur and Wilson 1967). This Theory in fact is exploited to explain the species-area relationship in other areas as well. This relationship will be referred to very often in this book

Spatial Patterns of Species Diversity

Complex spatial patterns of species biodiversity have often been recognised by dividing species richness into three major components, i.e. to characterise diversity on different scales (Whittaker 1972): **alpha richness, beta richness** and **gamma richness** (also often called respectively alpha diversity, beta diversity and gamma diversity). Alpha or point richness refers to the number of species in an area, i.e.,

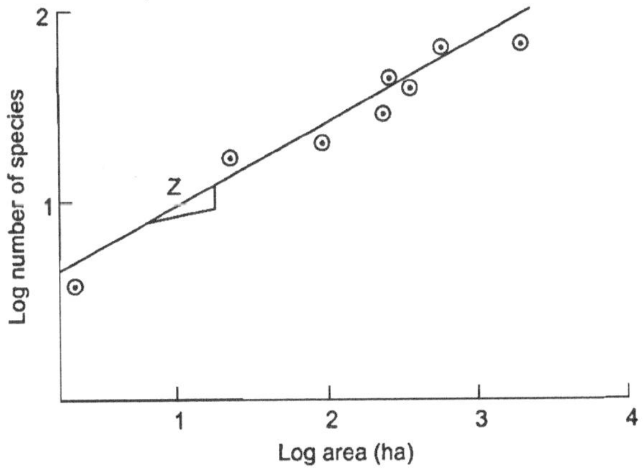

Fig. 3.3 Graphic representation of age-area relationship as per the Arrhenius relationship. Slope z indicates the rate at which the number of species changes with change in area.

[1]Some authors write this equation as $S = cA^z$ (see for example Meffe and Carroll 1994 and Crawley and Harral 2001). Others have slightly modified the equation (see May and Stumpf 2000) and write it as $S = cA^z \exp(-kA)$ where an additional parameter k is used; k is estimated from empirical a(A) curves, where a(A) represents the average fraction of species present in A that are found ('persist') in one-half A.

within-area diversity. Here we count the number of species using only their presence (and not abundance) in a given area of a given size. Beta richness or between-area richness refers to the changes in number of species between sites at local, small and homogeneous areas. Beta richness cannot be expressed in species numbers and can only be represented in terms of the similarity index between species diversity of different areas in the study region or of species turnover rates. Beta diversity 'is the ratio of gamma diversity of a region to the average alpha diversity of local areas within the region' (UNEP 1995). However, various authors have given different definitions for beta diversity. Gamma diversity refers to overall species richness within a large region, at the level of a landscape, i.e., biodiversity characterisation at a regional scale. Gamma diversity, therefore, does not have an upper limit.

The overall diversity of a given area is dependent on the range of habitats it includes as well as on the density of the habitats included. The greater the differences between the habitats included in terms of species diversity, the greater the overall diversity. This does not mean that an area with relatively low species richness will not make an important contribution to the overall diversity of the larger region of which it is a part, as long as this region contributes a significant number of species that do not occur elsewhere in the larger region.

Global Distribution of Species Richness

All analyses to date of trends in biodiversity world over almost always treat it in terms of species richness, as this is the only measure of diversity for which, to a large extent, adequate data are available on a global scale. Such analyses have indicated that biodiversity is not evenly distributed around the globe (Lugo and Brown 1996), but forms spatial patterns of diverse sorts.

Latitudinal gradients

Overall species richness decreases with increasing latitudes, i.e., there are far more species per unit area and in total in the equatorial region (tropical region) of the Earth than in the temperate regions, and far more diversity in temperate regions than in polar regions. This applies across many groups of plants (Stevens 1989).

Plant species diversity, however, varies markedly on small spatial scales within any one region, say Tropical or Temperate Zone. For example, there may be on average 70 tree taxa in one hectare of tropical moist forests, compared to 20 trees in one hectare of temperate forests. Even within tropics, species diversity may be strikingly lower in a desert than in a moist forest.

It should be understood clearly that latitude *per se* does not drive biodiversity gradient, but several abiotic and biotic factors and their intensity of interactions that might influence biological diversity at various scales co-vary with latitude (such as total land area, temperature, rainfall, energy flux, animal mutualists etc.).

It should also be understood that what was expressed above is only a broad trend; species richness in particular taxonomic groups or in particular habitats may show no significant latitudinal variation or may even increase towards temperate regions as in the case of polypore fungi, freshwater phytoplanktons etc. In some groups of plants, latitudinal patterns are expressed differently north and south of the Equator. For example, total tree species diversity decreases much more sharply from equatorial regions into the southern Temperate Zone than it does to the north. This is especially true of the African continent.

Altitudinal gradients

There are often gradients of species richness with altitude (Stevens 1992). Diversity generally decreases with increasing altitude in terrestrial ecosystems. This trend may be due to the result of species-area relationship, since available land area generally decreases with increasing altitude and since the number of species is closely related to the total area. Energy flux and temperature correlate inversely with elevation and may also account for the observed decline in diversity with increasing altitude. More striking examples for this gradient are the woody plant species in tropical montane areas (Gentry 1988).

Humped patterns of diversity on elevational gradients, with low diversity at the base and summit and high diversity in the midregion of a hill, are known for particular plant groups and in certain geographical areas. An example is the flowering plant diversity observed on the western slopes of the Andes in northern Chile (Arroyo *et al.* 1988).

Rainfall gradients

In terrestrial environments, biodiversity is higher in regions of greater rainfall and lower in drier areas. Therefore, the richest biodiversity areas are undoubtedly the tropical moist forests; these forests cover only about 7% of the world's area but contain more than 90% of all known species. This is true in all scales of measure: within habitats, between habitats and landscapes. Within a given altitudinal belt in the tropics, species diversity, especially of trees, increases with increasing precipitation.

Other factors

Species richness for most taxa correlates positively with structural complexity of the ecosystem. Structurally simple ecosystems such as the open ocean, grassland and the hot and cold deserts generally possess fewer species than structurally complex ecosystems such as a rain forest. This can be explained on the basis of energy cycle. Many investigators believe that species diversity and productivity (and consequently energy) are strongly correlated, although others question this correlation (see Lugo and Brown 1996; Meffe and Carroll 1994 and Odum 1983 for more detailed discussion on this and for the model proposed to explain this relationship). There are conspicuous exceptions to this generalisation. Some extremely productive ecosystems such as the sea-grass beds are very poor in species. The claim of a direct relationship between increase in species diversity with increase in productivity has led to very heated debates these last few years.

In aquatic environments, salinity is a major factor affecting species richness. In freshwater bodies, diversity decreases with increasing salinity.

Island ecosystems are generally poorer in species than comparable mainland ecosystems. The number of species found in islands correlates positively with their size, as per the Island Biogeography theory. Species diversity correlates negatively with distance of the island from the mainland. Low species richness in islands is attributed to low colonisation rates, higher extinction rates and lack of some specific resources.

Distribution of Higher Plant Species Diversity

A separate section on this topic was impelled by the fact that higher plants are not only the most dominant biota in the world, but also the most thoroughly studied.

Higher plants occur virtually in all ecosystems of the world. They occur even in the marine environment, as sea-grasses. In spite of their ubiquitous nature, their distribution is very uneven, with nearly 67% of the world's flowering plants occurring in tropical zones. Table. 3.2 provides a reasonable but provisional estimate of the number of higher plant species in various parts of the world. It is evident from this Table that the richest region is South America, which accounts for one-third of the world's

Table 3.2 Higher plant species diversity in terms of species number in different continents of the world (based on Akeroyd and Synge 1992)

Latin America	85,000
Tropical and Subtropical Africa	40,000 - 45,000
North Africa	10,000
Tropical Africa	21,000
Southern Africa	21,000
Tropical and Subtropical Asia	50,000
India	15,000
Malaysia (including E. Indies)	30,000
China	30,000
Australia	15,000
North America (including W. Indies & Pacific Islands)	17,000
Europe	12,500

higher plants. It should be emphasised that this continent is also the most poorly explored region in the world[2].

Centres of Diversity

Introduction

It has already been noted that there are broad gradients in biodiversity distribution from the geographic standpoint. Such gradients can be conveniently overlain with specific centres of diversity. The definition of centres of diversity is not very easy practically speaking due not only to scale problems, but also to choice of taxonomic groups (Mares 1992). Simply defined, a centre of diversity is merely an area with high species richness. Centres of diversity have been recognised on global, regional and local scales. Centres selected at one scale, say national, may have little relevance to another, say global (Heywood 1997b). Global and regional centres of diversity are recognised not only by species richness, but also by some other measures of biodiversity, such as the number of different phylogenetic lineages, number of endemics etc. On a local scale, the large number of species alone identifies such centres. Only in recent years have reliable analyses of centres of diversity covering all three geographical scales been attempted (Williams and Humphries 1994). On a global scale, four important analyses merit mentioning: (i) Myers' Hot Spots, (ii) IUCN's Centres of Plant Diversity (CPDs), (iii) Megadiversity Centres and (iv) Diversity zones of Barthlott *et al.* (1996).

Hot spots

Myers (1998) estimated that 25 places on the Earth could be classified as **'hot spots'** (a term borrowed from the physical sciences), based on endemic species richness. His earlier analysis and recognition of only 18 hot spots (Myers 1990) was limited because they were confined to higher plants, while the recent updating through Conservation International (CI) in Washington DC was based on other criteria as well. These hot spots support nearly 122,935 endemic or restricted-range plant species, which account for about 46% of the world's total vascular flora (Table 3.3). These hot spots are also rich in endemics of other groups of living organisms, in particular birds, reptiles, amphibians, butterflies and mammals. They harbour 60% of all species of organisms, including non-endemics, and at least 65% of all Red Data Book species. Although endemic species richness is the primary criterion for recognising hot spots according to Myers (1988a, 1990, 1998), the following other characteristics were also considered to qualify these areas as hot spots (Synge and Heywood 1987): (i) the site should already be threatened or under imminent threat; (ii) it should include

[2]The species concept used to arrive at the figures mentioned in Table. 3. 2 varies from place to place. For example, the species concept used in Latin America tended to recognise more species than the concept applied by taxonomists working in the Malaysian region. In other words, the species number reported for South America may come down to a lower figure if critical revision works are undertaken. This is suggested by the fact that the flora of Colombia was initially reported to contain 45,000 species (Prance 1977) but due to critical revision dropped to 35,000 (Forrero 1988).

Table. 3.3 The various hot spots of the world and their total number of endemic vascular plants (from Myers 1998)

Sl. No.	*Name of Hot Spot*	*Total Number of endemic vascular plants*
1.	Tropical Andes	20,000
2.	Mesoamerican forests	9000
3.	Western Sundas	8870
4.	Madagascar	8000
5.	Mediterranean Basin	7600
6.	Caribbean Islands	7000
7.	Philippines	6000
8.	Atlantic coast of Brazil	6000
9.	Cape region	5850
10.	Eastern Himalayas	5000
11.	Brazil's cerrado	4200
12.	Southwestern Australia	4170
13.	Western Ecuador & Darien-Choco	3760
14.	Pacific islands	3275
15.	Wallacea	3000
16.	Western African Forests	2960
17.	Western Ghats (India)/Sri Lanka	2690
18.	New Caledonia	2550
19.	California Floristic Province	2140
20.	Northern Indochina	2130
21.	Khasi-Manipur	2000
22.	New Zealand	1940
23.	Central Chile	1800
24.	Succulent Karro	1750
25.	Eastern Arc and Coastal Forests	1250

a diverse range of habitat types; (iii) it should have a significant percentage of specialist species, adapted to specific edapic conditions; (iv) it should contain important gene pools of plants of value to people or at least of potentially useful plants. The original 18 hot spots put together formed only 2% (i.e., 2,985,600 km^2) of the world's total area and differed from one another in size. Many of these hot spots occur in a tropical zone and very few in the Mediterranean biome (Table 3.3). Based on vegetation, these hot spots can be classified into 7 types: Tropical Rain forests, Tropical Montane forests, Mixture of Tropical Rain forests and Tropical Montane forests (Gentry 1991), Tropical Moist forest, Subtropical to Warm Temperate forests, a range of several Tropical habitats (Schneckenburger 1991) and Mediterranean Vegetation.

Myers proposed the hot spot idea primarily because of his belief that, at the global level, these are the areas of high conservation priority; if unique, restricted-range species are lost, they can never be replaced. These are also the areas, according to him, susceptible to unusually rapid rates of habitat modification or loss.

Megadiversity centres

The species richness and endemism in 25 selected countries of the world are provided in Table 3.4, in order to emphasise the geopolitical distribution of higher plant diversity. Of these countries, only a few, situated mainly in the tropics, possess a significant percentage of species diversity. Such countries with great species diversity have been termed **Megadiversity Countries or Centres**. These are the countries that should be considered seriously for priority conservation action (Mittermeier 1998; Mittermeier and Werner 1990). Only 12 Megadiversity countries have been identified (McNeely *et al.* 1990): Brazil, Colombia, China, Mexico, Indonesia, Ecuador, Australia, India, Peru, Malaysia, Zaire and Madagascar. It is interesting to note that these 12 countries are not the first 12 countries in terms of highest number of vascular plant species (compare Table 3.3); richness in other groups of organisms, especially animals, was also considered seriously for recognising these countries. Interesting countries, in terms of higher plant diversity that were left out include Venezuela, USA, Bolivia, Costa Rica, Thailand, Papua New Guinea, Tanzania, Argentina and Panama.

The 12 megadiversity centres together account for more than 70% of the world's vascular plants. Recognition of the 12 megadiversity countries is advantageous in one respect. Since these countries are distinct and

Table. 3.4 The 25 most plant-rich countries

Sl. No.	*Country*	*Total Number of Species*	*Total Number of Endemics (% in parentheses)*
1.	Brazil*	55,000	?
2.	Colombia*	35,000	1500(5%)
3.	China	30,000	18,000(60%)
4.	Mexico	25,000	3600(14%)
5.	Indonesia	20,000	15,000(67%)
6.	Venezuela*	20,000	8000(38%)
7.	Ecuador*	20,000	4000(20%)
8.	USA	19,500	4050(20%)
9.	Bolivia*	18,000	?
10.	Australia	15,000	(80%)
11.	India	15,000	5000(33%)**
12.	Peru*	14,000	?
13.	Costa Rica	12,000	1800(15%)
14.	Malaysia⁺	12,000	?
15.	Thailand	12,000	?
16.	Zaire	11,000	3200(30%)
17.	PNG (Papua New Guinea)	10,000	1120(55%)
18.	Tanzania	10,000	(11%)
19.	Argentina	9000	(25%)
20.	Madagascar	9000	(68.4%)
21.	Panama	9000	1225(13%)
22.	Turkey	8500	2650(31%)
23.	Cameroon	8000	160(2%)
24.	Guatemala	8000	1180(13.5%)
25.	Philippines	8000	3500(40%)
26.	Paraguay*	8000	?

*In these countries as well as Argentina, Chile, French Guiana, Guayana, Suriname and Uruguay 78.5 % of the higher plants are endemics (Heywood 1997b).

**Heywood (1997b) records a total number of 25,000 species of which 12,000 (i.e. 48%) are endemics.

⁺Heywood (1997b) records for South-east Asia including Malaysia, a total of 42,000-50,000 species of which 29,000-40,000 are endemics (i.e. 70-80%).

Question mark indicates either the absence or inadequacy of data.

definite geopolitical entities, conservation strategies can be planned and executed without much difficulty.

Centres of plant diversity

These are defined as places, particularly rich in plant species that, when protected, would safeguard the vast majority of wild plants of the world (Davis *et al.* 1994-1995). As per IUCN, the majority of these centres lie within the hot spots mentioned earlier, although some lie outside them. IUCN has recognised around 234 such CPDs the world over, and has classified them into three types: (i) botanically rich sites that can also be defined geographically (e.g. Mt. Kinabalu in Borneo); (ii) geographically defined areas with high species richness and/or endemism (e.g. The Atlas Mountains in N. Africa and the Cordillera Bética in Spain); and (iii) vegetation types and floristic provinces that are exceptionally species-rich (e.g. Amazon Rain Forest, South-western Australia and Atlantic Forests of Brazil).

The criteria for considering a region as a CPD are as follows: (i) The region must be species-rich, even if the actual number of species is not accurately estimated, and (ii) the region must have a large number of endemic species. At least one of these two criteria must be fulfilled. In addition to these two criteria, the following characteristics are also often considered in the selection of a region as a CPD: (i) the region should have important gene pools of plant taxa of value to human beings, or plants of potential use; (ii) it should have diverse habitat types; (iii) it should contain a significant number of specialist taxa that are exclusively adapted to specialised edaphic conditions; and (iv) the region should have threatened or potentially threatened habitats. It is evident from the foregoing that CPDs are selected largely on the basis of botanical importance rather than on degree of threat, although the latter is also taken into serious account. The site that is safe today might come under serious threat later, and therefore species richness is considered the most important criterion. In certain parts of the world, site selection was rather easy, in others very difficult; and in others has yet to be done. Standardised data sheets have been prepared for certain of the CPDs covering information relating to the vegetation, flora, useful taxa, threats and conservation requirement.

The value of the 'Centre of Diversity' concept can be very greatly strengthened, provided that ranking the functional roles of the species concerned and assessment of their phylogenetic status are done. For example, when these were done for the rain forests of southern South America (Chile region), it became evident that the diversity content of this forest with its many monotypic, endemic genera, one endemic family and a broad spectrum of woody life forms was much higher than that of the adjacent more species-rich areas of Mediterranean climate northwards in Chile.

Diversity zones

After a detailed evaluation of 1400 floras, floristic studies, bio-geographical surveys and vegetation analyses, Barthlott *et al.* (1996) mapped out the species richness in ten **diversity zones**, giving the numbers of species per 10,000 km^2 on a global scale. Six global diversity zones with more than 5000 species per 10,000 km^2 were identified. All were situated in humid tropical and subtropical regions: (i) Choco-Costa Rica centre (ii) Tropical Eastern Andes centre (iii) Atlantic Brazil centre, (iv) Eastern Himalaya-Yunnan centre (v) Northern Borneo centre, and (vi) New Guinea centre.

Future of Species Diversity Studies

Since our knowledge of the world's species is incomplete, our primary task is to make inventories and to catalogue species in several parts of the world and in several ecosystems. How to carry out this inventory is a topic of great discussion these days. Some have recommended the initiation of an intense global survey whereby all species believed to exist on the Earth are catalogued; an analysis and classification of all species should follow the survey. Others have drawn attention to the shortage of experts, funds, time and institutions to carry out the survey and analysis and, therefore, have recommended a realistic action program to rapidly recognise and preserve only those threatened ecosystems that contain the largest number of restricted- range and rare species. In other words, according to these people, the large-scale global inventory task should be relegated to a lower immediate priority.

The establishment of at least a small series of sites with an All-taxa Biodiversity Inventory (ATBI) was proposed by Janzen (1993a,b); such inventories could help the process of extrapolations. ATBI involves all sectors of society in the gathering, management and use of biodiversity information in sampled areas, especially in areas where species richness is known (hot spots) or expected to be high (Yoon 1993). This concept of ATBI has been criticised by some because an inventory covering all taxa is not presently feasible due to lack of expertise in many groups of organisms. Moreover, such ATBIs would be time-consuming and very expensive.

Wilson (1992) has strongly advocated for a mixed strategy of the above two: Worldwide surveys should be completed within 50 years and quicker attention should be focused on hot spots. He suggested a three-level strategy: The first is a **Rapid Assessment Program (RAP)** or **Rapid Biodiversity Assessment (RBA)** which would investigate within the next few years, poorly known ecosystems that might turn out to be unrecognised hot spots. A quick collection, analysis and dissemination of information is expected out of the RAP. The second level is the **BIOTROP** approach. This approach envisages the establishment of research stations in hot spots and conduction of detailed inventories. The third level, which to be completed within a time frame of 30 years, would combine the inventories from RAP and BIOTROP. Conservation International (CI) created RAP in 1989 and it works by assembling teams of expert tropical biologists and host country scientists.

The other important task that needs immediate attention is the creation of a synoptic or global master database for the presently known species in all groups, including synonymy (Bisby 1994). From this master database, the currently accepted name for each species, authorities and synonyms could be obtained and ascertained. The ILDIS database for Leguminosae is an example (Zarucchi *et al.* 1993a, b) (for more details, see Chapter 9).

4

AGROBIODIVERSITY AND CULTIVATED TAXA

Introduction

A strong reason prompted a separate chapter on the diversity of cultivated taxa. Cultivated taxa have influenced human life and civilisation more than wild species. Diversity of cultivated taxa is distinct from diversity of wild taxa and is now increasingly termed **agrobiodiversity**, to distinguish it from general biodiversity (Virchow 1998). Agrobiodiversity can be broadly defined as 'that part of biodiversity which nurtures people and which is nurtured by people' (FAO 1995; Virchow 1998). Although there are similarities between biodiversity and agrobiodiversity, the differences are greater (Table 4.1). A very important International Conference for the specific issue of agrobiodiversity was held in Leipzig in 1996, where the first report on the state of the World's Plant Genetic Resources for Food and Agriculture (PGRFA) was presented and the first global plan of action for the conservation of crop genetic resources was adopted by 150 countries (FAO 1996b).

Origin and Evolution of Cultivated Species Diversity

Introduction

Domestication or **cultivation** can be defined as the 'forceful' inclusion of wild populations of useful species, wholly or in part, into human society. Harris and Hillman (1989) defined domestication as 'human intervention in the reproductive system of the plant', which resulted in genetic and/or phenotypic changes. Cultivation grew out of food gathering by our ancestors, which imperceptibly led to elements of domestication (Harris 1969). Yen (1985) considers this transition from wild to cultivation as a 'form of intensification of plant gathering', involving a slow/gradual 'domestication of the environment'. Therefore, all cultivated plants were evolutionarily derived, directly or indirectly, from wild species through human-imposed unwitting/deliberate selections for desirable traits. The ancestors for many cultivated taxa have been identified, although those of some important crops, such as maize and common wheat, are still subject to research and argument.

In some cultivated taxa, the plants underwent more or less drastic morphological, physiological and other changes during the act of domestication, thus creating vast differences between the wild ancestors and the derived domesticates. In other words, they underwent changes related to the **Domestication syndrome** (Box 4.1). In yet others, the demarcating line between wild and cultivated taxa is extremely thin and diffuse. This is especially true of many forest and pasture

Table 4.1 Comparison of biodiversity and agrobiodiversity (from Virchow 1998)

Criteria	*Biodiversity*	*Agrobiodiversity*
Definition of diversity	Species diversity as variation of diversity between species	Genetic diversity as variation of diversity in one species
Common quantification of diversity	Number of species per defined area	Number of varieties of a given crop per defined area
Centres of diversity	Majority in tropical humid (forest) areas	Majority in tropical semi arid and arid areas
Driving force of diversity	Evolutionary process	Breeding process
Expansion	Unsystematically by humankind	Systematically by humankind
Extinction characterised by	Mainly species	Mainly varieties
Main causes of extinction	Human destruction, fragmentation, modification of habitats	Abandonment of land races
Prioritised conservation methods	In situ	Ex situ
General in situ conservation strategy	Reducing activities of humankind in areas of high diversity (habitat/ protected area approach)	Promoting the use of agrobiodiversity by farmers (on-farm management)
Polluter pays principle	Applicable	Non-applicable
Operational utility value	Pharmaceutically active compounds (genetically coded information)	Desirable traits (genetically coded function)
Supply for	All biotechnology industries including plant breeding industries	Conventional and biotechnology plant breeding industries
Companies involved	Mostly private companies	Public and private companies
Genetic sources for new products	Single component	Various components
Research and development system	Private	Public and private
Transaction chains between collection and ultimate producer:	Short	Long
Revenue-generating potential of products:	High	Low
Redistribution of sales as 'benefit sharing' to the providers of genetic resources:	Effective	Ineffective
Genetic resources exchange system:	Bilateral contracts	Multilateral system
Overlap	Wild relatives of plant genetic resources for food and agriculture	

species and tropical fruit trees such as oil palm. In many cases, there is an evolutionary continuum connecting the wild taxa with the present-day cultivars; there is also an ecological link connecting wild and cultivated taxa through semicultivated ones. The ecological and evolutionary continuum and relationships between wild progenitor and derived cultivar are very important in two respects: (i) Wild species are subjected to continual domestication, especially in a world where land use has intensified and pressure for new industrial raw materials increased. (ii) Wild relatives of domesticates are increasingly gaining importance as vital genetic resources for improvement of the derived cultivars, either for better yield and performance or for combating various stresses, including diseases.

Act of Domestication

Domestication is considered the foundation stone of crop plant diversity. This foundation first gave rise to **land races,** which are crop

Box 4.1 Domestication syndrome in crop plants (Harlan 1975a; Hawkes 1983; Schwanitz 1966; Simmonds 1979) [as cited by UNEP 1995]

Character change	*Significance*
Gigantism	Affects the part utilised—seed, fruit, root, stem, tuber—an almost universal characteristic of domesticated crop plants.
Suppression of dispersal mechanism	Results in retention of seed in the fruit or inflorescence; reduction of stolon length in potatoes, resulting in concentration of harvestable product and facilitation of collection. Another virtually universal character.
Suppression of sexual reproduction	Crops reproduced vegetatively by tubers, e.g. potatoes, show this. In the special case of banana, its culture as an edible fruit depends on this suppression as it serves to concentrate assimilates in the production of the harvested product.
Changed growth form	May be a consequence of gigantism—larger fruit structural support, allometric growth changes may result. Plant growth habits become more restrained and less rampant, facilitating crop husbandry.
Changed lifeform	Short-lived perennials (e.g. *Phaseolus* beans) may become biennial or annual, often in response to selection for higher yield and earlier maturity. The energy required for perennation can be diverted to production of biomass usable by humans.
Changed breeding system	Self-pollination has advantages in reducing weather-dependence for pollination, promoting yield stability in areas with unpredictable weather conditions.
Loss of seed dormancy	Promotes predictability in production of good crop stands and highly advantageous in cultivation. Short-term dormancy is useful in moist climates, inhibiting sprouting in the ear of cereals and seeds in pods of legumes.
Biochemical changes	Commonly involve loss of toxic or distasteful compounds, glucosinolates in brassicas, cucurbitacin in cucurbits, cynogenic glycosides in lima beans and cassava, lectins and protease inhibitors in some legume crops.
Changed ploidy level	May be auto-or allopolyploidy. Autopolyploid grassses (e.g. *Lolium* spp.) are cultivated. Allopolyploidy, more significant, has resulted in production of essentially new species such as the bread wheats and many soft fruit novelties (*Rubus* spp.).
Physiological changes	Photoperiod requirements can limit extension of the range of crops, which originate in low latitudes at high elevations. A change to day neutrality can enable this to occur as in *Phaseolus vulgaris*, the common bean.

varieties of peasant farming; they further diversified into modern **cultivars**[1] developed in the last one hundred years or so. In certain cultivated taxa, domestication took place very quickly, while in others it extended over several centuries. Much of what happened during domestication in certain instances is still obscure, although for many crops some elements of the domestication process are known. These elements concern (i) the wild progenitor, (ii) the site or region of initiation and intensification of domestication, (iii) the timing, and (iv) whether the domestication process was a single or repetitive event. For some domesticated species the origin will forever remain obscure (Frankel *et al.* 1995).

Harris and Hillman (1989) suggested three distinct, but not mutually exclusive, pathways of domestication:

(i) Very rapid genotypic changes took place within a very short span of 20 to 30 years in the progenitor, which eroded its ability to survive in the wild. This process has been shown in the domestication of cereals such as eincorn and emmer wheats and barley, and grain legumes such as pea, lentil and chickpea. Cultivated taxa belonging to this category are invariably inbreeders.

(ii) Very gradual genotypic changes took place, eventually rendering the progenitor incapable of survival in the wild. Wilkes (1989) described this process as 'incremental ennoblement'. Maize is a good example. Crops belonging to this category are invariably outbreeders.

(iii) Some 'plastic' phenotypic changes took place without immediate genotypic changes in the progenitor, leading to its domestication. In other words, domestication was conditioned by 'attuning the responses of the plants to specific environmental variables'. In many instances selective clonal domestication occurred in otherwise sexually propagating taxa. Examples of this category are banana, breadfruit, date palm, figs, grapes, pomegranate and olive as well as many root and tuber crops, e.g. yam. Cultivated taxa of this category are invariably vegetatively propagated.

Frankel *et al.* (1995) added a fourth category to domesticated plants.

(iv) Those cultivated species which do not fit into any of the aforesaid categories, mainly for want of information about them. Many vegetable and fruit species can be included here (e.g. winged bean, bitter gourd, mangosteen) which were maintained in peasant cultivation in the tropics for a long time. The wild progenitor of some is either not known or has disappeared.

The act of domestication may be a single or repetitive event in time and/or space. The type of domestication, according to this classification, depends on the breeding system of the taxon. Most seed-propagated species are considered to have had only one domestication event (Zohary 1989). Repetitive domestication was believed to have happened in cotton, *Capsicum*, *Chenopodium* (Pickersgill 1989) and

[1]Cultivar is the commonly accepted name for the individual precise race of a domesticated species; it was coined from the expression 'cultivated variety'. A cultivated variety may be an older variety, old cultivar or land race improved by breeding activities. Most cultivated species have many cultivars and the term used for their grouping is 'cultivar groups', although sometimes 'convar', 'provar' or 'cv' are substituted. Two sets of regulations govern a cultivar. The first is the commercialisation of a cultivar through registration in an official catalogue (governed by Public Law), while the other protects the rights of the breeders (governed by private law). In both instances the cultivars must follow the technical requirements of Distinctiveness, Homogeneity and Stability (DHS) or Distinctiveness, Uniformity and Stability (DUS).

many vegetatively propagated crops. Information about the type of domestication has both historical and evolutionary significance, especially in limiting or expanding the genetic base of the cultivars.

Two important genetic events are considered to be of great significance not only in initial domestication, but also in subsequent evolution of the crops. They are major gene mutations (as in cereals) in single or polygenes and polyploidy (as in wheat). The process of domestication has led not only to the evolution of new variations but in many taxa, to a proliferation of new forms far beyond the variation observed in the original ancestor, especially due to these genetic events. A typical example are the botanical varieties now available in *Brassica oleracea*, such as kale, cabbage, cauliflower, broccoli, kohlrabi and Brusselssprout.

Geography of Domestication

Vavilov (1926, 1949-50), an outstanding plant geographer and a renowned geneticist of the former Soviet Union, undertook a detailed exploration of the crop taxa in many less-developed and largely montane terrains of the world, considered by him areas of ancient agricultural civilisation. He believed that in these areas the indigenous crop varieties would not yet have given rise to cultivars selected by plant breeders. Based on detailed study of the geographic distribution of genetic diversity in various crop species, he identified many areas as **centres of genetic diversity**; he also believed that the centres of great genetic diversity are likewise the **centres of origin** of cultivated plants. He identified these centres on the bases of varietal diversity, homologous variation, endemism, dominant allele frequencies and disease resistance and recognised eight centres

Table. 4.2 Vavilov centres of origin of cultivated taxa and the most important cultivated taxa belonging to them

1. Chinese Centre	Millets, Sorghum, Buck wheat, Soybean, Kidney bean, Yam, Radish, Cannabis, Tea (Total cultivated taxa: 136)
2. (a) Indian centre	Rice, Finger Millet, Pigeon Pea, Green gram, Horse gram, Winged bean, Cluster-bean, *Amaranthus* sp., Brinjal, Bitter gourd, Bottle gourd, Snake gourd, Taro, Mango, Orange, Lemon, Myrobalan, Breadfruit, Tamarind, Sugar-cane, Mustard (Total: 117).
(b) Indo-Malayan centre	Coix, Velvet bean, Zinger, Pomelo, Litchi, Banana, Durio, Coconut, Cardamom, Black pepper, Turmeric, Gutta-percha (Total: 55).
3. Inner Asiatic centre	Peas, Lentils, Chick-pea, Flax, Safflower, Melon, Carrot, Onion, Basil, Pistachio, Pear, Grapevine (Total: 42).
4. Asia Minor	Eincorn type wheat, Soft wheat, Secabe, Cereals, Pea, Alfalfa, Vetch, Sesame, Castor, Figs, Poppy, Pumpkin (Total: 83).
5. Mediterranean centre	Artichoke, Emmer, Oats, Barley, Lentils, Horse bean, Chick-pea, Flax, Black mustard, Olive, Beetroot, Cabbage, Onions, Cumin, Fennel, Lavender (Total: 84).
6. Abyssinian centre	*Triticum durum*, Barley, Finger millet, Lentil, Fenugreek, Niger, Safflower, Sesame, Coriander, Coffee, Bhendi, Pearl millet, Sorghum (Total: 38).
7. South Mexican & Central American centre	Maize, Jack-bean, Squash, Chayote, Pumpkin, Sweet potato, Pepper Chilli, Upland cotton, Sisal, Papaya, Guava, Cherry, Tomato, Cocoa, Avocado (Total: 49).
8 (a). South American (Peru, Ecuador, Bolivia) centre	Potato, Lupine, Maize, Tomato, Peanut, Tobacco (Total: 45)
(b) Chilean centre	Strawberry (Total: 4).
(c) Brazil-Paraguay centre	Manihot, Peanut, Hevea rubber, Pineapple, Yam, Cassava. (Total: 13)

of origin of domesticated plants (Fig. 4.1), popularly called **Vavilov centres** or **Germplasm Treasures** later (Khohsoo 1991) (Table 4.2). The eight centres are East Asia (Chinese Centre), Tropical Asia (Indian and Indo-Malayan Centre), South-west Asia (Inner Asiatic Centre), the Near East (Asia Minor Centre), Mediterranean, Abyssinia, Andean, and Central America including Southern Mexico. Further studies established that these eight centres are undisputable centres of great genetic diversity in crop plants, but it was subsequently shown by others that sites of diversity and domestication did not and need not necessarily coincide.

Vavilov's concept underwent modifications due to very interesting further work on these eight centres. Harlan (1971, 1975a) recognised that some Vavilov centres do indeed fit the geographical concept of 'centre' but some do not. He therefore recognised 'centres' and 'non-centres', the latter covering vast areas of very great physical and cultural diversity. Thus he designated 12 areas in the world, including Vavilov's eight centres, as domestication areas (Table 4.3). Harlan's work showed that different species of the same crop were domesticated in different places and that independent domestication of the same species of the crop occurred in different places and at different times. Consequently one single centre of origin cannot be identified for all crops.

The 'centres of origin' concept sometimes gives rise to certain misconceptions because it ignores phenomena such as **Transdomestication** (Hymowitz 1972) and history; it also underestimates the contributions of peasant societies of areas outside the centres.

Dispersal and Diversification

Man has played a very important role in the diversification of gene pools of crop plants. Archaeological evidence from many parts of the world has shown that several cultivated species have had a fairly rapid spread through different parts of the world, mainly through human activity. In all places where they spread, they were subjected to modifications by the prevailing local environmental conditions as well as local cultivation practices. Secondary/tertiary centres of diversities evolved in those places where further genetic differentiation was favoured in the crop plants. Ethiopia, for

Table 4.3 Areas of the world where plants were domesticated along with the most common domesticated taxa of these areas (after Harlan 1976)

1. North America	:	Sunflower, Tepary bean
2. Mesoamerica	:	Maize, Tomato, Cotton*, Avocado, Papaya, Cocoa, Cassava*, Sweet Potato*, Common bean*
3. Lowland South America	:	Yam, Pineapple, Cassava*, Sweet Potato*, Cotton*
4. Highland South America	:	Potato, Peanut, Lima bean, Cotton*, bean*
5. Europe	:	Oats, Sugar-beet, Rye, Cabbage, Grapes*, Olive*
6. Africa	:	Sorghum, Pearl millet, Yam, Watermelon, Cowpea, African rice, Coffee, Cotton?*, Sesame?*
7. Near East	:	Wheat, Barley, Onion, Pea, Lentil, Chick-pea, Fig, Date, Flax, Pear, Pomegranate, Grapes*, Olive*, Apple?
8. Central Asia	:	Common millet, Buckwheat, Alfalfa, Hemp, Foxtail millet*, Grapes*, Broad-bean?
9. India	:	Pigeon-pea, Eggplant, Cucumber, Cotton*, Sesame*
10. China	:	Soybean, Cabbage, Onion, Peach, Foxtail millet*
11. South-east Asia	:	Oriental Rice, Banana, Citrus, Yam, Mango, Sugar-cane, Tea, Taro
12. South Pacific	:	Sugar-cane, Coconut, Breadfruit

*Taxa that were probably domesticated independently in different areas.

Fig. 4.1 The eight centres of origin of crop plants as proposed by N.I. Vavilov in 1926 (after Harlan 1971). The eight centres numbered in this figure correspond to the eight centres mentioned in Table 4.2.

example, became a secondary centre of crop plant diversification for Middle East crops such as barley, emmer wheat, peas, lentil and many others; Latin America formed secondary centres for beans, ground-nut, maize, cassava and rubber; and the Mediterranean for rye and oats.

During the processes of dispersal to and establishment in places other than the centres of origin, crop plant genomes underwent changes due to frequent or sporadic recombination within primary gene pools (Harlan and de Wet 1971) (see page 49) and due to mutations or polyploidy that were effected to confer resistance to parasites/predators or other biotic or abiotic stresses encountered in those places. Introgressive hybridisation from wild and weedy relatives of the secondary gene pools (see page 49) further enriched the existing genome of these crop plants, thus enhancing the scope for selection of superior adaptations. All these led to the evolution of distinct local land races or primitive cultivars, which can be distinguished from **advanced cultivars** produced from the former through scientific plant breeding programmes in the last one hundred years. As already stated, land races evolved through a possible combination of natural and deliberate selections under traditional conditions characterised by low-input cultivation. Advanced cultivars are the result of sophisticated breeding programmes for high performance/yield under intensive and high inputs that tend to level out environmental hurdles, if any; environmental levelling is invariably accompanied by genetic levelling which reduces the need for special local adaptations.

Diversity in Domesticated Species

The overall result of domestication was as follows: As the number of cultivated species increased, the number of wild species used for food (Flannery 1969) and other purposes decreased. Of the 511 plant families currently recognised (Brummitt 1992), only 173 have cultivated representatives. Of these 173 families, Poaceae has the largest number of domesticated species—380 (15.2% of all domesticated taxa), Leguminosae (*sensu lato*) follows with about 340 species (13.6%), Rosaceae ranks third with about 158 species (6.32%), followed by Solanaceae (155 species, 4.6%) Asteraceae (86 species, 3.44%), Cucurbitaceae (53 species, 2.12%), Lamiaceae (52 species, 2.08%), Rutaceae (44 species, 1.76%), Brassicaceae (43 species, 1.72%), Apiaceae (41 species, 1.64%), Chenopodiaceae (34 species, 1.36%), Zingiberaceae (31 species, 1.24%) and Arecaceae (30 species, 1.2 %). Many families, numbering about 50, have only one domesticated species. Although the majority of domesticated taxa are used as sources of food, some were selected for domestication for their fibres (cotton, hemp, flax, etc.). Among the 25,000 plants believed to be useful/used in various systems of medicine, only a very few have been domesticated [e.g. *Cephaelis ipecacuanta* of Rubiaceae, used as a source of Ipecac to treat amoebic dysentery, *Cinchona officinalis*, used as a source of quinine for Malarial fever, and *Valeriana mexicana*, used as a sedative (Oldfield 1992)].

The number of cultivated taxa increased as a result of diversification of human needs, only to be drastically reduced in the last one hundred years. Of the 400,000 plant species estimated and 300,000 species documented, 4000 species are edible. According to Mangelsdorf (1966), of these early man used at least 3000 plant species as food, but resorted to active cultivation of only 150-200 of them (see Boyle and Lenne 1997) to the extent that they formed items of world commerce. This was followed by a tendency to restrict cultivation to only the most efficient among them. Today the world is fed by about 15-20 plant species only (Boyle and Lenne 1997). Even among these, only 4 are major crop species (rice, wheat, maize and potato), providing for more than 50% of the food requirements of people. Thus the levels of genetic diversity of cultivated crop species have been subjected to profound changes at the species level from the time of their initial selection (i.e., in the preagricultural hunter-gatherer stage) up to about 100 years ago.

Although the total number of species of domesticated crops was reduced, efforts to increase the infraspecific diversity within the selected species were well underway i.e., variation within selected species increased immensely. For example, there are an estimated 130,000 distinct varieties within the rice species *Oryza sativa* (Chang 1995). In the last two to three decades, due to modern scientific breeding, there has been a very drastic reduction in use of infraspecific variations as a result of human preferences, cost of production, yield/performance potential, etc. There has also been a drastic reduction in area cultivated for local varieties, resulting in very few elite varieties in each crop possessing greater genetic vulnerability, defined as the genetic constitution of a crop. Such varieties are very uniform and homogeneous and replace the genetically diverse, albeit genetically less vulnerable varieties. This has substantially eroded the genetic diversity of crop species. Certain specific examples will illustrate this point. Philippino farmers were using several hundreds of rice varieties earlier, but in recent years two varieties alone account for 90% of the area planted (Friis-Hansen 1994; NRC 1993). In Argentina local varieties of *Amaranthus* have been almost totally replaced by modern varieties; in China, of the 10,000 wheat varieties used in 1949, only 1000 were in use by 1975; and in Sri Lanka, of the 2000 rice varieties in 1959, only five are presently cultivated (see Virchow 1998). In the USA, of the 7098 apple varieties documented in the US Department of Agriculture as having been in use up to 1904, approximately 86% have been lost. Similar losses have been recorded in cabbage varieties (95%), pea varieties (94%) and tomato varieties (81%) (Fowler 1994). The extinction rates of vegetable varieties of asparagus, beets, onions and others in the USA between 1903 and 1983 ranged between 87 and 98% (Fowler and Mooney 1990).

The present-day plant diversity of each crop thus ranges at the evolutionary level from wild ancestors (as in oil palm) to very advanced cultivars (as in wheat); at the ecological level, from components of a 'primeval ecosystem to those of high-input agriculture and horticulture'; and at the genetic level, 'from populations to genes' (Frankel *et al.* 1995). A functional classification of crop plant germplasms (or genetic resources) into the following categories was introduced by the International Biological Programme (IBP 1966; see also Frankel and Bennett 1970): Land races, Advanced cultivars, Wild relatives of cultivated plants and Wild species used by man. To these Frankel *et al.* (1995) have added Genetic stocks and Cloned Genes.

Land Races

Land races, as already defined, are crop varieties of peasant farming initially derived due to the domestication process, which constituted the base for further diversification into modern cultivars. The other attributes of land races are summarised by Harlan (1975b) as follows: (i) Land races have a 'certain genetic integrity' although they are genetically diverse and dynamic; (ii) They can be very well recognised morphologically; (iii) peasants have specific names for them; (iv) they are highly adapted to specific soil types and local environmental variables; (v) they have a specific time of seeding, date of maturity, nutritive properties and uses; (vi) they are balanced populations and not strictly varieties and are in equilibrium with prevailing local environmental conditions and pathogens/predators; and (vii) 'the genetic diversity of land races has two dimensions: among sites and populations and within sites and populations'. To this may be added the following: (viii) they are not registered in Official Seed Variety Lists (Virchow 1998).

The evolution of land races was subject to selection pressure for hardiness and dependability but not necessarily for performance and yield. The peasant was therefore happy if his land race could resist diseases and could grow under low-input adverse conditions, even if its yield was not very good. It is interesting to note that under very adverse conditions, however, land races of many crop species outyield the best of the modern cultivars of the same species. This has been very clearly demonstrated for local land races of wheat, which outyield the modern wheat cultivars in Madhya Pradesh of India (Rahul and Jacob Nellithanam 1998).

Land races of many major crops are now mainly used as donors of genes in order to enhance in some way the biological and/or economic adaptations of those crops; they are also used to improve the quality and quantity of yield of advanced cultivars. Under unfavourable marginal environments, land races have remained the predominant cultivars, making a substantial contribution to the food self-sufficiency of tribal/rural populations, especially in the remote parts of the third world. It is therefore argued that traditional farming with land races is 'not just a profession but a way of life that is dominated by the conservationist ethic as opposed to the consumerist culture that inspired the Green Revolution' (Rahul and Jacob Nellithanam 1998).

Land races should not be regarded merely as a source of superior genes or as important in subsistence agriculture, but also as sources of information on the origin, evolution and genetic differentiation of crop species. Very little, however, is known about the population structure of land races—within as well as between populations.

In many agricultural ecosystems, especially in those of developed countries, land races have almost disappeared. This has been due to the modification of traditional farming systems through the introduction of modern crop varieties and/or high-input technologies (Mooney 1988). In contrast, many rural/tribal communities still largely depend on land races of traditional crops. Land races of such crops as maize, potato, pepper, beans, tomato and squash are still prevalent, for example, on the hill sides of Mesoamerica, the Andes, and the low lands of South America (Ford-Lloyd and Jackson 1986). The following factors, although listed for Latin American staple crops, account for the maintenance of land race diversity in other parts of the world as well (Altieri and Anderson 1992): (i) Continuing cultivation of land races by traditional methods. (ii) Farming systems small scale. (iii) Environmental diversity available within fields. (iv) Local adaptation to environmental and biological stresses present. (v) Deliberate seed selection and maintenance by peasants. (vi) Geographic fragmentation that creates isolating mechanisms conducive to rapid differentiation. (vii) Seed networks and exchange among peasants within and between villages and cultural groups. (viii) Ethnic diversity leading to various classifications, uses and management of distinct crop varieties. (ix) Tolerance of weedy relatives within and around fields promoting hybridisation. (x) Ecological exchanges between the vegetation in a peasant's field and the wild vegetation surrounding the field.

Advanced Cultivars

Advanced cultivars or **Breeders' Lines** (= the **plus plants**) are recently evolved and invariably have a homogeneous population derived from careful individual selection and progeny testing, in preference to mass selection practised earlier. Once produced, these advanced cultivars form the genetic base for further improvement of the crop by the breeder who produced them as well as by others who use them (also see footnote in page 42). A survey of 100 American plant breeders showed that the majority used elite and advanced cultivars to expand the crop plant germplasms they were using (Duvick 1984). In crops other than wheat, they were also using land races to broaden their germplasm. In cotton and wheat, related wild species were also used for crop improvement.

Wild Relatives of Cultivated Plants

Wild relatives (also known as weedy crop relatives) and species contribute substantially to expansion of the genetic base of cultivated taxa; hence they are invariably used to breed and improve the latter (Harlan 1976; Prescott-Allen and Prescott-Allen 1988; Stalker 1980). They contribute in the following two ways: (i) Wild relatives often provide genes that are not available in domesticated plants. These are invariably genes affording resistance to diseases/pests and other environmental stresses. Such resistance genes have been acquired by the wild relatives through their long periods of coevolution with microbes/pests as well as survival for a long time under stressed environments of various sorts. A good example is the transfer of tobacco blue mould resistance gene from the wild *Nicotiana debreyi* and *N. goodspeedii* growing in Australia, to the cultivated American species, *N. tabacum*. Many of these resistance genes are major genes and control specific resistances, although minor genes for non-specific resistances may also be transferred from wild to cultivated taxa. It must be remembered, however, that resistance conferred by major and minor genes of wild relatives on transfer to cultivated taxa is finite and that ever-evolving relatives are continuously required for many future needs to improve the cultivated taxa. (ii) Wild relatives contribute a number of economic traits to the cultivated species. Good examples are the transfer of cytoplasmic male sterility from the wild *Helianthus petiolaris* to cultivated sunflower, and many fruit quality parameters from the wild *Lycopersicon cheesmanni* into cultivated tomato (Rick 1982).

It is evident from the foregoing that there is an urgent need to collect, study and conserve wild relatives of cultivated plants as they form part of the **primary, secondary** and **tertiary gene pools** (Harlan and de Wet 1971; see also Hawkes 1987). These gene pools are also known as **Genetic Resources Profiles** of crop species (Smartt 1990) (see Table 4.4). The Genetic Resources Profiles are ranked in terms of the accessibility of their germplasm through conventional hybridisation methods or through biotechnological intervention. Primary gene pools (GP1) represent the true biological species including all its cultivated (cultigen), wild and weedy forms; hybrids among these forms are fertile and gene transfer to the crop is simple,

Table 4.4 Genetic resource profiles of crop species (after Smartt 1990)

Harlan and de Wet Category	*Constituents*	*Order of recourse and accessibility to breeders*
GP1-primary gene pool	(a) Cultigen	1st order
	(b) Weedy form	2nd order
	(c) Wild prototype	
GP2-secondary gene pool	Cross-compatible species producing ± fertile hybrids	3rd order
GP3-tertiary gene pool	(a) Cross-compatible species producing viable but sterile hybrids	4th order
	(b) Cross-compatible species producing non-viable hybrids	5th order
Quaternary gene pool	Incompatible related species	6th order

direct and poses no problem. Most of the primary gene pools show at least 80% genetic closeness to the crop species. The secondary gene pools (GP2) represent the group of species that can be artificially hybridised with the crop but gene transfer may not be easy. The hybrids, if produced, are usually weak or partially sterile. The secondary gene pools show around 60% genetic closeness to the crop species. The tertiary gene pools (GP3) include all species that can be crossed to the crop species but with some difficulty. The hybrid zygote/embryo needs embryo rescue. Gene transfer is almost impossible or requires special techniques. These gene pools show around 40% closeness to the cultivated species. There is also a **quaternary gene pool**; its constituents are incompatible with related species. Gene transfer requires biotechnological intervention. All categories of gene pools may or may not exist in all crop species. In some legumes such as *Glycine max*, there is no third order (GP2) gene pool while in *Vicia faba*, there is neither a second nor third order gene pool.

Wild Plants

These represent taxa that live in the wild but are still used by man for various purposes. Many constitute new life-support and underutilised crops, often exploited by tribals and rural communities. Life-support taxa are plants that help to sustain man in stress-prone areas and under emergency situations (Paroda *et al.* 1988). Underutilized plants are predominantly tropical and subtropical and are restricted to specific environments. They include a wide array of species which are actually used by tribals/locals or have potential use, and range from vegetables and fruits to medicinal plants (NASC 1975). Many of them, though now used in a restricted environment, have the potential for expansion beyond their present area of occurrence and form good and attractive alternatives to the presently used taxa. The International Centre for Underutilised Crops (ICUC), an independent and autonomous body, seeks to promote the cultivation of underutilised plants in larger areas, as well as to identify the priority species at regional/national levels. This organization is also striving to foster the genetic and agronomic improvement of certain species, establish databases and organise training activities to popularise their cultivation.

Feral Plants

Feral plants can be defined as domesticated forms that have escaped and maintained themselves in the wild without human intervention. In southern India, the most notorious feral taxon is *Lantana camara*. This was first introduced as an ornamental domesticate in hill stations in the early years of the 19th century but has now established itself as a wild plant. Feral taxa are potential genetic resources and hence should not be exterminated without genetically characterising them. Where they are positively harmful to local taxa, feral taxa can be controlled.

Domesticated Microbes

It is important to give here at least a brief account of the microbes that have been domesticated and their diversity, in particular the microbial variants or strains employed either in producing important/novel products, or for specific processes of value to human society. Microbial strains have been developed for production of antibiotics, vaccines, vitamins, food, and fermented products such as alcohol and organic acids. Fermentation of milk and milk products (butter, cheese, yoghurt, buttermilk) as well as sugary substrates to produce alcohol and organic acids has been known since 5000 B.C. (Nisbet and Fox 1991). Over the years humans have evolved several specialised strains of microbial taxa, e.g. *Lactobacillus, Lactococcus, Leuconostoc, Streptococcus*, Yeasts (*Kluyveromyces, Candida, Saccharomyces,*

Schizosaccharomyces etc), *Penicillium, Rhizopus, Aspergillus, Clostridium, Acremonium chrysogenum, Bacillus thuringiensis* and members of Enterobacteriaceae for purposes ranging from fermentation to antibiotic production. Interesting additions to this list are the domesticated mushrooms, e.g. *Agaricus bisporus* and species of *Lentinus, Pleurotus, Volvariella, Auricularia, Flammulina* and *Tremella*. Mushroom domestication started in the early 1700s. The tendency for feral return to the wild has been retained by many of the domesticated microbial strains.

5

ECOSYSTEM DIVERSITY

Introduction

Assessment of biodiversity at the ecosystem level remains highly problematic and very difficult. This difficulty is compounded by the inclusion of abiotic components in ecosystem concepts. The two components—abiotic and biotic—exhibit very considerable dynamism individually as well as in their interactions; it is this dynamism that makes assessment of ecosystem diversity very problematical (Loreau *et al.* 1995; Primack 1992). There is also a lack of unique definition and classification of ecosystems at the global level and often at regional levels too. The earth possesses an enormous range of terrestrial and aquatic systems and to classify this variation into a meaningful and measurable system is not an easy task. In addition, several ecological terms are used interchangeably in ecosystem classifications, often unnecessarily complicating the issue. Words such as **community, habitat, niche, ecosystem, biome** etc. are used in a rather loose, arbitrary manner. Let us take a term, say community, on which classification is based versus one at the other extreme, a classification based on the physical characteristics and appearance of the area irrespective of species composition (such as forests, grasslands, deserts, wetlands etc,) (see Groombridge 1992). Both approaches pose difficulties. Consequently global ecosystem classification has become 'highly subjective like the classification of plants themselves'. Another stumbling block is that ecosystems are essentially dimensionless and lack boundaries.

The foregoing difficulties notwithstanding, the main advantage of assessing biodiversity at the ecosystem level is that it is much easier to record and to monitor changes and trends and the effects of human activities on ecosystems than on individual populations of species (Heywood 1997b). In the ecosystem approach, basically biodiversity is considered within areas (alpha diversity) and between areas. But, our knowledge of both within-area and between-area ecosystem diversities is highly inadequate, as already appraised in Chapter 3. Alpha diversity indicates the total number of species in a given area; it is the single central measure of ecosystem diversity. Some persons measure alpha diversity by assessing the extent of endemism. Assessment of between-area ecosystem diversity is faced with many practical problems, including the fact that in defining ecological units there is nothing equivalent to the species as a biotic concept.

Classification of Ecosystems

Four of the many global ecosystem classifications are presented here, although for reasons stated in the introduction, none is fully satisfactory.

Udvardy (1975) proposed the first of these four systems,which has the approval of IUCN and UNESCO. It recognizes 14 types of biogeographical biomes in the world. These biomes are further subdivided into 193 provinces. The 14 biomes are Tropical humid forests, Subtropical/Temperate rain forests/Woodlands, Temperate Needle forests/Woodlands, Tropical dry forests/Woodlands, Temperate Broadleaf forests, Evergreen Sclerophyllous forests, Warm deserts, Semideserts, Tropical grasslands/Savannas, Temperate grasslands, Mixed Island systems, Tundra Communities, Mixed Mountain systems, and Cold winter desert and Lake systems. Various countries in the world possess one or more of these fourteen biomes. Udvardy's system covers only terrestrial vegetation and does not take into account the marine ecosystems and the interface between land and sea and fresh water and sea; such interfaces are also very rich in biodiversity.

The second classification is the one proposed by Bailey (1989a,b). This system uses three hierarchical levels for characterising ecosystems: Domains, Divisions and Provinces. The four domains recognised are Polar, Humid Temperate, Dry, and Humid Tropical. The Polar domain has 6 divisions (icecap, icecap mountains, tundra, tundra mountains, subarctic and subarctic mountains). The Humid Temperate domain has 12 divisions (warm continental, warm continental mountains, hot continental, hot continental mountains, subtropical, subtropical mountains, marine, marine mountains, prairie, prairie mountains, Mediterranean and Mediterranean mountains). The Dry domain has eight divisions (tropical/subtropical steppe, tropical/subtropical steppe mountains, tropical/subtropical desert, tropical/subtropical desert mountains, temperate steppe, temperate steppe mountains, temperate desert and temperate desert mountains). The Humid Tropical domain has four divisions (savanna, savanna mountains, rain forest and rain forest mountains). All these 30 divisions under four domains are further subdivided into provinces. The number of these provinces depends on the division and the geographic location of the division. The total number of provinces is 96. Various countries have one or more of the four domains. The polar domain covers a world area of 38,038,000 km^2 (26% world area), Humid Temperate domain 22,455,000 km^2 (15.35%), Dry domain 46,806,000 km^2 (32%) and Humid Tropical domain 38,973,000 km^2 (26.64%).

The third classification was proposed by Olsen *et al.* (1983). This system contains eight broad groups subdivided into 46 ecosystems distributed across the world and resolved at the level of a 0.5° × 0.5° grid. One grid cell in the Equatorial region includes 3000 km^2. The per cent coverage of these eight groups for the entire world and the various continents is given in Table 5.1. The major feature of this system is

Table 5.1 Estimates of major vegetation cover percentage as per Olsen *et al.* (from Olsen, Watts and Allison 1983) Numbers indicate percentage of area occupied by the respective ecosystem by the respective ecosystem type

Countries	*I*	*II*	*III*	*IV*	*V*	*VI*	*VII*	*VIII*
Asia	4	1	16	9	24	17	10	18
Former USSR	3	2	5	26	10	8	21	26
Europe	6	0	—	9	4	35	22	23
North & Central America	5	2	3	33	9	10	17	21
South America	2	3	5	2	32	8	14	33
Oceania	4	1	18	0	18	5	38	16
Africa	2	2	30	0	28	7	14	17
Entire World	4	2	13	12	20	11	17	22
India	3	0	2	2	12	44	23	14

I. Other coastal aquatic vegetation II. Major wetlands III. Desert & Semidesert IV. Polar and Alpine V. Grasslands & shrub VI. Cropland & settlements VII. Interrupted wood VIII. Major forests

that it attempts to deal with transitional and disturbed systems as well.

The fourth classification of ecosystems is the Holdridge **Life Zone classification** (Holdridge 1967). It is based on the chief environmental variables such as temperature, rainfall and evapotranspiration, and their effect on vegetation. Holdridge recognised seven life zones and 25 ecosystem types within these seven zones (Fig. 5.1). This system is so designed as to predict disturbed and undisturbed ecosystems. It is also now used in models of global change.

Measuring Ecosystem Diversity

There is no authoritative index for measuring ecosystem diversity. As mentioned earlier, it is often evaluated through the estimation of component species richness and species abundance. If the component species of an ecosystem are more equally abundant, then the ecosystem is considered more diverse. Often weight is given to the numbers of species in different class sizes (e.g. herbs, shrubs, trees in a forest ecosystem, or planktons, multicellular filamentous plants and macrophytes in an aquatic system), at different trophic levels or in different taxonomic groups. For instance, if an ecosystem has higher hierarchical taxonomic groups even though fewer species, it is considered more diverse than an ecosystem containing fewer taxonomic groups with more numerous species. This is true for some marine ecosystems (see below).

Major Ecosystem Types of the World

As already stated, it is very difficult to design a classification of the world's ecosystems acceptable to all. For the sake of convenience, the following few major ecosystems were selected for discussion of their diversity, without reference to any one of the classification systems proposed above.

Tropical Moist Forests

Tropical moist forests are found between the Tropic of Cancer and the Tropic of Capricorn. In the New World, these forests extend from Mexico south to the east coast of Brazil. In the African continent, such forests are present only along the coast of Guinea, in the Zaire Basin and in the eastern region of Madagascar. In the Asian continent, they are found on the southern slopes of the Eastern Himalayas, in south-west India (Western ghats), Sri Lanka, Malaysia, Indonesia, Philippines, Thailand and New Guinea. In Australia, they are confined to a small narrow strip on its eastern coast.

Tropical moist forests cover only 6 to 7% of the Earth's surface. According to Sommer (1976), the total land area of Earth occupied by this type of forest is 9,350,000 km^2, whereas FAO/UNEP (1981) estimated the total area to be 12,007,990 km^2 (Table 5.2). The continent-wise distribution (excluding Australia) in terms of area occupied is also shown in this Table.

Tropical moist forests are often equated with rain forests. These forests contain a closed community of essentially, but not exclusively, broadleaf, evergreen trees. These trees occur in two or more strata ('Forests piled upon forests'). Ground vegetation is extensive and vines, lianas and epiphytes abundant. A great variety of microbial populations (viruses, bacteria, fungi and microalgae) as well as of Bryophytes and Pteridophytes is characteristic. Tropical moist forests account for more than 50% and possibly as much as 90% of all known plant species. For example, a one-hectare plot in Amazonian Ecuador reportedly contains 473 different tree species (Valencia *et al.* 1994). Furthermore, these forests are also rich in endemics. In fact, in 14 of the originally proposed 18 areas of the Earth with an unusually high degree of plant endemism (hot spots), Tropical moist forests constitute the major vegetation. These forests contain more than 37,000 endemic plant species,

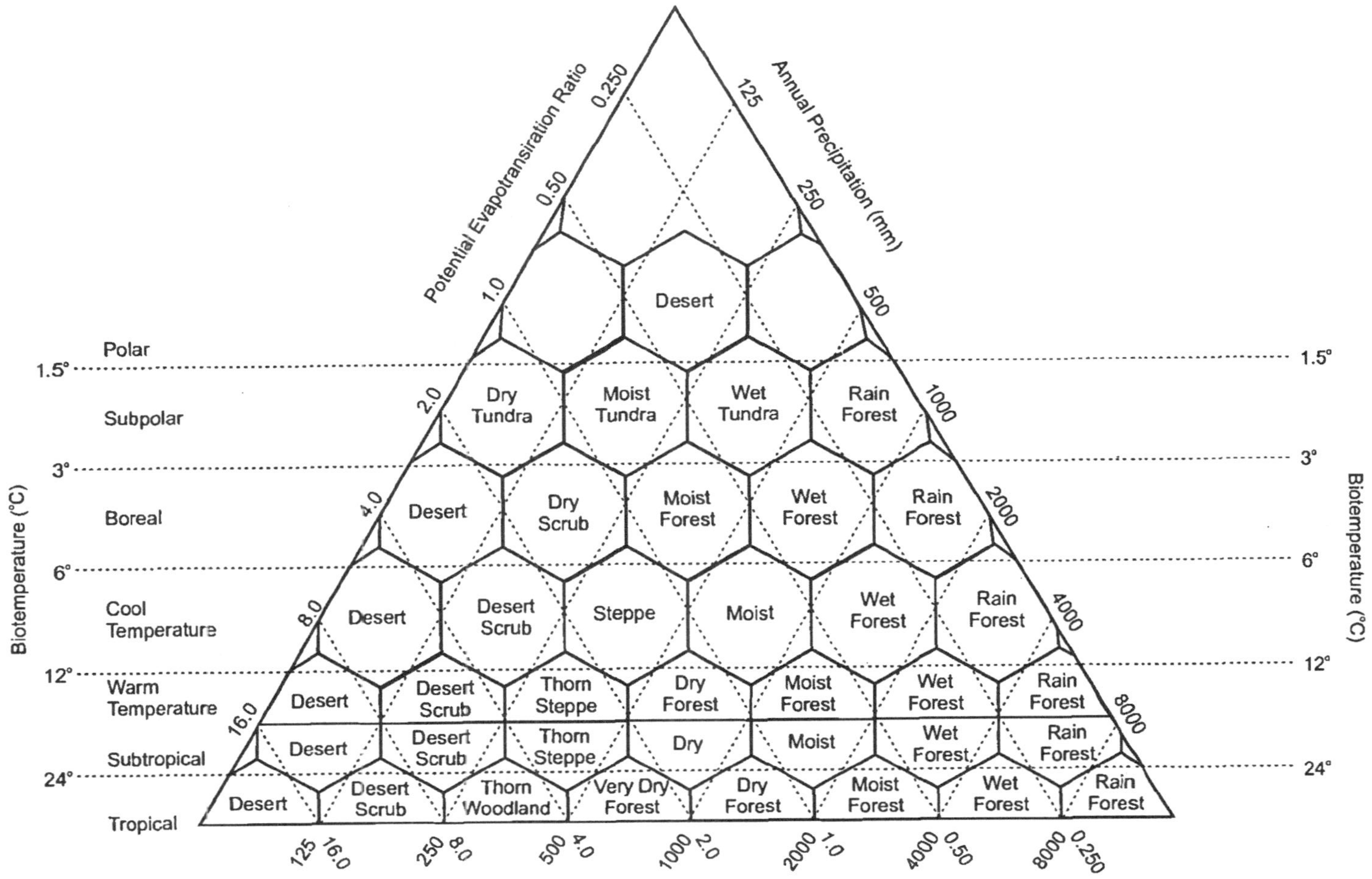

Fig. 5.1 The Life Zone Classification system of Holdridge (adapted from Groombridge 1992).

Table 5.2 Distribution of tropical moist forests

S.No.	Region		Total area (in km²) occupied by forests	Percentage of world extent of this forest type
1.	Africa			
	(Sommer 1976)			
		E. Africa	70,000	
		C. Africa	1,490,000	
		W. Africa	190,000	
		Total	1,750,000	18.72%
	Africa (FAO/UNEP 1981)		2,166,340	18.04%
2.	America			
	(Sommer 1976)			
		Latin America	4,720,000	
		Central America & Caribbean	340,000	
		Total	5,060,000	54.12%
	America (FAO/UNEP 1981)		6,786,550	56.52%
3.	Asia			
	(Sommer 1976)			
		Pacific	360,000	
		S.E.Asia	1,870,000	
		S.Asia	310,000	
		Total	2,540,000	27.16%
	Asia (FAO/UNEP 1981)		3,055,100	25.44%

i.e., 15% of all the plant species, in less than 30,000 km^2 or 0.2% of the Earth's land surface (Myers 1988a, 1990). Life-form classes are also extremely diverse in these forests.

Although known as rain forests, even the wettest among them have a dry season for at least one to a few months in a year and receive less than 100 mm rainfall per month. Only in Malaysia and Indonesia, large parts of these forests are ever wet. Biodiversity of these forests is influenced not only by total average rainfall, but also by local rainfall variations.

Temperate Forests

Temperate forests (or woodlands) occur mainly in the Northern Hemisphere. They occur in Europe, eastern Asia, eastern North America, as a narrow band along the Pacific coasts of North and South America, and on the islands of New Zealand and Tasmania. In the forests of western Europe, eastern USA and Canada, precipitation in the form of snow or rain is evenly distributed throughout the year. But in the forests of eastern Asia, most precipitation is seen as summer rains. In the forests of the Pacific coasts of North and South America, rain occurs during winter months.

Temperate forests are dominated by deciduous hardwood trees and to a lesser extent by evergreen broadleaf hardwood trees and conifers. Approximately more than 1200 species of trees are reported in this type of forest, although eastern Asia is the most diverse with the greatest number of species.

Arid and Semiarid Ecosystems

The biological diversity in these ecosystems is rather poor and is influenced by water availability. Precipitation is extremely unpredictable in time and space as well as total amount. These ecosystems comprise drought-evading, enduring and resisting taxa.

Boreal Forests

These are circumpolar biomes covering approximately 13×10^6 km^2 as upland entities and 2.6×10^6 km^2 in peatland in North America and Eurasia (Apps *et al.* 1993; Olsen *et al.* 1983; Shugart *et al.* 1992). Boreal forests are generally poor in terms of species richness, but the functional diversity of component species is very high.

Arctic and Alpine Systems

These are cold-dominated ecosystems devoid of trees. They occupy about 8% of the terrestrial surface of the Earth with 5% in the arctic region and 3% in the alpine. These ecosystems support only about 4% of the Earth's flora, with 1500 species in the arctic zone and 10,000 in the alpine regions. Species diversity in both regions is concentrated in areas of high vertical relief where a well-developed organic mat of peat is lacking. On the other hand, in the peat-covered lowland areas there are very few species, often less than 10 vascular plant species per m^2 area.

Grasslands

Grasslands often form a natural vegetation on the land surface of the Earth, occupying approximately 25% (33×10^6 km^2) (Shantz 1954). Before human impact, they were believed to have covered 40% of the Earth's surface. However, subsequent estimates range from about 47% (Williams *et al.* 1968), 16 to 18% (Whittaker and Likens 1975) and about 23 to 27% of land surface depending on whether Antarctica is included or not in the calculations. (Atlay *et al.* 1979) (Table 5.3). These form an open type vegetation, often subjected to periods of drought. They also constitute the largest category of generally non-tilled but extensively used land across the world (West 1996; Williams *et al.* 1968). The grassland ecosystem is dominated by grass and grasslike species, although in some areas shrubby and herbaceous elements (often called 'forbs') as well as trees may be present (Risser 1988). In the last instance, there are often fewer than 10-15 trees per hectare of grassland.

The definition of grassland is often muddled, but when simply expressed means: 'Non-tilled land used primarily for livestock grazing'(West 1996) (see also John and Tothill 1985). Grasslands are known by different vernacular names throughout the world: **Steppes** in Eurasia, **Prairies** in North America, **Ilanos, Cerrados** or **Pampas** in South America, **Savannas**[1] in Africa and **Rangelands** in Australia.

Table 5.3 Estimates of the area of the world's grasslands. Numbers without % mark indicate the area in million km^2 (after Hornby 1992)

	Wittaker and Likens (1975)	*Atlay et al. (1979)*	*Olsen et al. (1983)*
Savanna	15.0	22.5	24.6
Temperate grassland	9.0	12.5	6.7
Total grassland	24.0	35.0	31.3
Grassland as % of world land area	16.1%	23.7%	20.7%
Grassland as % of world land area (excluding Antarctica)	17.9%	26.5%	23.1%

[1]Sometimes spelled as Savannaha or Savanas.

Three important factors have been responsible for the evolution and maintenance of grassland ecosystems: drought, fire and grazing by large ungulate herbivores as well as smaller ones.. The last have been known to exert the most important selection pressure in the evolution of grasslands. The maximum grassland plant community is usually attained under moderate grazing (West 1993), especially where wet conditions and a long evolutionary history of herbivory exist.

Grasslands can be classified into three major categories: (i) **Natural grasslands:** These have developed in areas where tree growth has been evolutionarily prevented by edaphic and/ or climatic factors and where browsing by wild herbivores has taken place over a very long period. Natural grasslands are characterized by unsown vegetation whose species composition and turnover are not significantly affected by human activity. (ii) **Seminatural grasslands**: Vegetation is unsown but species composition and turnover are very strongly influenced by human pressures. Invariably these are subjected to grazing by domesticated livestock. Species diversity is intermediate between natural grasslands and artificial grasslands. (iii) **Artificial grasslands**: The vegetation of these grasslands is the result of seeds sown by humans. These grasslands are intensely managed by humans and consequently biodiversity in terms of species richness is extremely poor.

As noted above, the diversity of grasslands, in terms of species number, is highly variable. The most diverse grasslands are seen in northern South America (Menaut 1983) and in South Africa (Fig. 5.2), as these have not been subjected to human impact to the extent that other grasslands have. The Pampas of Argentina and Uruguay have over 400 species of grasses (Cabrera 1970), in spite of their secondary origin. The Cerrados of Brazil include more than 300 plant species but are species poor, although more natural (Table 5.4). The African grasslands, called Savannas, are defined as tropical ecosystems 'forming a continuum of physiognomic types ranging from closed woodlands with a heliophytic grass understorey, through open savannah woodlands to treeless edaphic grasslands' (Huntley and Walker 1982). They develop in soils of low to very low fertility. The grasses invariably show a C4 photosynthetic pathway. The number of species (approximately 1750) is not far below that of rain forests with about 2020 species, contrary to the situation in the Americas (Menaut 1983). In the Somali-Masai region, which forms the heart of African Savanna, the ecosystem contains 2500 plants species, of which about 50% are ecological endemics (Stuart and Adams 1990). The wild animals associated with the savannas are elephant, tiger, lion, giraffe, zebra and the rhino.

The grasslands of Australia, often called rangelands, are defined very broadly (Moore 1970), although Groves (1981) considered the only naturally occurring grasslands in Australia those dominated by hummock grasses, belonging to species of *Triodia* and *Plectrachne* (the hummock). These grasslands are distributed over large areas in south, west and northern Australia. Other than hummock grasses, grass species adapted to desert conditions and fire are common, followed by secondarily adapted grasses and introductions. In general, Australian grasslands are very poor in species richness, which is attributed to the lack of variety in native herbivorous animals.

The natural steppe grasslands of Asia extend from Manchuria westward as far as Bulgaria and Hungary. These grasslands are subjected to very hot and dry summers and very cold winters and lack trees. Local herbivorous animals such as horses, wild sheep, gazelles and a number of small mammals influence the species diversity of Asian steppes.

In Europe, the grasslands known are only extensions of Asiatic steppes; in addition, there are secondary grasslands, many which were

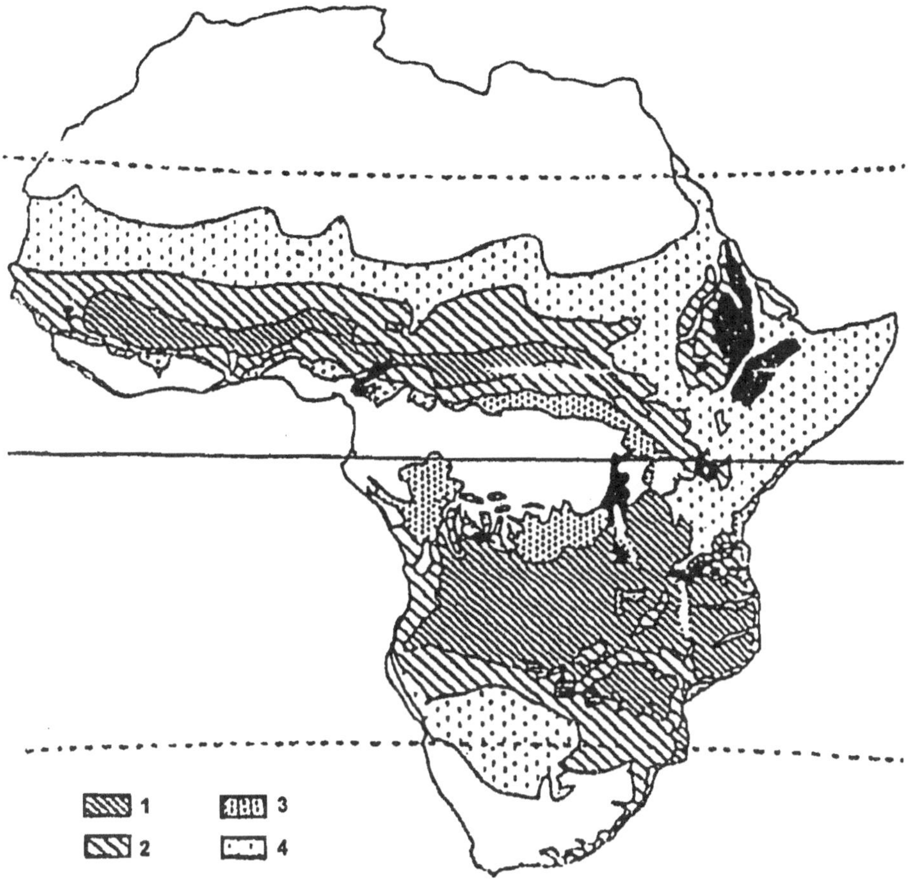

Fig. 5.2 Vegetation types of the African continent with special reference to grasslands (after Menaut 1983). 1. Woodland. 2. Tree/shrub savanna. 3. Forest/savanna mosaic. 4. Tree/shrub 'steppes'

Table 5.4 Floristic richness of various neotropical Savanna formations (from Hornby 1992)

Formation	*Area (km²)*	*No. of trees and shrubs*	*No. of subshrubs half-shrubs, herbs, vines, etc*	*No. of grass species*	*Total no. of species*
Cerrado in north-western Sao Paulo	50	45	175	17	237
Cerrado in western Minas Gorais	15,000	c.200	c.300	73	c.600
Whole cerrado region	2,000,000	429 (774)[1]	181	108	718 (1063)
Rio Branco savannas	40,000	40	87	9	136
Rupununi savannas	12,000	c.50	291	90	431
Northern Suriname savannas	c.3000	15	213	44	272 (445)[2]
Central Venezuelan llanos	3	69 (16)[3]	175	44	288
Venezuelan llanos	250,000	43	312	200	555
Colombian llanos	150,000	44	174	88	306

[1]Total flora including other plant formations; [2]Total flora including bushes; [3]Number of Savanna trees excluding groves.

created by humans. Natural grasslands occur in Bulgaria and Hungary. In fact, the most natural grasslands are those of Hungarian **Pusztas**. Seminatural grasslands occur in Spain and in the Atlantic regions of France and the British Isles.

There are no natural grasslands on oceanic islands. Those that occur in Madagascar and Sri Lanka are man-made or seminatural. However, natural grasslands are found on the islands of New Zealand and Tasmania, occurring in both at high altitudes and receiving very high rainfall. They are quite different from the grasslands that occur in the rest of the world. Cuba is very interesting in that it is the only island lacking native herbivorous animals but nonetheless supporting very high species-rich grasslands. The Cuban savannas are floristically comparable to the African.

Wetland Ecosystems

It is extremely difficult to define a wetland ecosystem as it covers a wide range of inland, coastal and marine habitats. The all-inclusive definition of wetlands is perhaps the one provided by the **Ramsar convention**, according to which wetlands are 'areas of marsh, fen, peatland or water, whether natural or artificial, permanent or temporary, with water that is static or flowing, fresh, brackish or salty, including areas of marine water, the depth of which at low tide does not exceed six metres'. When all the variables mentioned in this definition were considered, more than 30 categories of natural wetlands and 9 man-made ones could be recognised (Dugan 1990). The other problem in wetland characterisation is its highly dynamic nature, often changing with seasons and over longer periods of time, thus imposing problems in defining its boundaries with precision.

Freshwater Wetlands

Only 0.014% of Earth's water occurs in soils and freshwater wetlands (la Riviere 1989). Aselmann and Crutzen (1989) and Mathews and Fung (1987) have made the best estimates of freshwater wetlands. Both estimates were made when the authors tried to assess methane production world over, since only freshwater wetlands produce methane. Their studies have provided very useful information. Aselmann and Crutzen (1989) recorded about 5.3 million km^2 of wetlands and recognised the following categories of freshwater wetlands (Table 5.5).

Bogs: These can be defined as peat-producing wetlands of moist climatic regions. Organic matter accumulates over long periods without decomposing due to lack of /slow activity of microbes. Water and nutrient input are entirely through precipitation. Since peat is abundant, the soil is acidic and nutrient deficient. *Sphagnum* is the predominant taxon in bogs.

Fens: These are wetlands producing peat; in contrast to bogs, fens are influenced by soil nutrients flowing through the system. Fens are more prolific than bogs in terms of supporting vegetation. The dominant members of fens are grasses, sedges and mosses.

Swamps: These are forested wetlands on waterlogged or inundated soils. However, there is no accumulation of peat here. Unlike bogs and fens, tree representatives are dominant in bogs. The very characteristic swamps of India occur in the Western Ghats with species of *Myristica* predominant.

Marshes: These are usually herbaceous 'swamps' or 'mires'. Grasses, sedges, reeds and foxtails usually dominate the vegetation. Marshes may be seasonal or permanent.

Flood-plains: These are wetlands periodically flooded with water occurring alongside rivers or lakes. They vary considerably in vegetation.

Lakes: Lakes are open water bodies varying in depth from a few centimetres to several metres.

Lakes and rivers cover about 2.5×10^6 km^2, i.e., less than 2% of the Earth's surface (Wetzel

Table 5.5 Global freshwater wetland areas* (from Aselmann and Crutzen 1989)

Region	*Bogs*	*Fens*	*Swamps*	*Marshes*	*Floodplains*	*Lakes*	*Total*
USSR (former)	917	531	·25	39	-	-	1512
Europe	54	93	1	4	1	1	154
Near East	-	-	11	-	-	-	11
Far East	-	-	11	-	-	-	11
China	11	-	3	18	-	-	32
South-east Asia	197	-	44	-	-	-	241
Aust/NZ	2	3	1	-	9	-	15
Africa	-	-	85	57	174	39	355
Alaska	?	250-400	?	?	?	?	(325)
Canada	673	531	14	44	-	6	1268
USA**	13	-	80	40	95	-	18
C. America	-	-	15	2	1	-	18
S. America	-	-	851	62	543	68	1524
Total	1867	1483	1130	274	823	114	5691

*In 1000 km^2; **excluding Alaska; Aust. = Australia; NZ = New Zealand

1983). Freshwater lakes contain 1.25 x 10^5 km^2 water (Wetzel 1983).

Marine Ecosystems

Biodiversity of the marine environment in general and of marine sediments in particular is very poorly known. This is true both in descriptive terms of species richness and distribution along latitudinal and depth gradients, and the ecological and evolutionary processes regulating it (Heip 1996). Such ignorance or lack of interest in marine benthos in the coastal system and in pelagic or deep-sea communities of open seas is unwarranted since the marine environment occupies 71% of the Earth's surface (and about 51% of its surface by ocean over 3000 m in depth). It is also an important agent in major global biogeochemical cycles (Heip 1996). True assessment of marine ecosystem diversity is very difficult because of its very complicated biogeography. The sea is also more uninterrupted and internally moving, which pose great problems for monitoring and inventorying its biodiversity. The sea is three-dimensional and more dynamic than is land (Norse 1993). Thus, in principle every marine organism can reach all parts of the marine environment. There is also lack of a broadly accepted marine biogeographic scheme, analogous to that of terrestrial environment. However, 49 large marine ecosystems have been recognised in the world, accounting for 20,000 marine plant species and a number of marine viruses and microbes (May 1994).

Open sea and ocean

All environments on the Earth minus landmasses and the coastal-marine system (see below for details) are included in this ecosystem, comprising over 90% of the Earth's habitable volume. This system is completely devoid of higher plants. The dominant components of biodiversity are planktons, of which the phytoplanktons account for the great productivity of oceans. The diversity of open-sea phytoplanktons is very poorly known. The most conservative estimation of planktons, in terms of species number, is half a million (May 1992a), although some higher estimates place it around 10 million. There are no deep-sea plants, although the number of undescribed species of animals and microbes might be roughly comparable to the number in tropical forests (Grassle and Maciolek 1992).

Coastal-marine system

The coastal-marine ecosystem refers to the marine region extending from the 'upper tidal limits out across the continental shelf, slope and rise' (Brink 1993); it thus includes rocky shores, sandy beaches, kelp forests, subtidal benthos and the water column over the shelf, slope and rise (Fig. 5.3). According to Highet (1992), the coastal-marine system generally encompasses the Exclusive Economic Zones of Nations, and is approximately 200 nautical miles wide with a 440,000-km long continental profile.

The coastal-marine system remains largely neglected, despite its very huge productivity (Ray 1996; Ray and McCormick-Ray 1989). Hayden *et al.* (1984) described 21 types of oceanic and coastal marine realms and 45 coastal provinces. The inclusion of mid-water, deep-water and off-shelf benthic areas could substantially amplify marine biogeographic provinces to at least 300 (Ray *et al.* 1992).

There are perhaps 20,000 marine plant species and a much smaller number of viruses, bacteria, fungi and microalgae. Although the named terrestrial species outnumber those in marine environments by seven to one (May 1994), by any measure other than total described species, marine systems are globally about twice as diverse as land systems.

Mangrove ecosystem

Mangroves are intertidal forested wetlands characteristically located in littoral, sheltered and low-lying tropical and subtropical coast. They dominate river deltas, lagoons and estuarine complexes developed from terrigenous sediments (Thom 1982). Mangroves are also found on islands. They are highly salt-tolerant and various structural and physiological adaptations have been developed for this characteristic in different taxa of mangroves. Although mangroves have a diverse collection of trees and shrubs, there are exclusive species and non-exclusive species. The former are found only in the mangrove habitat while the latter are not restricted to mangroves. According to Saenger *et al.* (1983), about 60 species in 22 genera are exclusive, while 23 species in 16 genera are non-exclusive. According to Tomlinson (1986), 54 species of mangroves are trees. In addition, the mangrove habitat promotes the growth and establishment of a wide variety of other organisms (Table 5.6).

Mangroves do not exhibit a uniform appearance. They may vary from extremely closed forests of 40-50 m high trees in parts of South America and Sundarbans in Bangladesh, to open forests with sparsely distributed trees to stunted shrubs of less than 1 m high (Finlayson and Moses 1991).

The world extent of mangroves is shown in Table 5.7. They cover 240×10^3 km^2 area of the world's coastal line (Lugo *et al.* 1990; Twilley *et al.* 1992). The largest mangrove ecosystems are found in Indonesia, followed by Nigeria, Mexico and Australia in that order.

Agroecosystems

Agroecosystems are those in which man has deliberately selected specific crop plants to replace the natural vegetation. Therefore, it is an artificial ecosystem. The artificiality and biodiversity seen in these systems vary enormously depending on the intensity of human intervention.

Agroecosystems include shifting cultivation, home and kitchen gardens, pastures, mixed farming (all classified under low intense intervention), multiple cropping, horticluture, alley farming (all classified under middle intense intervention), cereal and pulse cropping, orchards and plantations (all classified under high intense intervention). All high intense agroecosystems are monospecific and usually composed of genetically uniform plant stands.

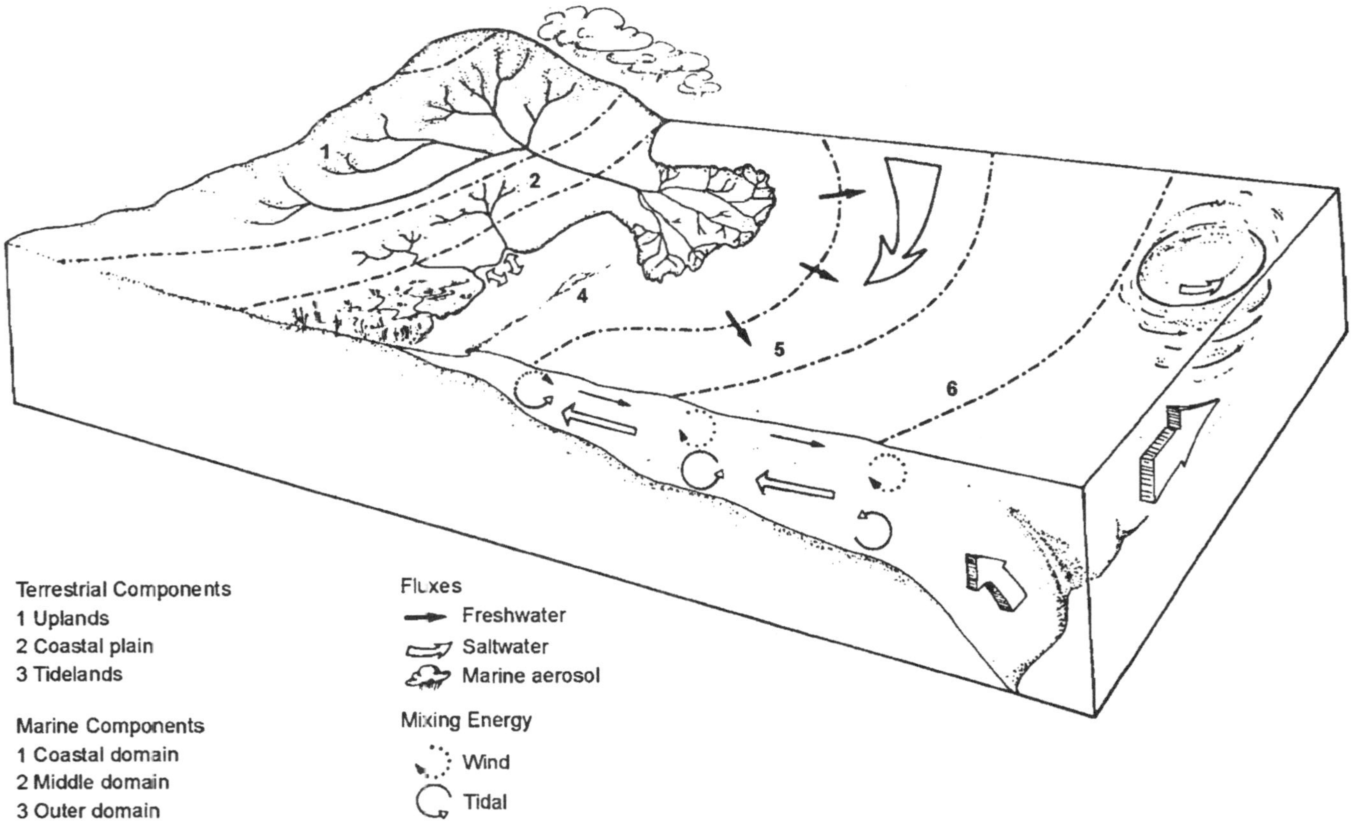

Fig. 5.3 The coastal-marine zone. This zone consists of both terrestrial and marine components linked together functionally by the processes shown here (from Ray and McCormick-Ray 1989).

Table 5.6 Species richness in biological groups associated with mangroves (from Saenger *et al.* 1983). Numbers indicate the number of species recorded in each group.

Taxonomic Group	*Asia*	*Caribbean W. and Atlantic*
Bacteria	10	-
Fungi	25	-
Algae	65	105
HIGHER PLANTS		
Bryophytes/Ferns	35	2
Monocotyledons	73	20
Dicotyledons	110	28
ANIMALS		
Protozoa	18	3
Sponges/Bryozoa	5	36
Coelenterata/Ctenophora	3	42
Non-polychaete worms	13	13
Polychaetes	11	33
Crustaceans	229	87
Insects/Arachnids	500	-
Molluscs	211	124
Echinoderms	1	29
Ascidians	0	30
Fish	283	212
Reptiles	22	3
Amphibians	2	2
Birds	177	138
Mammals	36	5

Table 5.7 Total Area of Mangroves (in hectares)

Countries	*Area (ha)*
1. Asia	12,900,345
2. N.& C. America	4,218,085
3. S. America	1,814,025
4. Africa	5,909,880

Although extremely poor in species diversity (and almost nil in high intense systems), agroecosystems also have weeds, escapes, parasitic microbes/pests and soil organisms.

Urban and Peri-urban diversity

Introduction

It has been estimated that by about 2025 A.D. about 80% of the people of developed countries will live in cities, and that throughout the world only 40% will live in rural areas (UN 1989). Consequently, **urban** and **periurban biodiversity** will come to play a major and increasing role in human activities. With more and more urbanisation, this will be the only kind of biodiversity to occupy a major part in most human life. There are two basic issues related to urban biodiversity. The first relates to the type of biodiversity that could be best linked to cities, while the second relates to the question of the role which cities and towns could play as centres for creation and maintenance of biodiversity (Folch 1996). In the light of these two issues, assessment of urban and periurban biodiversity will be one of the critical considerations in developing future conservation and biodiversity strategies.

Urban biodiversity conservation movement, although started in the mid-1970s (Goode 1993), has only now become widely accepted. The reasons for this are as follows (Heywood 1996): (i) increase in the amount of urban wastelands; (ii) rapid development of environmentalism among people; (iii) recognition and increasing popularity of the discipline of urban ecology; (iv) establishment of several Wildlife/Nature/Environment organisations, clubs and societies; and (v) more responsive attitude on the part of city administration authorities. Urban biodiversity covers both natural and seminatural habitats and artificially created habitats (Heywood 1996). Considerable numbers of wild species of plants, fungi and other microbes are found in natural urban habitats. Seminatural and artificially created biodiversity include the diversity of cultivated species, including ornamentals in parks, gardens, roadsides, houses and other created habitats; this diversity is also very high in many cities. It is a pity that many authors give very scant consideration to this biodiversity; some authors deliberately exclude it on the grounds that it is artificial and does not, therefore, form an integral 'part of biodiversity as such, which should concern itself with ecological integrity' (Angermeier 1994; Heywood 1996). However, several others have argued that although the city is an artifice, it is often constructed in the midst of nature and hence does contain sufficient natural biodiversity elements here and there.

Today, the literature on urban landscapes and diversity of many cities of the world has substantially increased (Adams 1994; Lopez-Moreno 1993; Platt *et al.* 1994; Smith and Hellmund 1993). Several books have been published; many journals deal with issues on urban diversity; hundreds of research and popular articles as well as reports on urban biodiversity have been published. Some cities, e.g. Edinburgh and Moscow, have prepared an Urban Nature Conservation Strategy. In newly organised cities, a considerable area is devoted to 'green space'.

Nature of Urban Biodiversity

In the USA, urban green areas range from 42.2% in Honolulu to 0.3% in Baton Rouge. It has been found that about one-fifth of cities is open space and a recent study in Brussels (Belgium) suggested that up to 50% is actually green and photosynthesising. Bangalore (Karnataka, India) is also estimated to have about 50% area under vegetation.

Urban green spaces comprise a range of habitats: natural, seminatural and artificial. The natural habitats in cities/towns represent the remnants of native original vegetation. As examples the following may be cited: (i) Extensions of the Mata Atlantic forests into Rio de Jeneiro. (ii) A goodly portion of the Singapore Botanic garden into Singapore. (iii) The National Park Ei Avila in the city of Caracas in Venezuela. (iv) York University Campus in Toronto. (v) The Ridge Forest of New Delhi, (vi) The Guindy Park and Tambaram (representing the native dry evergreen forests) in Chennai, Tamil Nadu, India. The seminatural habitats in cities represent vegetation of recent origin developed on their own on derelict land, disused industrial land or any other open field such as cemeteries, railway embankments, hospital and school grounds, gravel pits etc. These habitats were termed 'unofficial countryside' (Mabey 1973).

Artificial habitats in urban areas are extremely diverse. Man has created all these. Parks and gardens, canals, arboreta, domestic gardens, roadside trees, greenways, nurseries, streams and ponds are included under this category.

Species Diversity in Urban Habitats

Species diversity in urban habitats has not been estimated in the majority of places, although an enormous array of biodiversity is present. But much of this diversity falls in the introduced category, although native species may account for as much as 50% of total diversity in some cities. For example, the island of Singapore has lost over 99% of its original forest cover (including mangroves) in the past 175 years, but yet has lost only 26% of its plant species (Turner *et al.*1994). A survey in the Netherlands indicated that 50% of its flora is found on roadside verges; this includes rare species to the extent of 16.8% (Sykora *et al.*1994).

Many thousands of exotic species are cultivated in parks, gardens, arboreta, domestic gardens and roadsides. In fact, this category of taxa is extremely important as these plants have scientific, cultural and educational value that cannot be adequately estimated. However, their genetic diversity may not be very great, although new horticultural and garden cultivars are continuously produced. This is especially true for orchids, tulips, roses, begonias, narcissi, delphiniums, polyanthus, violas and pansies.

Importance of Urban Biodiversity

As already stated in the Introduction to this section, Urban biodiversity is going to play a very important role in the coming years since urbanisation is taking place at a very rapid rate throughout the world. Hence, adequate measures should be taken not only to assess such biodiversity, but also to plan conservation strategies (Nicholson-Lord 1987). This has been recognised in the bioregional approach to biodiversity conservation developed in the Global Biodiversity Strategy (WRI/IUCN/UNEP 1992) and efforts are ongoing to develop a framework within which such diversity can be developed and conserved. Urban diversity, without doubt, provides urban citizens social values and benefits that can be derived from contact with Nature, even if artificially created. Thus it provides urban children inheritance value (Heywood, 1996).

6

VALUES AND USES OF BIODIVERSITY

Introduction

This chapter is concerned with the values and uses of biodiversity to human society. Smith (1776), the greatest of classical Economists, stated that 'the origin of all wealth came from the bosom of earth', implying the existence of great bondage between Economics and the Earth's resources, especially biodiversity. The value placed on biodiversity depends on whether it is a private or social one. Private value is given by the 'price' of the resource itself, while social value refers to the value of the resource to society. The social value of biodiversity in reality refers to the manner in which bioresources are used or abused, but not actually estimated in terms of their 'worth' to society. Hence the social value of biodiversity tends to vary between countries and between cultures. Societal status, social preferences, degree of technological advances and the distribution of income and assets are some of the factors largely responsible for the differential perception of value of bioresources between countries and cultures. Although for certain resources the market prices will also reflect social value, for bioresources the private value differs from the social (Perrings 1997) and the former tend to be very poor approximations of social value. The components of biodiversity are a part of the wealth of society and their allocation between competing uses largely depends on their relative private value. Private value of biodiversity components needs to be understood in order to ascertain the driving forces behind biodiversity loss, while their social value has also to be understood in order to know how much biodiversity should be conserved (Perrings 1997).

There is also a debate on whether values are to be considered 'ethical judgements or equivalence measures' (UNEP 1995), i.e., whether values are statements of principle or a reflection of social opportunity costs (Turner and Pearce 1993; UNEP 1995). Ethical judgements influence people's preferences and therefore can be 'translated into a willingness to commit resources to biodiversity conservation' (Perrings 1997). Equivalence measures of value that are needed to fix the desirable level of conservation are the 'opportunities foregone in committing resources to conservation' (Perrings 1997). Equivalence measures of value in reality are not blind to ethical judgements (UNEP 1995) and *vice versa*.

Individuals may place the same value on biodiversity but commit very different resources to it due to differences in their monetary endowments. For example, a developed and a developing country may place the same value on biodiversity but commit very different amounts to, say, its conservation because their monetary backgrounds differ. In other words, valuation of biodiversity reflects the relative importance of both valuation of biodiversity in a society and the latter's ability to pay for it. A rich country willing to pay more for biodiversity than a poor country does not signify that the components of biodiversity in the former are more valuable than the components of biodiversity in the poor country. In other words, the most important point here is the distribution of income and assets (Perrings 1997).

Biodiversity Values

Valuing biodiversity is mandatory eventhough it is really beyond valuation. At least three important systems of classification of values of biodiversity have been proposed to date whose salient features are summarised in Box 6.1. The first system of classification of biodiversity values breaks the value down into a number of components: use or non-use, direct or indirect use, consumptive or productive use etc. This system looks at biodiversity from the point of view of a practising economist. The most important and recent formulations incorporating these components for calculating the total value of biodiversity are provided in Box 6.1. The component values involved in these formulations are defined and, in some cases, briefly explained below:

(i) **Total Environmental Value (TEnV):** UNEP (1995) defined this as a function of **primary value** and **total economic value.** A team of ecologists and economists working together arrived at a surrogate evaluation of all the environmental goods and services. It amounts to $33 trillion worldwide per year and thus is larger than the global economy of $29 trillion (1997 figures) (see Myers 2000). In other words, global natural resources are more valuable than global national products.

Box 6.1. Formulations of the components of biodiversity values as per various authors

II. First System of Classifications

1. Given (1996); McNeely (1988)

   DV (CUV, PUV) + IV (NUV, OV, EV)

2. Pearce (1990); Pearce and Moran (1994).

   TEV = UV + NV = (DUV + IUV + OV) +(EV + BV)

3. UNEP (1995)

   i. TEV = F [DUV, LUV, OV, QOV (Use Values), BV, EV (Passive or non-use value)]

   ii. TEnV = G [PV (Non-anthropocentric instrumental value), TEV (Anthropocentric value)]

4. Perrings (1997)

   V = F (DUV, IUV, OV, QOV, BV, EV)

5. Ravi and Pushpangadan (1997)

   TEV = DUV (CUV, PUV, NUV) + IUV + OV +QOV + NV (EV,AV)

II. Second System of Classification

6. Meffe and Carroll (1994)

   TEV = I_1V (G,S, I, PS) + I_2V

III. Third System of Classification

7. Norton (1987)

   TEV = DeV + I_2V + TrV

AV = Aesthetic Value; BV- Bequest Value; CUV- Consumptive Use Value; DeV- Demand Value; DUV- Direct Use Value; DV- Direct Value; EV-Existence Value; F - Function of; G - Goods; I - Information; IUV- Indirect Use Value; IV- Indirect Value; I_1V- Instrumental Value; I_2V- Intrinsic Value; NUV- Non-consumptive Use Value; NV- Non-use Value; OV- Option Value; PS- Psychospiritual value; PUV- Production Use Value; PV- Primary Value; QOV- Quasi-option Value; S - Services; TEV- Total Economic Value; TEnV- Total Environmental Value; TrV-Transformation Value; UV- Use Value; V- Value.

(ii) **Primary Value (PV)**: This may be defined as the value of the system characteristics upon which all ecosystem functions depend (UNEP 1995). Therefore, it represents the prior value of the ecosystem; it is called the primary value because the structured ecosystem produces functions which have secondary value. The secondary value will exist as long as the ecosystem retains its 'health', existence, homeostasis, operation and maintenance. The primary value is also known as 'glue' value since its notion is related to the fact that the system holds everything together and in principle, therefore, has economic value.

(iii) **Total Economic Value (TEV)**: Also called **Total Value** (TV) or simply **Value** (V). TEV denotes the sum total of all kinds of values attached to biodiversity minus the primary value. According to UNEP (1995), it is the function of use and non-use values, 'with due consideration of any trade-offs or mutually exclusive uses or functions of the resources/habitat in question'. Care must be taken to avoid simply adding up the resultant values to obtain TEV.

Total Economic Value by itself will underestimate the true value of ecosystems; it has to be considered along with the primary value for which details are priorly provided.

(iv) **Use Value (UV)**: This represents the value arising from an actual use made of a given component of biodiversity (MacArthur 1997). It is often a function of **Direct and Indirect Use Values**. Pearce (1990) and Pearce and Moran (1994) also include **Option Value** as a third function of use value, while UNEP (1995) includes **Quasi-option Value** also under use value.

(v) **Direct Use Value (DUV):** Also called **Direct Value.** MacArthur (1997) defines this as 'actual uses, especially in consumption'. According to UNEP (1995), it represents the economic values derived from direct use or interaction with a biological resource or resource system. The bioresource makes a direct contribution to human welfare in the form of either enjoyment or satisfaction. DUV is relatively easily observed and measured, often by assigning market prices. McNeely (1988) considers DUV a function of **Consumptive Use Value** (CUV) and **Productive Use Value** (PUV). Ravi and Pushpangadan (1997) consider DUV as a function of CUV, PUV and **Non-consumptive Use alue** (NUV). Some estimates of direct use values of selected wild resources are shown in Table 6.1.

(vi) **Consumptive Use Value (CUV)**: This is a type of direct use value and represents the value placed on a biodiversity component that is consumed/enjoyed directly, without passing through a market (Given 1996; Groombridge 1992; McNeely 1988). Recreation may be cited as an example. Ravi and Pushpangadan (1997) give an altogether different definition for CUV—'Consumption in physical form'—and include all types of biomass (food, fuel, fruit, fodder, medicine, industrial raw materials such as herbs for pharmaceutical preparations, wood for different uses and microbial products) as examples. Many of these in fact do pass through a market while some do not. Consumptive use values seldom appear in the GNP of countries but are nonetheless very important. Fuel wood is a consumptive value of great importance in rural areas.

(vii) **Productive Use Value (PUV)**: The value given to a component of biodiversity that is commercially harvested or is a source for a commercially harvestable product; such items pass through a market. Examples: minor forest produce, fruits

Table. 6.1 Estimates of direct use value of selected wild resources in developing countries (in US $)

Activity/Use	*Estimated Value*	*Source*
Preban ivory exports, Africa	$35-45 million/year	Barbier et al. (1990)
Viewing value of elephants, Kenya	$25 million/year	Brown and Henry (1993)
Tropical forest product exports	$11 billion/year	Barbier et al. (1994)
Fruit/latex forest harvesting, Peru	$6330/ha	Peters et al. (1989)
Sustainable timber harvesting, Peru	$490/ha	Peters et al. (1989)
Pharmaceutical prospecting, Costa Rica	$4.81 million/product	Aylward (1993)
Buffalo range ranching, Zimbabwe	$3.5-4.5/ha	Child (1990)
Wetlands fish & fuel wood, Nigeria	$38-59/ha	Barbier et al. (1991)
Ecotourism, Costa Rica	$1250/ha	Tobias and Mendelsohn (1991)
Tourism, Thailand	$385-860,000/year	Dixon and Sherman (1990)
Genetic value, Cameroon	$7/ha	Ruitenbeek (1989b)
Medicinal plants in Belize	36-166 ($ /ha/year)	Balick and Mendelsohn (1992)
Gross benefits from fruits, herbs, medicinal plants etc. in India	117 -144 ($ /ha/year)	Chopra (1993)
Brazil nuts only in Brazil	97 ($ /ha/year)	Mori (1992)
Fuel and fodder in Tamil Nadu, India	80 ($ /ha/year)	Appasamy (1993)

and seeds, latex, timber, pharmaceuticals, medicines, fibres, gums and resins, wild relatives of cultivated plants. Table. 6.2 provides some information on the value of productivity contributions of wild relatives of crop plants. The values of such items are usually estimated at the production end (landed value, harvested value, farm gate value etc.). PUV is included in national economic statements and budgets.

(viii) **Indirect Use Value (IUV)**: MacArthur (1997) defined this as 'benefits arising from an ecosystem function'. It represents the 'economic value derived from the role of resources and systems in supporting or protecting activities whose outputs have direct value in production or consumption' (MacArthur 1997; UNEP 1995). Indirect contributions of biodiversity to human welfare are said to have this value. As examples for indirect

Table 6.2 Examples of productivity contributions of wild relatives of crops (after UNEP 1995)

Crop	*Found in*	*Effect on production*
Wheat	Turkey	Genetic resistance to disease; valued at US $50 million/year
Rice	India	Wild strain proved resistant to the grassy stunt virus
Barley	Ethiopia	Protects California's US $160 million/year crop from yellow dwarf virus
Hops	N. Europe	Added US $15 million to British brewing industry in 1981 by lessening bitterness
Beans	Mexico	Genes from the wild Mexican bean used to improve resistance to the Mexican bean weevil which priorly destroyed as much as 25% of stored beans in Africa and 15% in South America
Grapes	Texas	Texas rootstock used to revitalise the European wine industry in the 1860s after a louse infection

contributions of biodiversity the following may be mentioned: biogeochemical cycles, photosynthesis, climate regulation, pollutant degradation, prevention of soil loss. Table. 6.3 gives an idea of the indirect use value of biodiversity.

(ix) **Non-Consumptive Use Value (NUV)**: This refers to the value which the components/systems of biodiversity possess in terms of functions or services offered. Some consider this a subcategory of IUV (Given 1996; McNeely 1988), while others (Ravi and Pushpangadan 1997) treat it as a category of DUV.

(x) **Non-Use Value (NV)**: Defined as the 'value relating to safeguarding the existence of assets, even though not related to their actual use in a foreseeable period' (MacArthur 1997). NV is also referred to as **Passive Use Value**, referring to the value of a biodiversity resource 'in production or consumption to someone/thing other than the user'. Such a value exists 'where individuals who do not intend to make use of such resources would nevertheless feel a "loss" if they were to disappear' (Brown 1990; Randall 1991). In view of this, people may like to conserve such biodiversity resources in their own right.

(xi) **Option Value (OV)**: Defined as 'willingness to pay to safeguard an asset for the option of using it in future' (MacArthur 1997). UNEP (1995) defined OV as follows: 'The potential value of the resource for future (direct and indirect) use'. The wild relatives of cultivated plants that are yet to be exploited may be cited as examples of biodiversity components possessing OV.

(xii) **Quasi-Option Value (QOV)**: According to UNEP (1995), QOV represents 'the value of the future information made available through the preservation of a resource' (also see Arrow and Fisher 1974). It should be mentioned that the distinction between option and quasi-option values is not always maintained.

Table 6.3 Estimates of the indirect use value of ecological functions (US $) (after UNEP 1995)

Resources/Function	*Estimated value*	*Source*
Cameroon		
Watershed protection of fisheries	$54/ha	
Control of flooding by forests	$23/ha	Ruitenbeek (1989a, b)
Soil fertility maintenance by forests	$8/ha	
Philippines		
Watershed protection of marine tourism	$13.9-19.2 million	Hodgson and Dixon (1988)
Watershed protection of fisheries	$6.2-8.1 million	
USA		
Water yield augmentation of managed forests	$232-388/acre	Bowes and Krutilla (1989)
Brazil		
Carbon storage by forests	$1300/ha/year	Pearce (1990)
Indonesia		
Support by mangroves of agriculture, fishing and cottage industries	$536 million	Ruitenbeek (1992)
Sweden		
Nitrogen reduction by restored wetlands	$18.7/kg N- reduction	Gren et al. (1994)

(xiii) **Existence Value (EV)**: Defined as the value 'deriving from the existence of a particular asset' (MacArthur 1997). UNEP (1995) defines EV as : 'The value of knowing that a particular species, habitat or ecosystem does and will continue to exist. It is independent of any use that the value may make of the resource'. EV notes the benefits derived by any one individual from the mere knowledge that the bioresource exists (see Pearce and Moran 1994). People who have donated money to a conservation organisation without expecting anything in return other than the satisfaction of knowing they have contributed something to the cause of biodiversity, may be said to have realised the existence value of biodiversity. Existence values of biodiversity generally generate sympathy and concern among people.

(xiv) **Bequest Value (BV)**: This is the 'value of knowing that others may benefit from the existence of an asset in future' (MacArthur 1997). According to UNEP (1995) it is a 'value defined by willingness to pay, to ensure that people's offspring or future generations inherit a particular environmental asset'. This value may thus be considered as the value of keeping a resource intact for one's heirs (Krutilla 1967). Some people, for example Aldred (1994), view BV as merely one of a number of types of Existence Value, and not warranting a separate category.

The recognition of a number of different categories of values by economists has created problems for lay biologists (who lack sufficient knowledge of economics) in understanding the real concept behind each of these subcategories of values. There is lack of clear-cut demarcation between at least some of these categories, and definitions and understandings overlap. The different classifications proposed by economists compound these problems, as different subcategories find different positions in different classifications (see Box 6.1). Lay biologists expect a simpler classification, which they can follow and practise without much difficulty. It is from this standpoint that the second classification system proposed by Meffe and Carroll (1994) captures our attention.

Meffe and Carroll (1994) detailed a second system of classification of values of Biodiversity. Value, as per this system, is classified into **Instrumental** (I_1V) and **Inherent Values**. The first is the value that something (in biodiversity) has as a means to another's ends. In other words, the components of biodiversity that are instrumental (i.e., absolutely necessary) in providing the material basis of human life come under this category. At one level these components maintain the biosphere as a functioning system and at another provide the basic materials for all utilitarian needs. Instrumental value, depending on the manner in which it constitutes the means to another's ends, can be divided into four categories of values: **Goods**, **Services**, **Information** and **Psychospiritual**. The components of biodiversity that provide the basis for food, forage, fuel, medicine, useful chemicals, fibre, ornamentals etc. (Hawkes 1987) come under **goods**. Pollination, dispersal of fruits and seeds, nutrient recycling, nitrogen fixation, biogeochemical cycles, role in maintaining a stable environment including soil stability, water purification, flood control, coastline stabilisation, waste treatment, disease regulation, maintenance of air quality etc., are some of the **services** rendered by biodiversity. Primary productivity through carbon sequestration is one of the most important services of plant biodiversity. The net primary productivity globally amounts to about 225 billion metric tons of organic matter annually (Ehrlich 1988). Biocontrol of pests/pathogens is another important service rendered by some components of biodiversity. If we accept the **Gaia Hypothesis** (Lovelock 1988a,b), the earth's temperature and the ocean's salinity are controlled by biodiversity. The **information** value of biodiversity is reflected in various aspects of basic sciences, applied biology, genetic engineering etc. It thereby emphasises the scientific value of plants, animals and microbes. The genetic information contained in elements of biodiversity is a very potential economic good and can be exploited for biotechnological applications. As Meadows (1990) states: '...biodiversity contains the accumulated wisdom of nature and the key to its future....Nature's knowledge is contained in the DNA within living cells. The variety of genetic information is the driving engine of evolution, the immune system for life, the source of adaptability'. The **Psychospiritual Value** is very difficult to define and can only be explained indirectly through examples. Meffe and Carroll (1994) cite aesthetic beauty, religious awe and scientific knowledge as some examples of the psychospiritual value category. The feeling of 'biophilia', according to Wilson (1984), can be equated to the special wonder, awe, and mystery in nature that one finds. It is the feeling of preference for Nature's variety instead of monotony (Soulé 1985). In other words, psychospiritual value is the realisation that an ordinary plant, however ordinary it may be, is as potentially beautiful as any work of craft or art.

The **Intrinsic value**, as per this second system of classification (Meffe and Carroll 1994), is the value that something (some component of or the entire biodiversity) "has as an end in itself" (Meffe and Carroll, 1994). The intrinsic value of biodiversity, as a whole, is a matter of great controversy; some question its recognition as a separate category. Unlike Instrumental value, intrinsic value cannot be divided into subcategories. It is also not clearly known whether intrinsic value exists objectively or is subjectively conferred. Which aspects of biodiversity could be considered to possess intrinsic value *per se* is another unresolved problem (Callicott 1986; Elliot 1992). If one values some component of biodiversity for its own sake, irrespective of its role, the intrinsic value of that component would appear to be subjectively conferred. Contrarily, if the value of some component of biodiversity is automatically recognised or felt by an individual, its intrinsic value can be said to exist objectively.

Intrinsic and instrumental values are not mutually exclusive. Many components of biodiversity may be valued not only for their utility, but also for themselves. Hence Norton (1991) has argued that by dividing biodiversity values into Instrumental and Intrinsic, one is doing more harm than good to the conservation of biodiversity. According to him, the intrinsic value issue has divided biodiversity scientists into two mutually suspicious fractions: Anthropocentrists and Non-anthropocentrists.

Norton (1987) advocated the third system of classification of the values of biodiversity. He recognised three kinds of values for Biodiversity: **Demand value**, **Intrinsic value** and **Transformative value**. Demand values occur when a component of biodiversity provides satisfaction for some felt preferences, commonly recognised in terms of Goods, Services and Information. Intrinsic value is defined in the sense already discussed in the second system of classification of values, i.e., a component of biodiversity can have value of and in itself , without reference to its usefulness to humanity. So, intrinsic value cannot be quantified. Transformative value exists where the object of biodiversity provides 'the occasion for examining or altering a felt preference rather than simply satisfying it'. It involves a transformation of, or change in a person's earlier set of felt preferences (for further discussion, see Meffe and Caroll 1994).

Ethical and Aesthetic Values

These two categories of intrinsic values (as per the system found in Meffe and Carroll 1994; Norton 1987, 1994) warrant special discussion. For some cultures, especially those in existence for several centuries, ethical benefits provide the strongest grounds for their deep concern for biodiversity. In India and certain East Asian countries, biodiversity is considered to have great value on cultural and religious grounds (see Chapter 8, for further details). Realising the importance of ethical value, the IUCN's Working Group on Ethics and Conservation produced a document on the ethical foundation for conservation of biodiversity (IUCN/UNEP/WWF 1990). The implications of this document were detailed by Engel and Engel (1990). Moral and ethical values differ from place to place, culture to culture, time to time and person to person. They also differ between different components of biodiversity. For example, the ethical value attached to sacred basil is not accorded to a cactus in India. In other words, most people value certain species more than others almost subconsciously.

The aesthetic value of biodiversity is very well known. Most people react more 'aesthetically' (often instinctively) towards plants that are appealing, visually or otherwise. Most cultures, irrespective of geographic location, have attested to the effect of plant and animal beauty on the human mind and emotions. Poets, writers and artists from various cultures have given expression to the aesthetic appeal of plants and animals. However, the relative aesthetic judgements differ from place to place, time to time and culture to culture. Roses, for example, kindle the aesthetic sense much more than cacti, succulents and carnivorous plants, although the latter have their own admirers. Such relative aesthetic judgements could presumably compel greater concern for certain biodiversity elements than for others, those deemed 'aesthetically not worthy'. Interestingly, the plant and animal species that particularly deserve protection, quite often have the least aesthetic appeal.

Precautionary Principle[1]

At present, only a relatively small percentage of biodiversity is actively exploited by man and valued. However, there are other elements of biodiversity that may be very important for the different reasons listed below:

(i) They may have values unused or unknown at present, but once discovered and exploited, could substantially enhance the well-being of humankind, and

(ii) They may become useful at some future time due to changing circumstances.

These reasons support a precautionary approach to maintenance of all biodiversity. Biodiversity elements with actual (yet unknown) or potential use should not be lost simply because we presently do not know their value.Further, it must be understood that biodiversity elements once lost cannot be recreated even with our best technologies. In addition to future options for humanity, future options for continuation of evolution of biodiversity elements (such as genes, species and ecosystems) should be borne in mind. The aforesaid are in essence the tenets of the precautionary principle.

Methodologies for Valuation of Biodiversity

Many methods for quantifying the benefits of biodiversity, i.e., for valuation, have been proposed and refined by the growing group of Environmental Economists. Some of these methods are listed in Table 6.4 from which it is evident that the terms used in the different approaches and methodologies vary considerably but do not actually reflect dissent regarding content and coverage.

[1]The 'Precautionary principle' is now widely applied in another context. The basis for this alternative usage emerged in the European Environmental Policies in the late 1970s and is now enshrined in numerous international treaties and declarations. The Precautionary principle is, accordingly, also applicable in developing the absolute safety precautions mandatory in the adoption of new biotechnologies (Foster *et al.* 2000)

Table 6.4 Methods for valuation of biodiversity provided by recent workers (based in part on MacArthur 1997)

1. Dixon and Sherman (1990)

- A. Techniques based on market prices
  - a. Change in productivity approach
  - b. Loss of earnings approach
- B. Techniques based on surrogate market prices
  - a. Hedonic pricing
    - (i) Property value
    - (ii) Wage differential
- C. Survey-based approaches
  - a. Contingent valuation methods
    - (i) Compensating variation
    - (ii) Equivalent variation
  - b. Delphi techniques
- D. Cost-based approaches
  - a. Opportunity cost of lost benefits
  - b. Alternative cost
  - c. Cost effectiveness
  - d. Expenditure-based approaches
    - (i) Preventive expenditure approaches
    - (ii) Mitigation cost approach
    - (iii) Replacement cost approach
    - (iv) Shadow project
    - (v) Relocation cost approach

2. Winpenny (1991)

- A. Effect on production approach
- B. Preventive expenditure and replacement cost.
  - a. Prevention expenditure
  - b. Replacement cost
- C. Human capital
- D. Hedonic methods
  - a. Property value
  - b. Wage differential
- E. Travel cost method
- F. Contingent valuation method
  - a. WTP
  - b. WTA

3. Pearce and Moran (1994)

- A. Direct methods
  - a. Experimentation (and Research)
  - b. Questionnaires
    - (i) Contingent ranking
    - (ii) Contingent valuation
- B. Indirect methods
  - a. Hedonic values
    - (i) Land values
    - (ii) Wage premia/penalties
  - b. Travel costs
  - c. Aversive method
- C. Contingent market approach
  - a. Dose response
  - b. Replacement cost.

4. UNEP (1995)

- A. Contingent valuation and Ranking
  - a. Contingent valuation
  - b. Contingent ranking
- B. Revealed preference methods
  - a. Travel cost method
  - b. Hedonic travel cost method
  - c. Random utility method
- C. Production function approaches
- D. Revealed preference and opportunity cost methods
  - a. Change in productivity
  - b. Change in earnings
  - c. Defensive or preventive expenditures
  - d. Replacement cost
  - e. Substitution or proxy
  - f. Shadow project
  - g. Compensation costs
  - h. Benefits transfer

It is not necessary to detail all the methodologies developed so far. Only the most important are described below:

Changes in Productivity Method

This is also known as the Production Function method. Changes in the supply of biodiversity resources result in changes in the economic value of their production. This principle underpins the Production Function method.

Changes in the supply of resources may occur for several reasons, including habitat changes, loss of biodiversity components etc. For example, soil degradation (due to continuous use of the land, use of chemical fertilisers and pesticides etc.) can affect agricultural productivity through the following: (i) increase in crop production costs for a certain level of output, and (ii) the resultant changes in quantities of output and prices change the benefits received by consumers and producers. Initial levels of surplus compared with resultant levels enable estimation of the value of changes in the supply of biodiversity resources.

The production function technique is a natural complement to **cost-benefit analysis.** This method has been used to estimate the value of soil conservation (Pimentel *et al.* 1994), of coastal wetlands (Aylward and Barbier 1992), of mangrove systems (Ruitenbeek 1992), flood-plains in northern Nigeria (Barbier *et al.* 1991) etc. For more details refer Dixon *et al.* (1988).

Contingent Valuation Method

The basis for this method is what people are willing to pay (WTP) for increment in biodiversity quality, or what people are willing to accept (WTA) in compensation for foregoing such benefits. It is generally assumed that WTP is equal to WTA, but empirical research has demonstrated that WTA often exceeds WTP. In spite of this problem, economists use WTP to calculate the value of biodiversity. The procedures involved are as follows:

(i) Elicit people's WTP for biodiversity goods or services directly through surveys/ questionnaires.

(ii) Create a hypothetical market situation

(iii) Use the respondent's replies to place value on biodiversity items that are not usually marketed.

The resultant valuation is 'contingent' because the value derived from this method depends on every individual's perception of number of background factors that influence the market under survey.

The advantages of this method are:

(i) It can be used to elicit values across the spectrum of total economic value.

(ii) This is the only method available for arriving at option prices and existence value.

Disadvantages include the following:

(i) A poorly designed/implemented survey will not give the true WTP.

(ii) There will tend to be differences between consumer intentions as expressed on the questionnaire and consumer preferences as revealed in the market.

The contingent valuation method was designed in North America and has been used there and in Europe; it has been little used in developing countries, except for Brazil. For more details on this method see Briscoe *et al.* (1990).

Hedonic Pricing Method

Certain biodiversity services are non-marketable. However, the values of these services are frequently incorporated into prices of marketable goods and services. Therefore, such market values must be desegregated to uncover the relative value of the two to human welfare. For example, soil fertility, scenic beauty or air quality of a land are not directly marketable, but the hedonic pricing method enables explicit valuation of these services that are implicit in the price of the land. The procedures involved in this method are as follows:

(i) Estimate the value of the land

(ii) Estimate econometrically the value contributed by the chosen 'service' attributes to the value of the land.

(iii) Work back from this hedonic price equation to the actual demand curve.

For more details see Prescott-Allen and Prescott-Allen (1982).

Travel Cost Method

This method can be used to find only the recreational value of a landscape. The methodology involved in this technique is as follows:

(i) Collect information on the expenditure incurred by visitors to the particular site/ landscape.

(ii) Aggregate the number of visitors by what it costs them to travel to and from the site. This will provide a surrogate market indicating what people are WTP for access to the site.

(iii) Estimate travel costs of visitors from distant places and nearby places. The WTP from these two groups of people will provide information on the relationship between distance and travel costs, based on which the benefits enjoyed by visitors to the site can be estimated.

The shortcomings of this method are:

(i) Unobserved travel cost is quite likely.

(ii) The method assumes that leisure time and travel to the site are necessarily a cost.

(iii) Travels to a site are often multipurpose.

For more details on the Travel cost method, readers may refer to Tobias and Mendelsohn (1991). This method was employed in evaluating the value of forests in the UK (Garrod and Willis 1991) and Achray forest in central Scotland (Hanley 1989).

Uses of Plants

Introduction

Species of plants provide an array of products used by people worldwide. Certain plants can be exploited directly from the wild, while others sustain humanity through cultivation. In spite of vast overall development, plant biodiversity as a global resource largely remains poorly understood, underexploited and poorly documented. Knowledge of plant use from indigenous people has not been translated into wider use largely because of poor documentation of ethnic information. However, plants have been a major source of food, medicine, horticultural and ornamental plants, timber, fibre, dyes and other chemicals, fuel and renewable energy, and a host of other products used in industry and commerce. A general outline of the major uses of plants is provided below.

Food

The most important contribution of plants to humanity is food. In the early years of man's evolution, plants were consumed raw and obtained from the wild; gathering food from the wild continues even today in tribal communities throughout the world. However, with the evolution of civilisation, man began to domesticate plants for food. Of the about 250,000 species of flowering plants, 75,000 species are edible but to date only about 3000 are regarded as a source of food. Of these, around 200 plant species have been domesticated with 15-20 constituting crops of major economic value. Species belonging to Poaceae, Papilionaceae, Brassicaceae, Rosaceae, Apiaceae, Solanaceae, Lamiaceae, Chenopodiaceae, Araceae, Cucurbitaceae and Asteraceae are the major sources of food.

The very high probability of global climatic changes is expected to cause large-scale shifts in natural vegetation and agricultural crops. Hence there is urgent need to protect genetic resources of food plants to maintain crop productivity in different climatic conditions. Wild species related to crop plants often provide this 'insurance value', as already indicated in Chapter 4. Heywood (1992) has suggested that there are several species of useful plants in the tropics alone whose uses could be extended from emergency sustenance in isolated locations or disaster areas to fully exploitable alternative sources of food. Future prospects are limitless and unforeseeable.

Fodder and Forage

Many species of plants are used as fodder. They are either used directly from the wild, as in pastures and rangelands, or domesticated. Grasses and legumes are the most important fodder sources.

Timber

Wood, the source of timber, is one of the most utilised plant commercial commodities throughout the world. Although predominantly harvested from the wild, monoculture plantations under agro-and social-forestry programmes are increasingly being raised as a source of timber. Wild sources of timber, especially from hardwoods, are predominantly tropical and, in fact, account for a very significant proportion of export earnings for developing countries in the tropics. The USA, some European countries and former USSR provinces account for the major supply of softwoods. Malaysia, Myanmar, Indonesia, Papua New Guinea and Gabon are the most important tropical countries involved in timber trade. Wood is exported as logs, sawn wood or plywood. It is difficult to assess the extent to which timber either for domestic consumption or for export is derived from plantations. Industrial timber plantations of temperate countries predominantly consist of coniferous species.

Ghana has 674 tree species of great timber potential but timber is exploited from only 60 species in the past. Peninsular Malaysia has about 3000 tree species, of which over 400 have been a source of good timber for national and international markets.

Because of continual exploitation and lack of adequate replantation, most timber tree species of tropical countries are now threatened; habitat loss, forest fragmentation, improper and inadequate management etc. have also contributed to this threat. More than 80 tree species of timber value are already listed as endangered the world over.

Rattans and Canes

Rattans and canes constitute the most important resources exported from tropical countries. Most of the 600 or so species, all belonging to Arecaceae (palms), are native to South and South-east Asia and the vast majority are endemics. The Philippines, China, Indonesia, India, Sri Lanka and Thailand are the most important rattan-exporting countries. Rattans and canes are used for cane furniture, mats, baskets, fish traps, dyes, medicines etc.

Rattans and canes are obtained almost exclusively from wild sources, although 10% of the supply comes from plantations in Central and South Kalimantan. They are mainly obtained from species of *Calamus* (15 species of this genus are more important sources).

Medicinal Plants

Plants are very important in health care. In less developed/ developing countries, 80% of the people still rely only on traditional medicines obtained from local plants and 85% of traditional medicine involves the use of plant extracts (Farnsworth 1988). Further, since adequate hospital facilities and allopathic doctors are absent in much of the tropics, any destruction of tropical forests would concomitantly destroy the primary healthcare network involving local plants and traditional 'doctors' (Balick 1990). Some 200 chemicals extracted in pure form from *circa* 90 plant species are used in medicine throughout the world, i.e., about half of the world's medicinal compounds are still derived or obtained from plant sources (Hamann 1991). Many of these chemicals cannot be synthesised. Therefore, medicinal plants are of great significance to both developed and developing countries.

At present only a very small percentage of the world's plants contributes on a global scale to health care. There is clearly a great range of higher plants from which to draw and there is also a great repository of traditional knowledge in the various cultures/societies of people using medicinal plants. WHO has listed over 21,000 plant species worldwide which are reportedly of medicinal value Heywood (1991a) estimated some 25,000 species of medicinal plants. More than 2500 species of plants are used in the Ayurveda, Siddha, Unani and other traditional health care systems.

Natural plant diversity might be increasingly valued for the 'blueprints' it provides for new synthetic drugs, in spite of an increasing technology to design and manufacture synthetic drugs. Principle (1991) estimated that the potential annual market value in OECD countries of the species of medicinal plants likely to vanish before the year 2050 is US $60 million. This figure is about 0.15% of the amount spent on plant-based medicines. It represents a benefit foregone rather than an actual loss. It is, however, only a market value and does not include other components of the total economic worth of the drugs, such as the cost to a society deprived of them and the benefits of good health. Therefore, the total economic value could be 5 to 50-fold higher.

Medicinal plants, especially those used in traditional medicine, are still largely harvested from the wild and relatively few cultivated. Cultivation has been attempted only for the last 25 years and a number of medicinal plants have reportedly lost/ become poor in medicinal properties upon cultivation. Yet species such as *Papaver somniferum, Cinchona officinalis, Mentha piperita, Ocimum sanctum, Digitalis purpurea, Gentiana lutea, Valeriana mexicana, Vinca rosea* and others have been effectively domesticated.

Because of constant exploitation, a number of medicinal taxa have become threatened in various parts of the world. Such taxa include species of *Dioscorea, Ephedra, Solanum, Rauvolfia, Parkia* and others.

Ornamentals

Ornamentals are domesticated wild plants and like food plants have a long history. In China, lilies have been cultivated for more than 2000 years and similarly in Rome, roses, violets, anemones, narcissi and lavender have a long history of cultivation. The number of ornamental and decorative plants under cultivation far exceeds the number of food plants. In the UK alone, *circa* 3000 species are ornamentals.

Ornamentals are important commercially and contribute significantly to international trade in countries such as the Netherlands, the USA and Japan. Both whole plants and cut organs such as flowers and leaves/twigs have ornamental value. Among whole plants of importance, the most important are orchids, succulents (cacti and euphorbias), cycads, insectivorous plants, bulbous species etc. Cut flowers of orchids, tulips, lilies, narcissi, violets, roses, anemones etc. are very important. More than 5000 species of orchids and their hybrids were recorded in the trade statistics of CITES during 1983-1989, a figure that must have increased substantially by now. Thailand, Malaysia and India account for major trade in tropical orchids. Although a number of these orchids are artificially propagated *in vivo* and *in vitro*, exploitation from natural habitats is still enormous, threatening endangerment of many orchid species. In Japan already 70 taxa of orchids have been entered in Red data lists. CITES appendix 1 (see page 157) includes species of *Paphiopedilum* and *Phragmipedium* as highly endangered.

The average international trade in cacti per annum is approximately 14 million plants as per CITES statistics, obviously a gross underestimate. One nursery in the Netherlands alone produces over 18 million cacti annually, the USA between 10-50 million, , while Mexico exports around 50,000 every year.

The most important succulent genera (other than cacti) in terms of commercial value are *Aloe, Euphorbia* and *Pachypodium*. Madagascar is among the chief exporters of succulents at around 135,000 plants annually according to CITES—all collected from the wild.

The bulbous plants form the next important group of ornamentals, albeit little information about them was available from CITES. While wild sources still contribute substantially, domesticated bulbous plants are very important commercially: *Galanthus, Cyclamen, Allium, Anemone, Arum, Crocus, Dracunculus, Eranthis,*

Fritillaria, Hyacinthus, Lilium, Muscari, Narcissus, Pancratium, Scilla, Tulipa and *Urginea*.

Other Uses

Plants have several other uses but only the most important are mentioned here. A number of species yield fibres of great value for cloth and other industrial purposes. Cotton, linen, jute, sisal, hemp, coconut, etc. are some of the fibres obtained. A number of fibre plants have been domesticated (cotton, linen, jute etc.) but fibres from wild taxa are still widely obtained, especially in tribal and rural areas.

Plants offer a good source of fuel, either as wood (firewood) or its transformed product, charcoal. Plant biomass from any source can also be converted into fuel. In fact, plants are very efficient sources of renewable energy.

Natural rubber, latex, gums, resins, dyes, essential oils and beverages are some of the other products of commercial value obtained from plants.

Uses of Microbes

Microbes are useful in many ways. One of their major users is the pharmaceutical industry. Their capacity to ferment various substrates has led to the production of a number of clinically and otherwise important antibiotics (see Table 6.5). Microbes are also good sources of various medicinally important enzymes (streptokinase and asparginase), toxins (botulin), immunomodulators (Cyclosporin A), hypolipidaemic (Lovastatin) etc. Vaccines such as BCG, typhoid, hepatitis B, hormones, and alkaloids such as ergot are also derived from some microbes. The ability of micro-organisms, especially transgenically modified ones, to transform steroids and antibiotics has also been successfully exploited by the pharmaceutical industry in recent years. Single cell proteins, microbicides, pesticides, insecticides and fungicides, flavouring agents, alcohol, acetone, butanol, glycerol and organic acids such as citric acid, fumaric, acetic and lactic, are also derived from the activity of microbes.

Table 6.5 Applications of microbially derived antibiotics

Antibiotics	*Applications*
Penicillins, Cephalosporins, Rifamycins	Antibacterial
Amphotericin B, Griseofulvin, Nystatin, Hamycin	Antifungal
Abikoviromycin, Kikumycin	Antiviral
Adriamycin, Belomycin	Antitumour
Avermectin, Hygromycin	Anthelminthic
Herbicidin	Herbicide
Milbemycin	Miticide
Tetranectin	Insecticide
Gibberellins	Plant hormone
Monascin	Food pigment
Detoxin	Detoxicant
Azalomycin	Antiprotozoal
Nisin	Food additive
Variginiamycin	Animal growth promoter
Colisan, Patulin	Antispasmodic
Monorden	Tranquillizer
Cyclosporin A, Alanosine	Immunosuppressive
Griseofulvin, Amicomycin	Anti-inflammatory
Dopastin, Spiramycin	Hypertensive
Filipin, Nogalomycin	Anticoagulant

Several micro-organisms are highly useful in the agricultural industry, either as biofertilizers due to their capacity to fix atmospheric nitrogen or in phosphate solubilisation. Some soil bacteria, such as *Pseudomonas putida*, enhance plant growth by producing very characteristic plant-growth regulators. Other important roles of micro-organisms include biomining (biologically extracted mineral ores), bioremediation (cleansing the environment through absorption or biotransformation of toxic chemicals), biosorption, biogas production, harnessing solar energy etc. Thus the uses of microbes are highly diverse and can be pithily summarised in this statement: **'Microbes can and will do anything; microbes are smarter, wiser and more energetic than microbiologists, chemists, engineers and others'** Naresh Kumar (1998), by quoting Perlman in Biodiplomacy.

7

LOSS OF BIODIVERSITY

Introduction

We already know that biodiversity can be expressed in terms of the different levels of biological organisation such as genes, species, ecosystems and landscapes. All these four forms of biodiversity can be subjected to loss, although the most easily recognisable form of loss is that of species. Since, as stated earlier, the different forms of diversity are intimately related to one another, biodiversity loss at any one level will lead to loss at other levels too. This should be kept in mind while reading this chapter. It is only for convenience that biodiversity loss is treated separately under each of the first three levels.

Loss of Genetic Diversity

Introduction

It must be evident to the readers that the unrestrained loss of species is almost always sure to be accompanied by a more subtle, but no less important process—the loss of genetic diversity. When a species is lost, all genetic information carried by that species is also lost. Species loss, as detailed below, is usually preceded by fragmentation of its contiguous populations to result in many small, isolated populations. Genetic diversity present in the whole contiguous population is also fragmented and slowly lost when the fragmented populations are also lost. In other words, the 'gene pools become gene puddles' and these 'gene puddles' are also lost subsequently. As noted earlier in Chapter 2, genetic diversity is analysed at the population level. Hence loss of genetic diversity is also studied in populations. New work suggests that the extinction rates of populations of species are far higher than has been estimated for species. The annual losses of populations are around 0.8 %, which is equivalent to about 1800 populations every hour (Hughes *et al.* 1997). It was further stated in the second Chapter of this book that genetic diversity is important for fitness and adaptive changes; loss of genetic diversity, therefore, becomes a serious matter for concern as it will affect the fitness and evolutionary adaptability of a species.

Factors Causing Loss of Genetic Diversity

Reduction in genetic diversity within populations of species may be caused by four factors, all of which are a function of genetically effective population size. These four factors are: **Founder effects, Demographic bottlenecks, Genetic drift** and **Inbreeding depression.**

Founder Effects

Founder effects occur when only a few individuals (called 'founders') of an originally larger population establish a new population. The genetic constitution of the newly established population will depend on the genetics of its founders. If the founders are not true representatives of the larger parent population in terms of their genetic constitution, or if only a few founders are involved in establishing a new population, then the newly established population is a biased (in terms of genetic diversity) representation of the original larger gene pool (of the parent population of founders) from which it came; thus it may have lower overall genetic diversity.

Demographic Bottlenecks

When a larger population suddenly experiences a severe, temporary reduction in size for whatever reasons, a demographic bottleneck results. The outcome of such a bottleneck is that the genetic diversity of all subsequent generations is contained in the few individuals (of the original population) that survive the bottleneck and reproduce. Expectedly, some genetic diversity will be lost in this process. The magnitude of loss is dependent on the size of the bottleneck, the reproductive ability of the surviving individuals, the new genetic changes caused in the surviving individuals (by mutation etc.) by the bottleneck and the growth rate of the derived population afterwards.

Genetic Drift

This represents a random change in gene (or allele) frequencies in small populations. In mathematical terms, it represents a chronic bottleneck that results in repeated erosion of heterozygosity (i.e., increase in homozygosity), loss of variability and eventual loss of genes or alleles; rare alleles are very often lost. A reduction in diversity of allelic combinations is also apparent in subsequent generations. In other words, genetic correlation among characters can also be altered as only a few pleiotropic alleles (alleles that affect two or more phenotypic characters) may remain. Genetic drift is believed to be a key factor in the loss of genetic diversity and therefore is important in conservation also. Since the magnitude of random genetic drift is a function of the population size (Wright 1938), it can be measured directly by the **effective size of a population.** The latter can be defined as the size of an ideal population whose genome is affected by random drift to the same extent, as is the genome of the real population under study. The ideal population is one in which all individual plants are equally likely to be the parent of any progeny, sexually reproducing, bisexual and producing both male and female gametes simultaneously. The gametes combine at random and self-fertilisation happens at a rate of $\mathbf{Ne^{-1}}$. The number of progeny per parent is a Poisson random variable. The generations are distinct and there is selection, migration or mutation. Such an ideal population is the proper yardstick for predicting the effects of genetic drift. The effective genetic size of a population is the effective population size, **Ne**, and is defined as $\mathbf{Ne = N/\ (1+ F_{IS})}$ where **N**= population size as per census and F_{IS} = deficiency of heterozygotes within that population size.

Inbreeding Depression

Inbreeding can be defined as mating of individuals related by common ancestry. There is greater probability of inbreeding occurring in smaller populations. The most important consequence of inbreeding is inbreeding depression, which may be defined as a 'decrease in the mean of a character upon inbreeding' (Lande 1996). Inbreeding depression causes decrease in growth rate, reduction in vigour and fertility, decreased survival rate, physical deformities etc.; all these individually or in combination will be evident in any component of fitness, under a specific environment. The other important genetic outcome of inbreeding is a reduction in heterozygosity and an increase in homozygosity 'through increased identity by descent'.

Two hypotheses have been posited so far to account for changes associated with inbreeding depression (see Charlesworth and Charlesworth 1987): (i) According to the Overdominance hypothesis, loss of beneficial heterozygosity has been responsible for inbreeding depression. (ii) According to the Partial dominance hypothesis, an increased homozygosity for deleterious recessive or partially recessive alleles has been responsible for inbreeding depression. It must be understood that these two mechanisms of inbreeding depression are not mutually exclusive and that they may occur at a single locus or multiloci levels. A critical perusal of the literature on these two hypotheses clearly indicated that the majority of careful experimentations carried out to date support the Partial dominance theory, although some role from overdominance cannot be totally ruled out.

The magnitude of inbreeding depression can be very great. In plants, the magnitude is measured by comparing the performance of progeny derived after self-pollination, with that of progeny obtained from deliberate crossing. The values for the ratios of fitness components in progenies obtained after selfing and crossing are then computed. These experiments were almost always done under greenhouse conditions. As a result, the measurements often resulted in underestimates of fitness reduction that would otherwise have accompanied inbreeding in nature. In other words, the magnitude of inbreeding depression calculated under greenhouse experimentations may not be comparable to that resulting from inbreeding in nature. This conclusion is reflected, for example, in the results obtained by Dudash (1990) in *Sabatia angularis*. It is, however, unfortunate that hardly any studies exist on the effect of inbreeding depression on progeny performance (i.e., fitness) conducted throughout the entire life cycle of plants. A more complete estimate of fitness can be obtained only if it is assessed at as many points in the life cycle of a plant as possible. This is possible in the case of annuals but not in perennials. Ritland (1990) developed a method to estimate fitness from electrophoretic data alone for long-lived organisms.

Founder effects, genetic drift, demographic bottlenecks and inbreeding depressions will all reduce the original genetic diversity of small, isolated populations over a period of time. This loss in genetic diversity is at the rate of ½ **Ne**. For example, a population of 1000 individuals will retain 99.95% of its genetic diversity in the first generation, while a population of 50 individuals will retain only 99% of its genetic variability in the first generation. But over many generations, this small loss may be magnified. For instance, after 20 generations, the population of 1000 will still retain over 99% of its original genetic variability, while the population of 50 will retain less than 82%.

Loss of Species Diversity

Introduction

The loss of species is a natural process. We know from fossil and historical data that all species have a definite life span. We also know those forces that led to loss of species as well as those that allowed certain species to survive. The theoretical calculations from fossil data suggest that as much as a quarter of Earth's species become extinct each million years. The actual reasons for this loss are not known. The explanations offered thus far range from interspecific competitions (Darwin 1859), climatic changes (Florention *et al.* 1991; Stanley 1988), accumulation of deleterious genes (Raup 1991a), result of inbreeding (Jimenez *et al.* 1994) or extraterrestrial impacts such as those of asteroids (Alvarez *et al.* 1980). Well over 95% of all species that have evolved on this Earth thus far have become extinct. We also know that extinct species outnumber living ones by a factor of perhaps a thousand to one.

A species is said to have become extinct when all its individuals are lost without producing progeny. Such a loss of species is called **true extinction**. There have also been **pseudoextinctions**, wherein a species disappears when its lineage is transformed over evolutionary time or divides into two or more separate lineages. The relative frequencies of these two types of extinctions in evolutionary history are not clearly known. Extinctions of the true type generally occur when a natural or man-made environmental change or challenge exceeds the adaptive capacity of the individuals of a species, and there is no safe place to which the species can retreat. Species extinction without the intervention of man is often called **background extinction** (Raup 1978). The background extinction rate, on average, is calculated to be 4 million years for each species (Raup 1991b). This may appear to be an incredibly long time to humans but is remarkably short with reference to the nearly 4000 million years of history of life on the Earth. Background extinction rates are considerably lower in land plants (Niklas *et al.* 1985) since they tend to belong to evolutionary clades that are more extinction-prone than others.

We are now aware that the history of life on the Earth has been punctuated with **mass extinctions** (Balmford *et al.* 1998; May and Tregonning 1998; Pimm *et al.* 1995). A mass extinction can be defined as an exceptional loss in biodiversity that is substantial in size and global in extent; it should also affect a broad range of taxonomic groups over very short periods of geologic time (Jablonski 1986; Meffe and Carroll 1994; Sepkoski 1988). Some of these mass extinction episodes were of short duration while others extended for at least several million years. Eight extinctions have been identified and grouped into five major mass extinctions: (i) Ashgillian (end Ordovician), (ii) Givetian, (iii) Frasnian and (iv) Famennian (all three, late Devonian), (v) Guadalupian and (vi) Dzhulfian (both, end Permian), (vii) Norian (end Triassic) and (viii) Maastrichtian (end Cretaceous) (Raup and Sepkoski 1982). Most species extinctions have been due to these mass extinctions. For example, during the mass extinction of the Permian-Triassic boundary 96% of all Earth's species at that time reportedly perished (Raup 1979b). The Norian (end Triassic) and Maastrichtian (late Cretaceous) extinctions were also substantial (Sepkoski and Raup 1986).

Processes Responsible for Species Extinction

Two types of processes are fundamentally responsible for species extinction:

(i) **Deterministic processes** or cause-and-effect relationships: Examples: glaciation, deforestation, habitat fragmentation etc. In these processes, some essential components of ecosystems are removed while others, lethal to the ecosystem, are added. Deterministic events are either readily observed or easily detected (more details on these processes are given later).

(ii) **Stochastic processes** or chance events: Four types of stochastic processes are distinguished (Gilpin and Soulé 1986; Shaffer 1981, 1987).

(a) **Demographic uncertainty**: This resulted from the effect of random events on the survival and reproduction of individuals in a finite population. In a small population of, say, less than 100 individuals, demographic uncertainty can be seen. As an example, let us cite the instance of an extremely skewed sex ratio in a small population of a dioecious taxon and its effect on the overall reproductive effort of the population. Menges (1991) found that for *Astrocaryum mexicanum*, a population size of at least 50 individuals is needed so that the probability of its survival is greater than 95%. The greater the population size of a species, the better its chance of survival.

(b) **Environmental uncertainty**: This is due to unpredictable environmental events such as sudden changes in weather, food supply, disease incidence, extent of competitors, predators and parasites etc. (Shaffer 1987). Unlike in demographic uncertainty, there is no critical population size that once reached guarantees a high level of long-term security from environmental uncertainty.

(c) **Natural catastrophies**: These are extreme cases of environmental uncertainty, e.g. floods, hurricanes, fire, drought etc. These catastrophies are usually short in duration but massive in impact.

(d) **Genetic uncertainty**: This refers to random changes in the genome, mutations etc. The already described founder effects, genetic drift, inbreeding depression etc. are also included under genetic uncertainty.

Deterministic and stochastic processes may either act independently or in combination (Gilpin and Soulé 1986; Schaffer 1987). In the latter case, their effects compound each other.

Population Size as a Critical Factor in Species Extinction

Introduction

The effect of population size on the loss of genetic diversity due to founder effects, demographic bottlenecks, genetic drift and inbreeding depression was detailed above. Here, more details are given on the importance of population size in species loss/retention.

We now have increasing evidence for the fact that the size of the population of a species is a major feature determining its survival (Lande and Barrowclough, 1987). Smaller populations are generally more prone to extinction than larger ones for reasons mentioned in the last section. The dwindling of species size (i.e., reduction in population size) has been described as **'collapse'** by Soulé *et al.* (1979) and as **'relaxation'** by Diamond (1972). The size of the population of a species required to avoid extinction is termed the **minimum viable population** (MVP) and varies for each and every species temporally as well as spatially. All the four stochastic processes as well as the deterministic factors mentioned above affect smaller populations. Moreover, they can interact to reinforce one another's negative impacts and draw smaller populations into 'extinction vortices' (Gilpin and Soulé1986).

MVP and Population Viability Analysis

The MVP of any species can be determined by **population viability analysis** (PVA), also termed **population vulnerability analysis** (Shaffer 1990; Soulé 1987a,b). PVA involves the study of ways in which all factors interact to determine extinction probabilities for populations of individual species. According to Gilpin and Soulé (1986), PVA takes into account the following three constantly changing and interacting aspects: phenotype of population, environment, and population structure and fitness. The components of each of the three aspects are presented in Table 7.1.

From data compiled on these components and their interactions, mathematical models for PVA can be developed. However, a word of caution; the theoretical basis for PVA is still developing and there is no single mathematical model sufficiently sophisticated to simultaneously incorporate all classes of chance events (= stochastic events) (Shaffer 1995). Neither are we sure that the components of the aforesaid three major components, even if known completely for any species, suffice for designing an effective model of PVA. These problems notwithstanding, considerable attention has been given in recent years to the MVP concept and the search for a 'magic number' for a general and universal minimum population size. Most work on MVP and PVA has been done on animals and it is extremely difficult for many reasons to extrapolate the methodology and the results derived therefrom to plants (for details, see Given 1996).

PVA is a powerful tool for predicting the probability of persistence of a species (or its

Table 7.1 Components of phenotype, environment and population that help in PVA (from Gilpin and Soulé 1986)

(i)	Phenotype of population	
	a. Morphology	— Variations in size, shape and patterns including geographic and temporal variations.
	b. Physiology	— Metabolism and its efficiency, reproduction and disease resistance.
	c. Interspecific interactions	— Mutualism, predation, parasitism, competition etc
	d. Distribution	— Dispersal, migration (including genetic), habitat selection.
(ii)	Environment	
	a. Habitat quantity	
	b. Habitat quality	— Abundance of resources and interacting organisms, patterns of disturbance (duration, frequency, severity and spatial scale).
(iii)	Population structure and fitness	
	a. Dynamics of spatial distribution	
	b. Metapopulation structure and fragmentation	
	c. Age structure	
	d. Size structure	
	e. Sex structure	
	f. Saturation density	
	g. Growth rate	— Including variation in growth rate between individuals, within patches and between patches.

populations) over a specified period of time with a specified probability. As a theoretical instance we can say that a population is found to have 95% probability of surviving for 100 years, on the basis of PVA. In this light, it can be further stated that if an ecosystem of a particular size is fragmented into smaller units (the phenomenon referred to as **Fragmentation** earlier—see details below), the population size of constituent species of that ecosystem is also fragmented into smaller units. Then extinction of species in each of these fragmented areas will be faster than in the non-fragmented habitat. The latter is due to the fact that the population of species in a fragmented system shows a number of ecological traits that individually or in combination hasten the process of extinction. The combination of these is often referred to as 'ecological correlates of vulnerability to population extinction'. The most important of these traits are Rarity, Dispersal ability, Degree of ecological specialisation, Niche location and Population variability. Rarer populations are likely to become extinct than larger ones. Poor dispersers are more prone to extinction than good dispersers. Ecological specialists are more vulnerable to extinction than generalists. Species adapted to or able to tolerate the conditions at the interface between fragments (i.e., **edge effects**) will survive for a longer period. Stable populations of species are less vulnerable to extinction than fluctuating populations.

Metapopulation Concept

The MVP model described above considers all individuals as belonging to a single isolated population. In reality, most species are distributed in patches, thus consisting of subpopulations of a population. Such subpopulations are termed **metapopulations**. As per the metapopulation concept, subpopulations although geographically isolated and spatially discrete, are interconnected through gene flow; limited migration also occurs between them. Each subpopulation has a finite life expectancy. Species whose metapopulations are suddenly reduced in size often fail to establish new populations concomitant with the disappearance of existing ones. The stochastic processes with which they had coped earlier, now push them to extinction; on the contrary, in dynamically stable species new populations arise as established ones disappear, because conditions are favourable. The metapopulation model focuses on the dynamics of the demographic and genetic processes that affect individual species rather than simple changes in their number. This concept also explicitly recognises the effects of environmental heterogeneity both within and between subpopulations (Merriam 1991).

How do we determine whether we have at hand a population of a species with an unbroken genetic and temporal continuity or a species with a metapopulation structure of ephemeral subpopulations? According to Soulé (1987c), by measuring the rate of species turnover with reference to the different populations in habitat patches of varying size.

Current and Future Species Extinction Rates

It was earlier remarked that species extinction is a natural process and that during the history of the Earth several mass extinction events took place. Natural extinctions are distinct from those triggered by human intervention. According to Wilson (1988b), extinctions caused directly or indirectly by human beings are occurring at a rate that far exceeds any estimates of background extinction rate (i.e., 4 million years); the human-induced extinction rate may be 1000 to 10,000 times greater than the average background extinction rate (Balmford *et al.* 1998; May *et al.* 1995; Pimm *et al.* 1995). Many studies made recently indicate that species extinctions are occurring today at very high rates on both local and global scales and that we are now in the opening phase of another mass extinction—this time triggered by human intervention alone. The present mass extinction, if it remains unchecked, will purportedly rival and even conceivably surpass in extent any of the previous great mass extinction episodes (Given 1996).

How can one predict the present and future extinction rates? Quantifying present rates of species extinction is extremely difficult, while predicting future extinction rates with precision is almost impossible. Actual counting of lost species is not only difficult, but problem-laden. Furthermore, recorded species extinctions are not entirely reliable (see discussion later on page 90). Therefore, estimates of extinction are not based on observed or recorded species extinctions but rather on extrapolations from estimates of habitat loss and of species richness in different habitats which bear in mind the basic assumption from biogeography pertaining to species-area relationships. The chosen habitat for the purpose of extrapolation is the tropical forest ecosystem, where estimates of species richness and of actual and projected deforestation rates are taken into consideration. The selection of this habitat and the parameters are justified not only because the majority of global species live in tropical forests, but also because outright destruction, gross degradation and fragmentation of habitats are more rampant in this ecosystem. Of the conservative estimate of 10 million species, 7.2 million species are believed to live in the tropics, with 5 million in different types of tropical forests. According to recent calculations (Myers 1992), tropical forests are being destroyed at the rate minimally of 150,000 km^2 per year. FAO (1993) estimated an annual loss of 134,000 km^2 of tropical forests, approximately 11% lower than Myers (1992). But besides outright destruction, additional loss caused by logging (from selective to intensive) (Jacobs 1988; Whitmore 1980), slash-and-burn conversion to pasture and cropland (especially plantation crops), fragmentation, edge effect, habitat alteration and environmental deterioration should be taken into account (Ledig 1992). Myers' estimated loss of tropical forests to the extent of 150,000 km^2 per year represents 2% of the remaining forest area, a figure which by now on our entering the 21st century has doubled. In terms of hot spot analysis (recall that most hot spots occur in the tropical belt), 5 of the 18 hot spots originally designated have already lost 90% or more of their original habitats, and the rest are expected to lose 90% of their habitats within the next two decades. These will account for 50% of the species.

The current annual 2% rate of forest loss does not mean that 2% of the forest species are lost as well. This 2% annual rate of forest destruction is translated into annual species loss the using Arrhenius equation (Chapter 3, Fig. 3.3). The number of species lost is estimated from average values of z and rates of deforestation (Ehrlich and Ehrlich 1981; Lovejoy 1980; Simberloff 1986). How this method can provide estimates of species extinction for different z values and rates of deforestation is illustrated in Fig. 7.1 (30% to 45% loss of species with z = 0.15 and 60% to 80% loss with z = 0.40 and so on; Lugo and Brown 1996; Wilson 1992 see, however, the discussion below). The most important generalisation made from this type of analysis is that a 10-fold reduction in area results in the loss of half the species present. Using this extrapolation method, it is estimated that at present the annual species loss in tropical forests alone is around 27,000 or about 75 per day. Using the same method, many others have estimated the extinction rates of species in tropical forests (Lovejoy 1980; Myers 1988b; Raven 1987, 1988b; Reid and Miller 1989; Simberloff 1986; Wilson 1988b; see also the review by Reid 1992). Estimated rates range from 2 to 25% depending on the group of organisms, different assumptions made in relation to the various components of the Arrhenius equation, different forest regions sampled etc. As already indicated, recent work suggests that the extinction rates of populations of species are far higher, with 1800 populations lost every hour (Hughes *et al.* 1997).

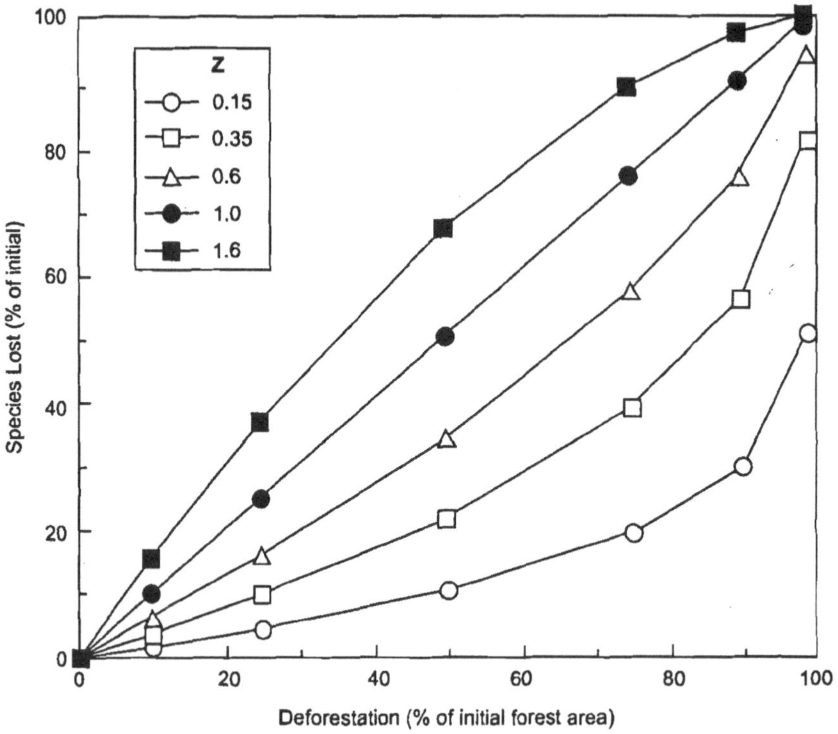

Fig. 7.1 Estimates of species lost due to deforestation using different z values (from Lugo and Brown 1996).

Lugo (1988b) and Lugo *et al.* (1993) were of the opinion that this extrapolation methodology overestimates the rates of species extinction for the following reasons:

(i) It assumes that deforestation causes the complete loss of species in a particular site.

(ii) It ignores the possibility of resiliency or persistence of a species in the site through dispersal or survival, even after habitat change.

(iii) The predicted extinction rates for certain groups of plants, for example trees, are higher than actual empirically observed extinctions in areas where deforestation has taken place.

(iv) This method does not take into account the efforts of man to restore species through afforestation, land rehabilitation and other programmes.

Other problems in the methodology may be mentioned (Jenkins 1992):

(v) The principal assumption that species richness and habitat destruction within tropical forests are evenly distributed is not really true since richness is known to vary greatly between different sites of tropical forests at all scales of composition.

(vi) The inadequacy of data on the poorly known groups of living organisms, which constitute a major chunk of the world's total number of species, makes an unrealistic assessment of species loss due to loss of habitat in any area.

(vii) Areas differ radically in number of endemic species they contain. It is therefore possible that the complete destruction of even a small area with a large number of endemics will experience more species extinction than a same-size area with only a few or no local endemics, even if the latter has greater species richness.

(viii) The available or cited figures for rates of habitat loss are also doubtful.

(ix) The model does not also take into account the presumed 'residual' extinctions that will occur through fragmentation of the remaining forests. It is known by the Island Biogeography Theory that fragmented habitats will have greater rates of extinction through various stochastic processes. Under such a situation, we shall not be able to distinguish those species which might have already been committed to extinction without direct human intervention from those that will suffer extinction due to natural processes.

(x) Certain habitats are capable of self-recovery, which can deter extinction that may have started (Heywood and Stuart 1992; Reid 1992). As an instance, we can cite the Atlantic forests of Brazil, which despite their reduction in size to 10% of their original extent in the last one hundred years, have lost only a few species (Brown and Brown 1992) thanks to self-recovery.

(xi) The species/area relationship may not be the same for all groups of organisms.

(xii) This model does not consider future changes in human behaviour or resource management that could 'ameliorate or exacerbate' these effects (Heywood and Stuart 1992; Reid 1992).

Recent studies using the Arrhenius equation have suggested that there is no exponential rise in species number with increase in area sampled (Crawley and Harral 2001). It has been shown that z is greater than 0.25 in islands on larger areas and smaller than 0.25 in smaller area samples. It has also been shown that the relation between species richness and area with slope $z = 0.25$ cannot be global. Values of z were low (0.1 to 0.2) at small scales (< 100 m^2), high (0.4 to 0.5) at intermediate scales (1 hectare to 10 km^2), and low again (0.1 to 0.2) for the largest scales. There are

250,000 species of vascular plants on the Earth and the total land suface is 1.4×10^8 km^2. The law predicts that $C = S/A^Z = 250{,}000\ (1.4 \times 10^8 \times 10^6)^{0.25} = 72.7$ species at a scale of 1 m^2, which is not always true for all regions of the world (Crawley and Harral 2001).

How well do the estimates of species extinction calculated through the extrapolation method tally with the actual figures compiled? Compilations made by Reid and Miller (1989) and Reid (1992) indicate only 724 recorded cases of species extinctions since 1600 A.D. worldwide, of which only 384 are plants. The data compiled by WCMC and presented in Groombrige (1992) records the loss of 599 species of plants (Table 7.2). The World Conservation Monitoring Centre. (WCMC 1992; figure cited from WCMC pers. comm. 15 Feb. 1995 and cited in UNEP 1995) recorded 654 plant species (mostly flowering plants) as extinct since 1600. These figures, if taken as correct, indicate almost no significant loss of species. But as Gentry (1996) opined, these figures are likely to be gross underestimates. Gentry (1996) further believes that many additional, unrecorded extinctions of plants must have occurred in recent years. The current figures of 172 extinct plants species from the USA-118 from Hawai, 51 from the continental USA, and 3 from Puerto Rico— itself amounts to 45% of the world's total extinct plant taxa (of 384) counted by Reid and Miller (1989). An earlier count of 213-228 extinct plant species in the United States made by CPC (Centre for Plant Conservation) and cited in Wilson (1992) included 39 taxonomically questionable species, and a couple of species that were subsequently rediscovered; this count would represent 55% of the world's lost plant species. That half of the recorded extinct plant species of the entire world should come from the United States, which contains only 7% of the world's plants, is far more likely 'to reflect profound ignorance of plant extinctions than exceptional levels of extinction in the United States'. Similarly, the Netherlands, with an area of only 35,000 km^2, had lost 37 of its 1410 species of higher plants by 1940 and another 33 species by 1990. These two instances clearly reveal that effective evaluation of species extinction rates cannot be done by simply counting historically recorded cases.

Table 7. 2 Extinct higher plant taxa of the world (as of March 1992) (from information available with WCMC and cited by Groombridge 1992)

Ser.No.	*Group*	*No. of extinct species*
1.	Fern allies	004
2.	True ferns	012
3.	Gymnosperms	002
4.	Dicots	461
5.	Monocots	120
	Total	**599**

What about future extinction rates of species? It has been estimated that natural extinction climbed to a rate of about one species every ten years between 1600 and 1950. It is also estimated that by the end of the 20th century, the rate of extinction may have risen to as high as 40 - 400 times the background rate, or at least to an average of 100 times (Ehrlich 1986). However, one of the major problems in predicting future extinction rates, is that such predictions depend on the size of the extant biota, which itself is a matter of controversy. The other problem lies in estimation of the degree of deforestation in the tropics. Based on already available data and mathematical models, it has been surmised that we may lose 20% of all species within the next 30 years and 50% or more species thereafter (Wilson 1992). Raven (1990) estimated that one-fourth of all plant species are likely to be eliminated during the next 30 years and half of the total species might disappear before the close of the 21st century. The Club of Earth (1990) has endorsed these estimates. According to another estimate, there will be a one-third reduction in species number within the next three decades (see Reid 1992).

Threatened Species

A threatened species can be defined as one believed to be at significant risk of extinction in the foreseeable future due to stochastic or deterministic factors or a combination of both affecting its population, or by virtue of its inherent rarity. Two aspects in this definition are difficult to define: (i) what level of risk is significant and (ii) what part of the future is foreseeable. Threatened species pose scientific, economic and moral challenges. Scientific, because their loss would disconnect evolutionary links that contribute to an understanding of plant life. Economic, because such species—or at least some of their genes—might prove useful in future, if not now. Moral, because human beings are to be blamed for having caused the extinction or endangerment of a component of Nature and a fellow being.

The problems of threat, depletion and extinction of plant taxa came to be better known by the general public largely through the United Nations Conference on the Human Environment held at Stockholm in 1992 and through the International Union for Conservation of Nature and Natural Resources (IUCN). The latter was instrumental in the establishment of the Threatened Plants Committee (TPC) with the objective of enlisting worldwide participation of plant scientists in collecting data on threatened plant species (Lucas 1976), their location and their preservation. Sir Peter Scott independently developed formulation of the Red Data Book (RDB) concept during the 1960s. The RDB categorises species at the threshold of risk according to the severity of the threats facing them and the estimated imminence of their loss. The RDBs were initially compiled on a global basis by the IUCN based on available information, but this concept was subsequently adopted at national and regional levels in several countries, resulting in local RDBs. Hilton-Taylor compiled the more recent one by IUCN in 2000.

Each species covered in the RDB is assigned a threat category largely based on analysis of the factors affecting its existence and the extent of the effect of these factors throughout the distributional range of the concerned species. Key factors include changes in distribution pattern and range, degree and type of threat, population biology etc. IUCN Red Categories are applied to species on a global scale and should not be confused with the threat categories assigned to species at the national level.

IUCN Threatened Categories and 'Unknown' Categories

Threatened categories

The major IUCN threatened categories (IUCN Red List Categories, 1995) currently recognised, together with their definitions (as given in the Red List) are as follows:

(i) **Extinct (EX)**: Species not definitely located in the wild during the past 50 years but which may survive in cultivation. The category **Ex?** has been assigned in a few cases, denoting that IUCN is virtually certain that the species has recently become extinct. Some authors (Koopowitz and Kaye 1990) suggest that 'Extinct' should denote those taxa that have been totally lost and that the terminology 'Extinct in the wild' should be used to refer to species lost in the wild, while living under cultivation. IUCN dubs the latter categories **EW.**

(ii) **Endangered (EN)**: Species in danger (i.e., at serious risk of disappearing from the wild state within a few decades) of extinction and whose survival is unlikely if the causal factors continue to operate. In this category are included those taxa whose numbers have been reduced to a critical level or whose habitats have been so drastically reduced that they are deemed to be in immediate danger of extinction. Also included are taxa that may now be extinct even though seen in the wild in the past 50 years. The other criteria are 50% decline in the last 10 years; < 5000 km^2 area of occupancy or < 500 km^2 in fragmented areas; 2500 individuals or subpopulation of 250 mature individuals. The category **Critically Endangered (CR)** includes species that face an extremely high risk of extinction in the wild in the immediate future. These are characterised by 80% decline in the last 10 years, 100 km^2 occupancy or 10 km^2 in fragmented areas; estimated 250 mature individuals or subpopulation of not more than 50 individuals.

(iii) **Vulnerable (VU)**: Taxa likely to move into the Endangered category in the near future if the causal factors continue to operate. Included in this category are taxa in which most or all populations decrease in size because of overexploitation, extensive destruction of habitat or other environmental disturbances. Also included are taxa with populations that are still abundant but under threat from severe adverse factors throughout their distributional range. The other criteria include 50% decline in the last 20 years; < 20,000 km^2 occupancy or < 2000 km^2 in fragmented populations; 10,000 individuals or subpopulation of 1000 mature individuals.

(iv) **Rare (R)**: Taxa with small populations that are not endangered or vulnerable at present but are at risk are included under this category. A species may be rare because of restricted geographical range, high habitat specificity, and small local population size, or thinly scattered over a more extensive range, or due to a combination of two or more of these characteristics. Human-caused rarity may be more devastating than rarity due to natural causes. Drury (1974) defined a rare species as one that 'either occurs in widely separated, small subpopulations or is restricted to a single population'.

(v) **Indeterminate (I)**: Species considered definitely to be endangered, vulnerable or rare, but for which information is insufficient to categorically assign them to any of these three categories.

'Unknown' categories

(i) **Insufficiently known (K)**: Taxa that are *suspected* but not definitely known to belong to any of the above categories, due to lack of information.

(ii) **Status unknown (?)**: No information.

(iii) **Candidate (C)**: Taxa whose status is being assessed and which are suspected but not yet definitely known to belong to any of the above categories.

Not-threatened category

(i) **Safe or not threatened (nt)**: Neither rare nor threatened.

Not-evaluated category

A taxon that has not been assessed against the criteria comes under this category.

It is to be noted that some combinations are permitted within the threatened categories. For example, the following combinations indicate that a species belongs to one of the two named categories in the combination: Extinct/Endangered (EX/EN), Endangered/Vulnerable (EN/VU), Endangered/Rare (EN/R), Vulnerable/Rare (VU/R.) etc. Combinations between the threatened and 'unknown' categories do not, however, indicate that a species could be anywhere on the scale covered by the two categories; if such is the case, the category **Unknown** should be used. Instead, combinations between threatened and not-threatened categories indicate that a species is on the borderline between the two categories, or that the species is in one category in part of its range and in another category elsewhere. For example, Vulnerable/Not Threatened (VU/nt) indicates that a species is vulnerable in one part of its range but safe elsewhere. The same holds true for Rare/Not Threatened (R /nt).

Categories must be assigned to entities in relation to a predefined area, which can be as large or small as one likes. Consequently, a species may be assigned to different categories of threat depending on the area considered. A species that is Extinct in one place may be Endangered in another and Rare in yet another, but might be only rare or even safe at the world level.

Census of Threatened Species

Raven (1987) mentioned that of the 250,000 species of vascular plants, 170,000 are tropical and subtropical. Of the 130,000 tropical plant species, some 60,000 are threatened and at risk of extinction within the next half-century. Of the 80,000 temperate plant species, about 8000 are threatened. In a more recent estimate, a total of 26,106 species of plants were considered threatened by the World Conservation Monitoring Centre (WCMC 1994); this number accounts for 8% of the world's plant species. Of these, 3632 species of plants belong to the category Endangered, 5687 to Vulnerable, 11,485 to Rare and 5302 to Indeterminate (WCMC 1994).

Table 7.3 provides data on the total number of threatened species in the major countries of the world, as per data presented by Groombridge (1992). A statistical analysis of threatened taxa of various countries indicated that the Maldives, Indonesia, India, Brazil, China and Australia have more threatened species in relation to the country's area.

Common Features of Threatened Species

As already indicated, the term 'threatened' is used to refer to a species which may be assigned to any one of the different categories described earlier. But the one common feature of all threatened categories is **rarity** (this word should not be confused with the 'Rare' category of threatened species). Rarity, according to Fiedler and Ahouse (1992), describes three different situations: (i) Taxa whose distribution is broad but population sizes are never large where found. (ii) Taxa whose distributions are clumped or narrow, yet whose populations are represented by many individuals where they are found. Finally, (iii) Taxa whose distribution is clumped and whose individual abundance is low where found (Fig.7.2). Incorporating a temporal component can extend the geographic definitions of rarity and by doing so, Fiedler and Ahouse (1992) obtained four categories of rarity (Fig. 7.3). A species may become rare for a large number of reasons, some at least operating in concert, to maintain a rare plant's distribution, abundance or either (Fiedler and Ahouse 1992) (Box 7.1). In addition to the reproductive biological causes mentioned by Fiedler (1986) and Fiedler and Ahouse (1992) (see Box 7.1), the

Table 7.3 Threatened species of plants, continent-wise (data from Groombridge 1992)

1. Asia	6608 [contributed mainly by Turkey (1994), India (1336), Jordan (752), Malaysia (522), China (350), Vietnam (338), Iran (301), Sri Lanka (220) and Philippines (159)]
2. USSR	No data.
3. Europe	2677 [contributed mainly by Spain (936) Greece (526), Portugal (240), Italy (210), Yugoslavia (190) and France (143)]
4. North and central America	5747 [contributed mainly by USA (2262), Mexico (883), Cuba (860), Panama (549), Costa Rica (419) and Guatemala (282)
5. South America	2061 [contributed mainly by Peru (360), Colombia (327), Brazil (318), Chile (284), Ecuador (256), Argentina (159) and Venezuela (106)]
6. Oceania	2673 [contributed mainly by Australia (2024), New Zealand (232) and New Caledonia (168)]
7. Africa	3308 [contributed mainly by South Africa (1016), Mauritius (209), Madagascar and Morocco (194 each), Tanzania (158), Algeria (145) and Kenya (144)]

Box 7.1 List of probable causes of rarity in vascular plant species (modified from Fiedler 1986).

I. AGE OF TAXON
1. Old and senescent
2. Young and incipient
3. Intermediate age

II. COEVOLUTION

III. EARTH HISTORY
1. Plate tectonics
2. Vicariance

V. ECOLOGY
1. Habitat specificity
2. Effects of present climate
3. Effects of edaphic conditions
4. Effects of specific predators
5. Effects of generalist predators
6. Effects of specific pathogens
7. Effects of generalist pathogens
8. Competitive ability

V. EVOLUTIONARY HISTORY
1. Effects of past climatic changes or stasis
2. Mode of origin of species
3. Phyletic momentum or inertia

VI. LAND-USE HISTORY
1. Effects of land management
2. Habitat conversion
3. Interaction with exotic plant competitors
4. Interaction with exotic flower visitors that do not enact pollination
5. Interaction with exotic herbivores
6. Effects of fire suppression and wildfires

VII. LIFE HISTORY 'STRATEGIES'
1. Herbivory escape
2. Allelopathy

VIII. POPULATION DYNAMICS
1. Schedule of births and factors influencing recruitment
2. Schedule of death and factors influencing mortality
3. Assessment of status of populations (e.g., e' declining stable, increasing)

IX. HUMAN USES
1. Horticultural trade
2. Aboriginal uses
3. Role in ancient and /or modern medicine
4. Role in past and /or current industry

X. REPRODUCTIVE BIOLOGY
1. Average number and range of flower production per reproductive individual
2. Average number and range of fruit production per reproductive individual
3. Average number and range of fruit set per reproductive individual
4. Average number and range of seed produced per reproductive individual
5. Average number and range of seeds set per fruit
6. Average number and range of seed set per reproductive individual
7. Pollination biology
8. Seed dispersal methods, agents and distance
9. Seed germination dynamics
10. Seedling establishment

XI. STOCHASTICITY
1. Demographic
2. Environmental

XII. TAXON GENETICS
1. Individual genetics (e.g. depauperate/ depleted genotype, detectable heterogeneity comparable to closely related common taxon)
2. Population genetics
3. Product of hybrid speciation

XIII. TAXONOMIC HISTORY
1. Meaningful taxonomic level (i.e., is rarity a taxonomic artefact?)

type of breeding system of the species may also play an important role in rarity (Weller 1996). Breeding systems by themselves may not contribute to a plant's rarity, but they operate through their profound effects on the genetic variations of the taxon, i.e., they have a pervasive indirect effect.

Rarity was analysed by Rabinowitz (1981) and Rabinowitz *et al.* (1986) and three dimensions applied in the analyses: (i) geographic range—whether large or small, (ii) habitat specificity—whether broad or restricted and (iii) local population size—whether large somewhere or always small. The authors

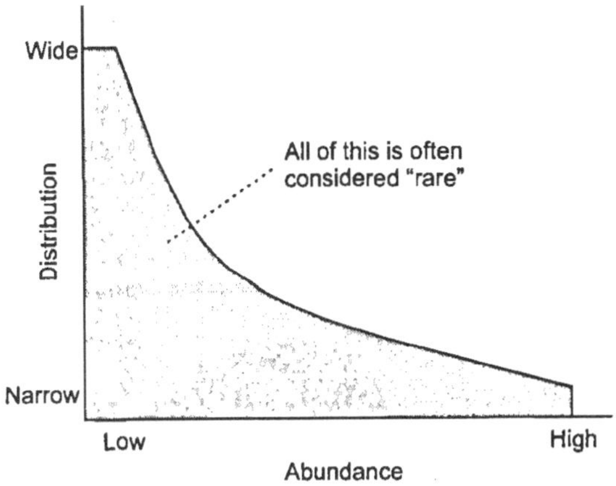

Fig. 7.2 Spectrum of combination of abundance and distribution values, from very narrow distribution with great abundance to very broad distribution with very low abundance, causing rarity of a species (from Fiedler and Ahouse 1992).

SPATIAL DISTRIBUTION	TEMPORAL PERSISTENCE: SHORT	TEMPORAL PERSISTENCE: LONG
WIDE	SHORT/WIDE (e.g. *Isotria medeoloides, Aconitum novoboracense, Agrostis hiemalis*)	LONG/WIDE (e.g. *Helianthemum canum, Torreya californica, Simmondsia chinensis*)
NARROW	SHORT/NARROW (e.g. *Betula murrayana, Clarkia franciscana, Spiranthes parksii*)	LONG/NARROW (e.g. *Lodoica maldivica, Cyathea dryopteroides, Rhododendron chapmanii*)

Fig. 7.3 Four categories of rarity in vascular plants, as defined by temporal persistence and spatial distribution (adapted from Fiedler and Ahouse 1992).

indicated that although the habitats of rare species should be studied and conserved as a priority, since all these three ecological characteristics are independent of one another, special attention must be paid to all three in any conservation effort.

Loss of Ecosystem Diversity

Loss of Ecosystem Diversity may be considered as the ultimate cause for loss of both species and genetic diversities. This has been amply indicated by fossil data as well as information presently available. Both deterministic and stochastic processes, described as responsible for species extinctions, are also responsible for loss of ecosystem diversity. This section delineates the threat factors affecting ecosystems in general as well as the magnitude of loss estimated for the major ecosystems of the world and the major causes for such a loss.

Factors Affecting Ecosystem Degradation and Loss

The various mechanisms involved in the loss of ecosystem diversity fall into five major categories (Diamond 1986; Fitter 1986): overkill, habitat destruction, impacts of introduced animals and weeds that later become invasive, pollution and secondary losses. Overkill denotes the uncontrolled organised collection (for scientific and industrial purposes) and killing of plants. Habitat destruction can be brought about by an array of organised land conversion causes, such as agriculture, housing, construction of roads and dams, industrial development, gravel and sand quarrying, wetland draining and filling, slash-and- burn (shifting) cultivation, tourism etc. Desertification can also be responsible for habitat destruction. Introduced animals, pests and invasive weeds cause impacts on ecosystems by displacing local taxa and by affecting community structure, biogeochemistry, fire regimes, erosion, geomorphology, hydrological cycle etc. (see Mooney 1996). Water table changes, trampling and overgrazing by animals, herbivory by smaller animals, unwanted competition between the introduced organisms and native ones, diseases and predation and disappearance of symbionts, pollinators and dispersers are other changes introduced directly or indirectly by exotic invasive organisms. Pollution can be caused by a number of factors, mostly human-generated. Land, water and air may all become polluted, markedly affecting the ecosystem components. Secondary losses may be induced by a combination of two or more of the aforesaid factors.

The most important phenomenon that causes ecosystem loss is **fragmentation;** it ranks among the most serious causes of erosion of biodiversity (Harris and Silva-Lopez 1992; Wilcox and Murphy 1985). Fragmentation may be defined as an 'unnatural detaching or separation of expansive tracts of habitats into spatially segregated fragments' that are too limited to maintain their different species for an infinite future. This phenomenon was observed as early as 1855 when de Candolle (see Browne 1983, p. 44) noticed that 'the break-up of a landmass into smaller units would necessarily lead to the extinction or local extermination of one or more species and the differential preservation of others'. Based on the operational mode of forces, fragmentation can be divided into five categories: Regressive, Enveloping, Divisive, Intrusive and Encroaching (Fig. 7.4).

Fragmentation leads to artificially created 'terrestrial islands' (Frankel *et al.* 1995). Such fragments experience microclimatic effects markedly different from those that existed in the large tracks of habitats before fragmentation (Groom and Schumaker 1993; Saunders *et al.* 1991). Air temperature at the edges of fragments can be significantly higher than that found in the interior (Kapos 1989); light can penetrate deep into the edge, thereby affecting the growth of existing species (Lovejoy *et al.* 1986). Fragmentation promotes the migration and colonisation of alien species and such colonisation is often substantial and continuous, profoundly affecting the survival of native species. The most serious effect of fragmentation is segregation of the larger populations of a species into more than one smaller population. The serious consequences of smaller populations on erosion of biodiversity have already been indicated. There is considerable evidence that the number of species in a fragmented habitat will decrease over time, although the probable rates at which it will happen are variable. In fact, actual data on rain forests show that forest fragments have lower species richness and fewer populations compared with continuous undisturbed forests.

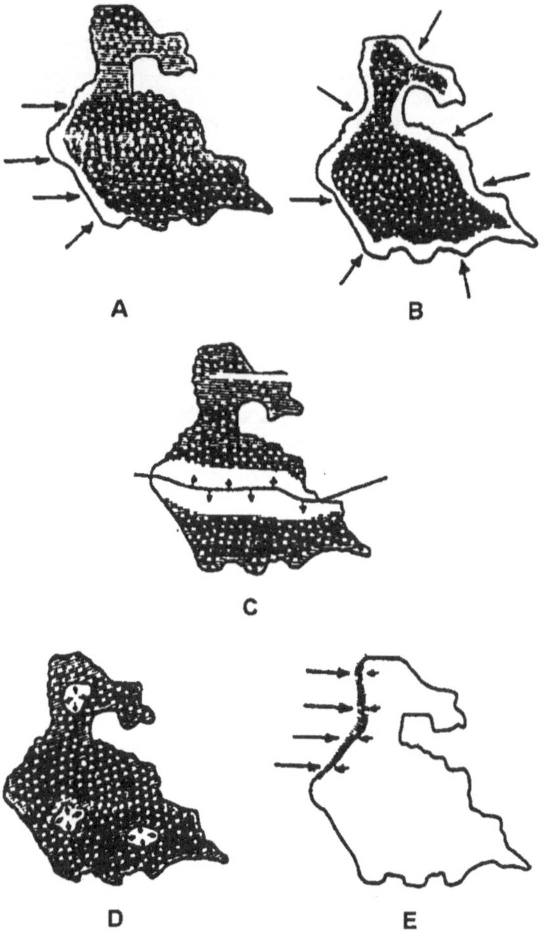

Fig. 7.4 Different types of fragmentation processes (from Harris and Silva-Lopez 1992). Arrows indicate the direction of extension of fragmentation processes while shaded regions indicate forest. Clear zones near the arrows indicate regions of edge effect.

Loss in Diversity of Major Ecosystems of the World

Tropical Forests

Factors responsible for loss

Logging, shifting cultivation and agriculture, mining and drilling, laying roads and railway lines, tourism, developmental activities (including development of townships and industries), dams and hydroelectric installations, flooding due to dams, development of cattle ranches (especially in Brazil and Central America), plantations (rubber, coffee, tea, sugar-cane, banana, oil palm, fruit trees, teak and other timber trees etc.), forest fire, collection of minor forest produce and fuel sources (firewood, charcoal etc), pollution, scientific activity, transmigration etc.

Estimates of loss

The true extent of loss is still not known, as it is very difficult to determine. Satellite data are incomplete and ground data are known only for a relatively few areas (Myers 1994). However, the rate of loss is currently very high (Lanly *et al.* 1991; Myers 1989; Whitmore and Sayer 1992). There are two important problems in the estimation of loss of tropical forest ecosystems. The first is the problem of definition of 'forest' and the second the problem of definition of 'deforestation'. There is great confusion in recognising what a forest is. Deforestation, according to some, refers to complete destruction, while according to others it includes degradation and fragmentation as well. Hence the estimated rates will tend to differ. Rates of loss estimated for various countries and continents also vary greatly (Figs. 7.5 and 7.6; Tables 7.4 to 7.6).

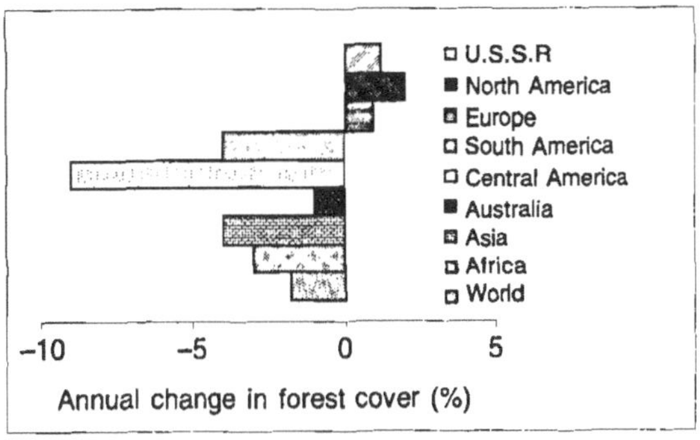

Fig. 7.5 Estimates of average annual rates of change in forest cover around the world for the period 1977-1989. Deforestation is estimated as the amount of forest converted to non-forest use (adapted from WRI 1992).

Grasslands

Factors responsible for loss

Fire (natural and man-made), pests and diseases, herbivores (grazing and browsing), inundation by water, conversion into agricultural land, introduction of alien grasses and legumes, weeds, human settlement and development of ranches.

Estimates of loss

There are fairly enough data on the loss of grassland ecosystem. A study by Denisiuk (1990) in Poland indicated that of the total 40, 400 km^2 area of grasslands, about 10,000 km^2 have been abandoned. The area of wet grassland was reduced by drainage from 36% to 23% between 1973 and 1978. In Paraguay, the Chaco grasslands originally occupying an area of 320,000 km^2, had been reduced to just 57% of this territory by *circa* 1990 (Redford *et al.* 1990). In Canada, there were 360,000-400,000 km^2 of open grasslands by the 19th century, of which only *circa* 80,000 km^2 remained by 1982; the rest of the area had been converted to other forms at a rate of 500 km^2 per year (Mondor and Kun 1982). In South Africa, the Fynbos habitat had an original extent of 75,000 km^2, but by about 1988 only 67% of this ecosystem remained (Mooney 1988). In North America, the tall prairie grasslands had an original area of 143,000 km^2 but by about 1990, only one percent of this grassland remained (WRI 1991). In Latin America, a study by Houghton *et al.* (1991) indicated a reduction of grasslands from about 340 x 10^6 ha in 1850 to about 125 x 10^6 ha in 1987 (interpreted from their graph).

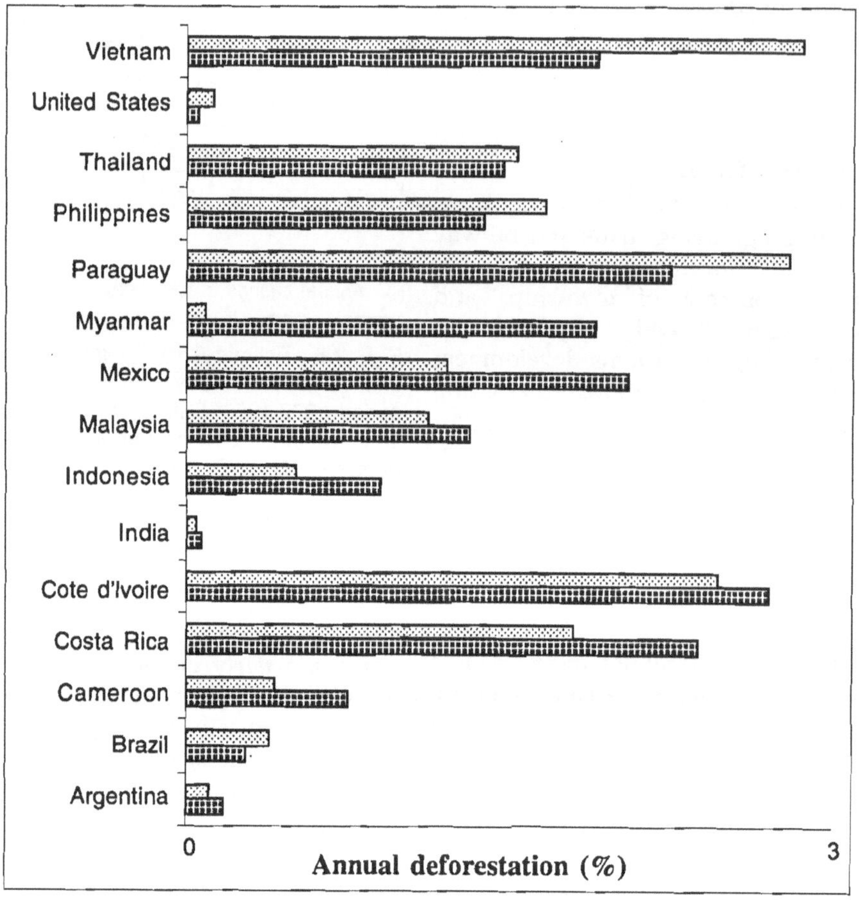

Fig. 7.6 Two different estimates of annual deforestation rates for 15 countries. Dark bar indicates data from FAO 1980 and 1990 estimates, while stippled bar indicates data from several independent estimates made between 1980 and 1991 (adapted from Groom and Schumaker 1993).

Inland Wetlands

Factors responsible for loss

A number of factors are responsible for the loss of inland wetlands. These include the following: pollution, hunting, fishing and associated disturbances, drainage for agriculture and ranching, disturbance due to recreation, reclamation for urban and industrial development, human settlement and encroachment, siltation, diversion of water supply, conversion to aquaculture ponds, alien species invasions (especially by species of *Salvinia* and *Eicchornia*) and introductions, increase in temperature and salinity, drought, navigation, transport, waste disposal, dam construction, groundwater abstraction, etc. (Allan and Flecker 1993; Dahl 1990; Dugan 1990; Finlayson and Moses1991; Moyle and Leidy 1992; NRC 1992; Scott and Carbonell 1991; Scott and Poole 1989).

Table 7.4 Some estimates of loss of tropical forests

Ser. No.	*Estimated deforestation*	*Reference*
1.	110,000 km^2 per year (1.176% loss)	Sommer 1976
2.	245,000 km^2 per year	Myers 1979
3.	200,000 km^2 & possibly more per year	Myers 1980b
4.	12.3% deforestation between 1980 & 2000	Lovejoy 1980; Lugo 1988b
5.	> 200, 000 km^2 per year	NRC 1980
6.	33% of forests destroyed by 2000	Myers 1985a
7.	0.7% loss per year	Wilson 1988b
8.	2% loss per year	Raven 1988a
9.	72,670 km^2 of closed forests per year (0.60% loss) between 1976 & 1980, 74,960 km^2 of closed forests per year (0.62% loss) between 1981 & 1985; 38,070 km^2 of open forests per year between 1981 & 1985 (0.52% loss)	FAO 1988
10.	0.8% loss per year as per one estimate and 1.6% loss as per another estimate	Reid and Miller 1989
11.	1.2% loss between 1981 & 1990 in 62 Tropical Countries (Total loss 168,000 km^2 per year)	FAO 1991a
12.	0.9% loss between 1981 & 1990 in 81 Tropical Countries (Total loss 170,000 km^2)	FAO 1991b
13.	1.8% loss per year	Wilson 1992
14.	About 2.5% loss per year between 1977 & 1989	WRI 1992

Table 7.5 Some estimates of global species loss due to tropical deforestation and their assumptions

Extinction estimate	*Total no. species (millions)/ per cent tropical*	*Tropical forest loss*	*Extinction/ area lost*	*Source*
1 species/h by 2000	5-10/40%-70%	245,000 km^2/y	50% species extinct when 10% area left	Myers 1979
33%-50% of all species between 1970 and 2000	3-10/25%	12.3% deforestation between 1980 and 2000	Species-area concave curve	Lovejoy 1980
1 million species by 2000	3-10/70%	> 200,000 km^2/y	50% species extinct when 10% area left	NRC 1980 Meyers 1980a
1 million species by 2000	5/40%	33% of remaining forests destroyed by 2000	50% species in area will go extinct	Myers 1985a
10% of all species by 2000; 25% of all species by 2015	4-5/50%	2% deforestation/y	50% species in area will go extinct	Raven 1988a
17,500 species/y	10/50%	0.7% deforestation/y	50% species in area will go extinct	Wilson 1988b

(Table 7.5 Contd.)

(Table 7.5 Contd.)

8.8% of all species by 2000	3-10/25%	12.3% deforestation between 1980 and 2000	Same as Lovejoy 1980	Lugo 1988a,b
5%-15% of all species between 1990 and 2020	10/ > 50%	0.8% deforestation/y	Species-area curve; z = 0.15, 0.35	Reid and Miller 1989
10%-38% of all species between 1990 and 2020	10/ > 50%	1.6% deforestation/y	Species-area curve; z = 0.15, 0.35	Reid and Miller 1989
27,000 species/y	10 in tropical rain forests	1.8% deforestation/y	Species-area curve; z = 0.15	Wilson 1992

Table 7.6 Tropical forest area and reported deforestation rates listed by country (from Skole and Tucker 1993)

			1970s		*Late 1980s*			
Country	*Total forest area (km²)*	*% of world total*	*Deforestation rate (km²) (FAO)*	*% of world total*	*Deforestation rate (km²) (Myers)*	*% of world total*	*Deforestation rate (km²) (WRI)*	*% of world total*
Brazil	3,562,800	30.7	13,600	19.7	50,000	36.1	80,000	48.4
Indonesia	1,135,750	9.8	5500	8.0	12,000	8.7	9000	5.4
Zaire	1,056,500	9.1	1700	2.5	4000	2.9	1820	1.1
Peru	693,100	6.0	2450	3.6	3500	2.5	2700	1.6
Colombia	464,000	4.0	8000	11.6	6500	4.7	8200	5.0
India	460,440	4.0	1320	1.9	4000	2.9	15,000	9.1
Bolivia	440,100	3.8	650	1.0	1500	1.1	870	0.5
Papua New Guinea	337,100	2.9	210	0.3	3500	2.5	220	0.1
Venezuela	318,700	2.7	1250	1.8	1500	1.1	1250	0.8
Burma	311,930	2.7'	920	1.3	8000	5.8	6770	4.1
Others*	2,829,930	24.4	33,300	48.3	44,100	31.8	39,610	23.9
Total	**11,610,350**	**100.0**	**68,900**	**100.0**	**138,600**	**100.0**	**165,400**	100.0

* Sixty-three other countries

Estimates of loss

It is extremely difficult to assess the loss of inland wetland ecosystem for most countries. In New Zealand, over 90% of natural wetlands have been destroyed since European settlement (Dugan 1990). Table 7.7 provides details about the severity of threats to wetlands of Asia. It is evident from this Table that the percentage of sites with moderate to high threat may reach as high as 86%, but on average 36% of fully protected wetlands and 47% of unprotected wetlands of Asia are subjected to moderate and high threats (Scott and Poole 1989). The sites considered already too degraded to merit any conservation effort are 23 of 93 major wetlands in Asia. In the USA, of a total 895,000 km^2 area of wetlands that existed in the 1780s, only 422,397 km^2 of wetlands remained in the 1980s,

Table 7.7 Severity of threats to wetlands of international importance in Asia (from Scott and Poole 1989)

	Number of Sites Known	*Degree of Threat*				*% Sites with Moderatre to High Treat*
		None	*Low*	*Mod*	*High*	
Bangladesh	11	1	1	5	4	82
Bhutan	5	3	—	1	1	40
Brunei	3	—	2	1	—	33
China	105	30	34	36	5	39
Hong Kong	3	1	—	2	—	67
India	88	4	44	22	18	45
Indonesia	129	1	54	66	8	57
Japan	38	8	11	17	2	50
Cambodia	3	—	1	2	—	67
Korea, DPR	5	5	—	—	—	0
Korea, Rep.	19	5	3	6	5	58
Laos	3	—	1	2	—	67
Malaysia	37	—	5	22	10	86
Mongolia	30	23	5	2	—	7
Myanmar	16	—	7	8	1	56
Nepal	14	2	7	4	1	36
Pakistan	42	1	20	15	6	50
Papua New Guinea	27	14	8	4	—	15
Philippines	49	2	13	24	10	69
Singapore	6	—	2	3	1	67
Sri Lanka	31	2	8	13	8	68
Taiwan	12	1	4	5	2	58
Thailand	36	1	18	14	3	47
Vietnam	23	3	14	4	2	26
Total	**734**	**107**	**262**	**278**	**87**	**50**

constituting a 53% loss; in Alaska, however, there was only a 0.1% loss of wetlands during this period (Dahl 1990).

Degradation of wetlands is a very critical problem when we consider their rates of recovery from damage. These rates are variable, but as a rule of thumb are proportional to the water renewal time and the generation times of the largest-living keystone species of the wetlands—riparian trees. Water renewal time for rivers is about 0.05 year but for lakes may range up to 100 years (Wetzel 1983); generation time for riparian trees may be as high as a decade or more.

Coastal Ecosystems Including Mangrove System

Factors responsible for loss

These ecosystems are subjected to a number of natural and man-made impacts, which directly or indirectly cause their loss (Aronson 1990; Barbier *et al.* 1994; Folke *et al.* 1991; Hammer *et al.* 1993; Keller and Jackson 1993a, b; Norse 1993; NRC 1995; Perrings *et al.* 1994; Saenger *et al.* 1983; Saenger 1986). The threats include: overexploitation of resources for short-term economic gain for purposes such as food, feed, fuel, pulp, tannins, glues, chemicals etc., coastal aquaculture, trampling, trawling, dredging, drilling, building, waste dumping, pollution (noise, thermal, chemical), oil spills, agriculture, urban and industrial development, diversion of fresh water etc.

Estimates of loss

Data on the degree of loss of coastal ecosystems are inadequate although it is generally agreed that these are among the most degraded ecosystems of the world. The mangroves in Nigeria have lost 50% of their original area of occupancy, i.e. 24,440 km^2 (Mackinnon and Mackinnon 1986).

Arctic and Alpine Systems

Factors responsible for loss

Most, if not all, factors known to affect the arctic and alpine ecosystems are not local but contributed from outside them. The most important is the CO_2-induced global warming (greenhouse effect), which is already causing upward migration of alpine plant and animal species (Grabherr *et al.* 1994) and is predicted to cause northward and upslope migration of tree-lines as well (Briffa *et al.* 1990; D'Arrigo *et al.* 1987). Arctic haze is again an effect caused by pollution originating at lower latitudes; this reduces the cover of mosses and lichens (Lechowicz 1987), the insulating layer that protects permafrost integrity (Tenhunen *et al.* 1992). In addition, intense agricultural activities in alpine areas and industrial development in arctic regions have substantially altered biodiversity. Inouye and McGuire (1991) indicated that human-induced global warming could alter the competitive balance and diversity of plant species in the arctic environment and 'could decouple flowering phenology and pollinators in such a way as to result in the elimination' of many plant species.

Estimates of loss

The exact extent of loss has not been estimated in the arctic and alpine ecosystems.

Boreal Forests

Factors responsible for loss

Greenhouse gases and consequent global warming, logging, plantations of conifers and other factors cause an alteration first of all in tree species diversity. There is also a gradual migration of species northwards because of increasing temperature (Pastor and Johnston 1992; Pastor and Post 1988).

Estimates of loss

Data highly inadequate.

Temperate Forest Systems

Factors responsible for loss

Almost all factors mentioned for tropical forest ecosystems are also involved here. For centuries these forests have experienced very intense human impacts. The temperate forests of eastern China were the first to be cleared on a large scale for agricultural development before 3200 B.P. (Teng 1927). Clearing of these forests was done much later in parts of Europe, North America and some areas of the Southern Hemisphere (Perlin 1989). Conversion of native forests into plantations, especially of conifers, was also a factor in the loss of temperate forest systems. Pests/pathogens (introduced newly), water and air pollution, use of fertilisers, development of roads/railway lines, surface mining and urbanisation were the other important factors involved in deforestation.

Estimates of loss

Information on the extent of loss of temperate forests is scant. Mathews (1983) indicated that more than 50% of temperate forests (of 18.6 x 10^6 km^2) along the Pacific coasts of North and South America have been converted to cultivation.

Arid and Semi-arid Lands

Factors responsible for loss

Grazing by livestock, introduction of domestic and game animals, creation of new water points, introduction of alien plants, removal of predators, intensive cultivation etc. (Jackson *et al.* 1991; Milchunas and Lauenroth 1993; Westoby *et al.* 1989).

Estimates of loss

Data highly inadequate.

Open Oceans

Factors responsible for loss

Increase in UV radiation incidence due to ozone holes, temperature rise due to global warming (an average increase of 0.32°C over the last 35 years—Parilla *et al.* 1994), pollution (air pollution due to land-based activities; water pollution due to oil spills, noise pollution due to ships, boats etc.), resource extraction and associated activities such as drilling, mining etc. All these factors individually or in combination affect the pelagic organisms of the open sea.

Estimates of loss

Data highly inadequate.

Loss of Agrobiodiversity

Agrobiodiversity is very often equated with richness in crop varieties, i.e., the more the number of varieties within a crop species, the greater the agrobiodiversity of that crop species. Agrobiodiversity richness is also the economic unit of benefit valuation. With loss of agrobiodiversity, narrowing of species richness has occurred on a global level. Of the 400,000 and more plant species reported, by conservative estimates, for the Earth, only *circa* 300,000 have been documented. Of the latter, just 10% are edible (i.e., 30,000 species) and among these 30,000 species, 7000 are either cultivated or collected by humans for food, feed or other agricultural purposes at one time or another, with just 200 constituting major domesticates. Of these, only 30 are of paramount importance and among these just four constitute primary staple foods, namely rice (26%), wheat (23%), maize (7%) and potato (3%). These four crops alone account for approximately 25-28% of all 6.2 million *ex situ* stored crop accessions (WIEWS 1996). Furthermore, most of the money spent on genetic resources activities is expended on these four crops.

Besides the narrowing down of species richness, a decrease in varietal diversity in agrobiodiversity has also been noted. Some examples were mentioned in Chapter 4. A conspicuous example, however, may be cited here: 75% of the area under rice cultivation which once accommodated 30,000 rice varieties is today sown with only 10 varieties.

It must be understood that every new breeding activity adversely narrows the genetic base of the source variety. On the other hand, incorporating new and favourable traits by introducing genes broadens the genetic base of the variety. Loss of genetically coded information from the agricultural field is on the increase due to the rapid disappearance of traditional varieties and land races.

Projected Scenario for Biodiversity Loss

Global biodiversity loss is taking place at an unprecedented rate as a complex response to several human-induced changes. The magnitude of this loss is so great and so strongly linked to ecosystem processes and society's use of biodiversity resources that it is now considered an important global change in its own right (Sala *et al.* 2000). There are, however, currently very few projected scenarios for biodiversity change in 10 terrestrial biomes and in freshwater ecosystems for the year 2100. Sala and colleagues (2000) identified the five important determinants of changes in biodiversity at the global scale: changes in land use, atmospheric CO_2 concentration, nitrogen deposition and acid rain, climate, and biotic exchanges (i.e. deliberate or accidental introduction of organisms into an ecosystem). Next they calculated the expected change of these drivers in each biome followed by estimation of the impact in each biome that a unit change in each driver exerts on biodiversity. They found that for terrestrial biomes, land-use changes would probably have the largest effect, followed by changes in climate, nitrogen deposition, biotic exchanges and elevated CO_2 concentration. Biotic exchange is much more important for a freshwater ecosystem. Mediterranean and grassland ecosystems are likely to experience the greatest losses in biodiversity. Changes in Northern Temperate ecosystems are likely to be the least because major land-use changes have already occurred there (Sala *et al.* 2000).

Loss of Biodiversity as an Economic Process

Biodiversity is the major natural resource of great economic value. Therefore, it is natural to expect that application of economic processes to bioresources will lead to significant losses of biodiversity (McNeely 1996). Loss of biodiversity is largely the result of these 'value-maximising' economic processes. Three important economic processes cause biodiversity loss: Conversion, Specialisation and Globlisation.

Biodiversity resources constitute a 'natural capital' and humans make the choice of whether to hold these resources in their original form or to modify them in some manner (Solow 1974). In the course of human history several instances are apparent wherein natural resources have been converted into more productive and usable forms acceptable to humans. One typical instance of the conversion process is the transformation of virgin land with several native plant species into a monoculture field. Economics has forced man to choose the single most productive plant species for channelling the solar energy in a field where earlier, several plant species grew and shared this energy; thus, the conversion process eliminates diversity in any given area.

The law of specialisation was one of the first laws of economics proposed by Smith as early as the 18th century. This law states: 'there tends to be increased productivity with increasing homogeneity in production methods and processes'. The prevailing methods of agricultural production, like the application of chemical fertilisers, pesticides, agricultural machinery etc., are biased against the maintenance of biodiversity of soil flora and fauna. Specialisation has also resulted in the use of an increasingly narrow spectrum of plant species to meet the needs of humankind. As mentioned earlier, of the several hundreds of plant species known to be edible, the most used are rice, wheat, maize and potato. Even within these four taxa specialised varieties are preferred. This specialisation process has eliminated other edible plants that were fairly widely used earlier, such as millets, thus leading to loss of agrobiodiversity.

Globlisation has come about in all aspects of human life, including trade in bioresources themselves or resource-based products. Globlisation and specialisation have gone 'hand-in-hand' to cause biodiversity loss worldwide in the process of 'furtherance of agricultural productivity'.

It may be concluded that most of the biodiversity loss is due to the exploitation of a relatively few plant species in agriculture at the cost of several others and to the specific methods of agricultural productivity. From an economic perspective this is inevitable. However, the moot point is whether biodiversity loss is presently taking place while optimally maximising benefits to humankind, or is it 'overshooting', i.e., is it very excessive. Certain indications support the second contention. Firstly, bioresources are presently traded off very readily for immediate gains, thus fulfilling Keynesian economic concepts. In the not so distant future, man will be searching for alternative sources of resources from natural biodiversity, but by then most of such biodiversity will have been wiped from this Earth. Secondly, economic specialisation in the production and use of specific elements of biodiversity is very risky because diversity provides insurance against unforeseen events while specialisation does not. The same can be said of conversions. It is imperative that man take adequate steps now not only to conserve biodiversity, but also to ensure sustainable use of the existing bioresources.

Conclusions

The causes and mechanisms of biodiversity loss and impoverishment have been detailed in this chapter. However, in concluding, attention is drawn to the six fundamental and complex causes often noticed within our policies, laws and management arrangements by various governments and people (UNEP 1995; WRI/IUCN/UNEP 1992):

(a) High rates of human population growth and biodiversity consumption;

(b) Greater and greater specialisation of traded products of agriculture/forestry leading to a very narrow spectrum of used products;

(c) Failure of economic systems and policies to adequately value biodiversity resources;

(d) Inequity in ownership and access to bioresources, including the benefits from their use and conservation;

(e) Inadequate knowledge and inefficient use of biodiversity information; and

(f) Poor or misused legal and institutional systems that promote an unsustainable use of biodiversity.

8

CONSERVATION OF BIODIVERSITY

Why Conservation and Conservation Biology?

Distinguishing between **'Preservation'** and **'Conservation'** is often problematical. Simply stated, 'Preservation' implies the protection of biodiversity from any kind of human activity or interference, while 'Conservation' implies the protection of biodiversity for sustainable utilisation (Virchow 1998). However, most of the earlier workers used the two terms casually, without differentiation, and even the present author has occasionally done so.

The idea that biodiversity is worth preserving rests on several fundamental arguments including nostalgia and human benefits and needs. The innate desire we all have for our children to experience the great pleasure or curious excitation biodiversity has given us is part of the nostalgic argument for biodiversity conservation. This argument goes so far as to say we were not bequeathed this Earth and its biodiversity by our ancestors, but have borrowed it from our children and must return it to them in the manner it was received. This nostalgic argument should not push us into construing conservation, however, as an act aimed at considering biodiversity an untouchable entity. There are arguments for 'explicit inclusion' of the biodiversity dimension within development planning (Weil 1986) and the argument of human benefit and need as more powerful 'in terms of the Descartian scientific tradition'. In essence, this means that unless there is biodiversity conservation, we may lose something of direct or indirect value to human society already known or yet to be discovered. This is amply reflected in the definition of conservation provided in the first World Conservation Strategy: 'The management of human use of the biosphere so that it may yield the greatest sustainable benefit to present generations, while maintaining its potential to meet the needs and aspirations of future generations. Thus, conservation is positive, embracing preservation, maintenance, sustainable utilisation, restoration and enhancement of the natural environment' (IUCN 1980). In other words, 'Conservation is a philosophy of managing the environment in such a way that it does not despoil, exhaust, or extinguish it or the resources and values it contains'.

Modern conservation biology traces its roots to the first International Conference on Conservation Biology at San Diego Wild Animal Park in 1973. A separate field of conservation biology emerged in the late 1970s and early 1980s as a response by the scientific community to the biodiversity crisis (Soulé and Wilcox 1980). Therefore, conservation biology was also called a 'crisis discipline' (Soulé 1985). The Society for Conservation Biology was founded in 1985 and has rapidly grown to include more than 3600 professional members (Duthie 1997). Many, however, would be tempted to ask the question, 'what is so new about conservation biology?', since people have been practising conservation for centuries (Meffe and Carroll 1994). But the 'new' and 'rejuvenated' conservation biology differs from the old in at least three respects (Meffe and Carroll 1994): (i) It now includes and, in fact, has been led in part by important contributions from theoretical academicians whose ecological/genetical models are increasingly applied to real-world situations. (ii) Most, if not all, traditional conservation efforts were based on an economic and utilitarian philosophy, which considers the maintenance of high yields of selected species for harvest as the principal motivating factor. On the other hand 'new' conservation biology views all biodiversity as equally important and as having inherent value[1]. In the light of this, 'new' conservation biology emphasises that management should be redirected towards the overall conservation of world biodiversity and ecosystems rather than to single target species. The rationale here is that the conservation of ecosystems can take care of individual species but not vice versa and that the ecosystems functioning in dynamic equilibrium not only serve as life-support systems for the Earth , but are also critical to the continuing survival of human kind. (iii) 'New' conservation biology emphasises that non-biologists also play a very important role in conservation. Conservation biology has thus become a new, mission-oriented and synthetic discipline (Gibbons 1992; Soulé 1986; Temple *et al.*, 1988). It applies the principles and results of diverse disciplines such as Ecology, Biogeography, Population Genetics, Economics, Sociology, Anthropology, Geology, Philosophy and many others to the maintenance and conservation of biodiversity across the world. It has therefore become a synthetic field 'to develop scientific principles and then apply them to developing technologies for the maintenance of biological diversity' (U.S. Congress Office of Technology Development, cited in Tangley 1988). Biodiversity conservation has thus been changed from an idealistic philosophy to a very serious technology. Conservation biology has also become a very challenging discipline in that it has 'to counter by human forces all that humans themselves have been responsible for'.

Current Practice in Conservation

Conservation of biodiversity can be attempted at three levels: Genes, Species and Ecosystems. As we have already seen in the first chapter of this book, the three constitute different levels of biological organisation and are interlinked. Maintenance of ecosystem diversity implies conservation of species which constitute that ecosystem, although it is feasible to conserve a species independent of the ecosystem of which it is a normal component. Maintenance of genetic diversity within a species implies maintenance of that species. Conservation of species diversity will take care to some extent of both ecosystem maintenance and genetic maintenance and hence many people consider it pivotal to the conservation of biodiversity. Moreover, loss of species diversity is more obvious and quantifiable than genetic or ecosystem diversity loss. Species in danger of loss can thus be readily identified and suitable conservation strategies planned. Conservation strategies based on species maintenance are labelled 'species-based' approaches. Some people, however, lay emphasis on 'Habitat- or ecosystem-based' approaches for conservation. The relative merits and demerits of either approach are discussed below.

[1]However, in very recent years certain biodiversity institutions, such as the Botanic Gardens Conservation International (BGCI), have instructed Botanic Gardens to give emphasis and priority to the *ex-situ* conservation of economically important plants that are threatened (see discussion later in this chapter).

Conservation of Genetic Diversity

It should be emphasised at the very outset that the unit for genetic conservation is a population of a species, i.e., genetic conservation strategies must be planned and executed at the population level.

All genetic conservation strategies and actions should be compatible with three conservation goals, on three time-scales of concern: (i) maintenance of viable population in the short term in order to avoid extinction, (ii) maintenance of the ability of the population to continue to undergo adaptive changes and (iii) maintenance of the ability of the population for continuing speciation. The first level of concern has a time-scale of a few days to a few decades and if not met, the other two goals are automatically denied.Maintenance of genetic diversity must also be planned in such a way that the population and species are able to genetically adapt and evolve in an unhindered manner. This second level of concern has a time-scale of several decades to millennia. The third level of concern, the capacity for continuing speciation, is considered the creative part of biodiversity, just as extinction is the annihilating part. The potential of a population (thereby the species) for continuing speciation is to be maintained at any cost and should be the ultimate goal of conservation, although its time-scale, tens of thousands or more years, makes appreciation of it difficult for humans. In any case, mere short-term conservation goals should be discouraged in the interest of biodiversity.

It is clear from the foregoing that genetic variation is important for short-term fitness, continuing adaptation and speciation. This realisation forces every conservationist to confront very difficult and practical questions: What are the units of genetic conservation and what, *de facto*, should we attempt to conserve? Salvation of every population of a species, every phenotypic variant and every unique allele is impossible. Therefore, as mentioned in the first para of this section, population seems to be the most reasonable level at which genetic conservation can be attempted. The reasons are as follows: (i) The population, and not the species, is the ecologically and evolutionarily significant (i.e. functional) unit (ESU). (ii) Genetic changes take place in the population over generations. (iii) Local adaptive changes likewise occur in the population. (iv) Geographically and genetically isolated populations offer the greatest potential for speciation. (v) Conservation at the species level will overlook the dynamics and attributes of individual populations within it as well as their ecological functions. (vi) Lastly, conservation at a level below the population, say at the allelic level, on the other hand, is impractical.

The acceptance of the population as the unit of genetic conservation strategies leads naturally to the next question: How many individuals in a population are needed for conservation of genetic diversity? Most research on this aspect has been done on animals, although the results are equally applicable to plants. It is suggested by the so-called '50/500 Rule' that a genetically effective population size (**Ne**) of at least 50 individuals is necessary for conservation of genetic diversity in the short term and to avoid inbreeding depression. A **Ne** of 500 is needed to avoid serious genetic drift in the long term (Franklin 1980; Soulé 1980). Although these figures, arrived at by simple genetic models, may appear to be reasonable estimates of minimum numbers needed, may be misleading in many cases (Lande 1988) for the following reasons: (i) the 50/500 rule very often ignores demographic, ecological, and behavioural considerations and (ii) this rule may also discourage conservation attempts in situations where less than the prescribed population size is available. Therefore, several authors feel that strict quantitative rules regarding genetic diversity conservation should be generally avoided, or at least be applied with great caution. A number of guidelines have emerged based on our current knowledge of genetic diversity conservation (see Box 8.1).

Genetic conservation has its own limitations; it will not be the saviour of biodiversity. Application of genetics to conservation is a very young science still in the developmental stage; several aspects have yet to be understood and such limitations impel a realistic approach. Genetic diversity of species and ecosystems in isolation may be nearly worthless, for reasons given below. Furthermore, logistic limitations in effecting genetic conservation have to be considered. Many genetic techniques useful in assessing genetic diversity are not cheap, are not easily learned and can be misused or misapplied; the techniques require experience, expertise and sophistication. There is also the danger of genetic conservation taking the focus off larger issues and 'sinking into esoteric details of genetic analysis'. Genetic conservation should be equated to 'fine tuning' management procedures, once the 'coarse tuning' of ecosystem/species conservation has been satisfactorily done.

Conservation of Species Diversity

It has already been proved that one of the main players in conservation conceptually, biologically and legally, is the species (Meffe and Carroll 1994). Most people are aware of the biodiversity entity called species, i.e., 'species' is familiar to many. Moreover, many powerful legislations on conservation at the world and national levels are focused on species. These include, for instance, CITES, Endangered Species Act of USA (ESA) etc. Loss of species diversity is also very obvious and more easily detectable and quantifiable than either loss of genetic or habitat diversity. Even conservation approaches based on habitats or ecosystems depend on an intimate understanding of the biology of their constituent species. The design and management of bioreserves are also often based on a knowledge of species-area relationships, life-history requirements of the species that are more focused for conservation in such reserves and the minimum of individuals of a species necessary to avoid major loss of genetic diversity.

Box 8.1 Qualitative Guidelines for Genetically Based Conservation Practices (from Meffe and Carroll 1994)

1. Large genetically effective population sizes are better than small ones because they will lose genetic variation more slowly.
2. The negative effects of genetic drift and inbreeding are inversely proportional to population size. Thus, avoid managing for unnaturally small populations.
3. Management of wild populations should be consistent with the history of their genetic patterns and processes. For example, historically isolated populations should remain isolated unless other concerns that dictate gene flow must occur. Gene flow among historically connected populations should continue at historical rates, even if that calls for assisted movement of individuals.
4. Low genetic diversity *per se* is not cause for alarm, because some species historically have low diversity. However, sudden and large losses of diversity in natural or captive populations are always cause for concern.
5. Avoid artificial selection in captivity. This is best done by keeping breeding populations in captivity for as few generations as possible, and also by simulating wild conditions as nearly as possible.
6. After a population crash, encourage rapid population growth to avoid a prolonged bottleneck.
7. Avoid possible outbreeding depression caused by breeding distantly related populations if other choices are available.
8. Avoid inadvertent introductions of exotic alleles into wild or captive populations.
9. Harvesting of wild stocks (hunting, fishing) can select for genetic changes, which can affect the future evolution of the population or species. For example, culling the largest individuals can select for earlier maturity at smaller body sizes. Thus, avoid selection in harvesting wild stocks.
10. Maintenance of genetic diversity in captive stocks is no substitute for genetic diversity in the wild. Technological mastery over the genome should not be used as an excuse to overexploit or destroy a species or population in the wild.

The species-based approach first identifies those species considered to be of high priority for conservation. This will ensure preservation of the highest level of biodiversity within limits of available resources. Most often, threatened species and those of actual or potential resource value or keystone, dominant and crucial species required for the well-being of an ecosystem are selected on a priority basis for conservation. The first category of species to be given priority are the Threatened species, exhibiting rarity (Fuller 1987; Furtado 1987). However, before doing so, the species selected for conservation should be broadly assessed for a range of factors leading to rarity (Whitson and Massey 1981). Details are given in Table 8.1. The preference for endemic taxa is one possible and positive approach. The other approach is through a method called **Cladistic Prioritisation**. This method is designed to assess, and to some extent quantify, the distinctiveness between taxa considered for conservation. It is based on phylogenetic relationship between species, expressed as divergence since their most recent common ancestor. Any character ranging from morphological to molecular can be used to establish this phylogenetic relationship. Vane-Wright *et al.* (1991) introduced a taxic diversity measure which in turn will define the degree of geneological differences (i.e., divergences) among species. The method has been shown diagrammatically in Chapter 3 (Fig. 3.2). Taxonomic weighting is based on hierarchical relationships, expressed in this figure in the form of a hierarchial tree. For each taxon, number of nodes to the root of the tree is used to determine its distinctiveness. The inverse of the node count provides a priority measure for conservation.

Table 8.1 A method for assigning conservation priorities to threatened species (from Given 1984)

Field	*Higher Priority*	*Lower Priority*
Geographic	Small range	Wide range
	Endemic to region	Not endemic to region
Taxonomy[1]	High level taxon	Low level taxon
	Small genus/family	Large genus/family
	Probably relict	Not relict
Habitat	Under threat	Not under threat
	Fragile	Resistant
	Specificity narrow	Wide habitat range
	Successional	Climax
Life form	Annual or short-lived perennial	Long-lived perennial
Populations	Small	Large
	Few	Many
Biology[2]	Rarely flowering	Often flowering
	Specific pollinator	Non-specific pollinator
	Dioecious	Monoecious
	Obligate outcrossing	Selfing readily
	Seed short-lived	Seed long lived
	Poor class structure	Good class structure
	Poor vegetative reproduction	Good vegetative reproduction
Miscellaneous	Harvested	Not harvested
	Region of high endemism	Region of low endemism

[1] High level taxa = family, genus; low level taxa = subspecies, variety.

[2] Poor class structure = disproportionate representation of one age class; good class structure = spread of population through age classes.

This method of Vane-Wright *et al.* (1991) is therefore based on the branching order, or topology of the phylogenetic tree. But, as Crozier (1992) pointed out, this method does not involve the length of branches, which could result in differences in lengths of independent evolution being obscured. Crozier (1992), therfore, advocated the use of genetic distance to obtain an estimate of branch length through the RFLP method.

The second category of species to be considered for conservation include directly harvested plants (e.g. forest trees, medicinal taxa, spices, ornamentals, food and forage plants etc.), plants which are a source of propagating materials for planting elsewhere, or plants that are sources of genetic variation useful for breeding and inprovement programmes.

In the third category, Given (1996) included the following: (i) **indicator species**, known to be particularly sensitive to pollutants, human interference, ecological instability or other distrurbances. (ii) **Umbrella species**, that unsually require larger space and that provide protection for other species within the ecosystem; they also require only very scarce resources. Such species are usually absent in smaller areas with residual vegetation. (iii) **Keystone species**, that are important to maintain the ecological integrity of the community and are essential to survival of other species. (iv) **Charismatic species**, that are significant from social, cultural or anthropomorphic standpoints, and/or usually attractive. (v) **Recreational species**, which are popular for collection, growing or observation.

Once the species are identified for protection, conservation strategies can be worked out for them either through *in-situ* or *ex-situ* methods or through a judicious combination of both (see details below). Species-based approach has resulted in identification, at the national and international levels, of taxa that need conservation and which are then listed by IUCN or national agencies in Red Data Books or detailed in the IUCN Species Survival Commission (SSC). The first IUCN Red Data Book on plants was published in 1978. The SSC has been divided into specialist groups based on geographic and/or taxonomic criteria. Some taxonomic groups whose species have been taken for serious conservation action include cycads, carnivorous plants and orchids. The SSC subgroups also periodically review the conservation status and needs of species within its remit and suggest recommendations for conservation action which will ensure their long-term survival.

The major advantage of the species-based approach is that it allows resource allocation to the most urgent cases, i.e., to species, which are in danger of immediate extinction on the basis of priority analysis. There are, however, disadvantages as well: (i) With our present knowledge and limited resources, only an extremely small proportion of the world's species can be adequately surveyed to set priorities for conservation. (ii) Even for those species which have been surveyed and identified for conservation, adequate resources are available to conserve only those which need the highest priority; such highest priorities are often based on individual prejudices. (iii) Benefits that could be obtained out of the taxa highly prioritised for conservation as per the above method can also play an important role for taking actual conservation action. Mostly those species that have actual or potential demand values, support values for other species or spiritual or traditional value (see Chapter 6) are preferred over an endangered taxon with no obvious value. At this stage, the question of actual cost required to conserve the useful taxa assumes importance.

Further details on species conservation are provided later.

Conservation of Ecosystem Diversity

Some people have suggested ecosystem conservation in place of either genetic or species-based approaches. This method attempts to ensure that representative areas of ecosystems or important habitat sites are maintained through a network of protected areas or through other controls on land use. A decision on which habitat should be selected for conservation is dictated by criteria such as species richness and degree of endemism (hot spots). Dinerstein and Wikramanayake (1993) developed a new approach (based on ecosystems) to conservation planning called **Conservation Potential/Threat Index**, which forecasts how current deforestation rates would affect the conservation and establishment of protected areas. Initially this approach was tested in 23 Indo-Pacific countries and subsequently was adapted for the Indo-Pacific region (Dinerstein *et al.*, 1995) and Latin America and West Indies (Dinerstein *et al.*, 1995b).

It is argued that if ecosystems are allowed to be kept intact and materials and energy flow unhindered through the ecosystem, then conservation of the diversity of its species and the latter's genes is automatically done. Without suitable ecosystems and dynamic ecological processes, genetic or species diversity in isolation may be nearly worthless. As already stated, ecosystem conservation is coarse tuning; only after it does fine tuning at species and gene levels become possible and efficient. In other words, ecosystem diversity conservation is the cheapest and most effective way of conserving both genetic and species diversities. The other important advantage of the ecosystem approach in conservation is that it requires no detailed knowledge of the status and distribution of its constituent species. This method is particularly useful for conserving the tropical rain forest ecosystems, whose species diversity has not yet been adequately studied or quantified.

The major difficulty of the ecosystem approach to conservation is the problem of devising satisfactory habitat or ecosystem classification (see Chapter 5) on which to base protected area networks. Another problem is that populations of threatened species which need urgent conservation steps are likely not be included in a network of protected areas. Thus there is risk that if attention is focused on the ecosystem, important biodiversity at the species and population levels may be overlooked. A third problem is that many ecosystems of the world are poorly known and understood. Also, the size, composition and complexity of an ecosystem can vary considerably in time and space.

Attempts to reconcile both species- and ecosystem-based approaches are the needs of the time. It is better to identify areas or ecosystems of high diversity, especially in threatened species, and undertake efforts to conserve such areas.

Relevance of Ecosystem Diversity as well as Services in Conservation

Almost all the recent initiatives of conservation, especially in the tropics, have been biased towards ecosystem (and species) diversity, ignoring ecosystem services. This is often justified on the ground that conservation of ecosystem diversity contributes to ecosystem services, but the question of whether diversity improves ecosystem services still remains unresolved (Singh 2002). The biodiversity-centred approach to conservation is primarily based on giving priority to areas of high diversity such as hot spots, centres of plant diversity, megadiversity centres etc., all of which are located only in the tropics. But many taxa belonging to different threatened categories occur outside these areas and their ecosystems may generate a significant amount of services for humans. The arid desert ecosystems, for example, may be very poor in biodiversity, but may generate services of value. Similarly, the seagrass ecosystem is extremely poor in species diversity but highly significant in terms of productivity. Thus it is evident that ecosystem services are not related to biodiversity. It should also be emphasised that 'even to justify the conservation of biodiversity-rich areas' from an economic perspective 'there may be need to take the support of ecosystem services as incentives' (Singh 2002). It is suggested that spending time and resources on biodiversity-rich areas alone for conservation may have to be reconsidered according to people who strongly advocate the concept of having ecosystem services as the basis for all conservation programs. Further details on ecosystem conservation are provided below.

Top-down and Bottom-up Protocols for Conservation

There are advantages as well as disadvantages in both species- and ecosystem-based approaches to conservation. Hence it is suggested that the two approaches be combined so as to effectively conserve both the desired species that need immediate attention and the ecosystems. Ganeshaiah *et al.* (2001) proposed two complementary approaches, viz. Top-down and Bottom-up protocols that address not only the immediate needs, but also the whole genetic resources of an ecosystem.

In the Top-down approach the major landscape elements that are likely to harbour the diversity of genetic pools of known and unknown species are identified first. Then the large landscape patches with greatest genetic diversity of a set of species and with a high level of biodiversity are identified for conservation actions. In the Bottom-up approach, pyramiding the species that need conservation attention on a priority basis is done. The various steps involved in the two approaches are shown in Boxes 8.2 and 8.3 (see also Fig. 8.1).

A combination of the two approaches is efficient and cost-effective in conserving species of large areas, especially in the tropics.

In-situ and *ex-situ* Conservations

Preservation of a species as a component of the functioning ecosystem, i.e., in an environment in which the species is subjected to continuing selection pressures and adaptive evolution, is called ***in-situ*** or **on-site conservation**. Hence

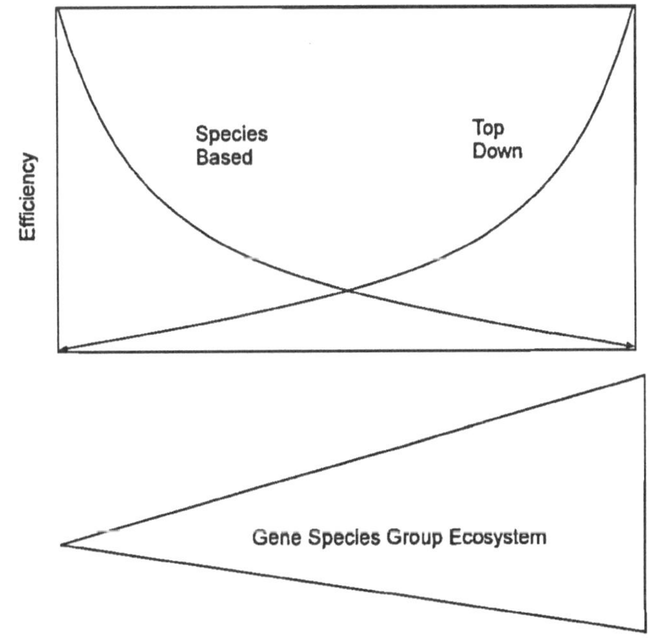

Fig. 8.1 Efficiency of bottom-up and top-down approaches in conserving genetic resources. While the top-down approach compromises details of target species, it addresses the conservation of genetic resources of the non-target species. The bottom-up approach specifically addreses conservation of the target species (adapted from Ganeshaiah *et al.*, 2001).

in-situ conservation is described as 'dynamic' (Frankel *et al.* 1995). Strictly speaking, it is the conservation of the whole ecosystem. This type of conservation has two aims: (i) to maintain economic production and (ii) to replant the ecosystem with local sources of seeds/ propagules. This is the most important form of biodiversity conservation for a variety of reasons. Among domesticates, land races cultivated in their native environment are subjected to *in-situ* dynamic conservation.

Maintenance of species away from their normal ecosystem/habitat is known as ***ex-situ*** or **off- site conservation**. It is considered a means of 'static conservation' (as opposed to dynamic *in-situ* conservation) (Frankel *et al.* 1995). *Ex-situ* conservation is slowly becoming more important as a back-up, and sometimes as a temporary replacement, for *in-situ* conservation (Heywood 1990). The distinction between *in-situ* and *ex-situ* conservations is not always absolute and the two are in no way mutually exclusive; rather they are complementary.

In-situ *conservation*

As already stated, *in-situ* conservation is the conservation of biodiversity where it currently exists. Governmental as well as non-governmental organizations are involved in this type of conservation. Both individual species as well as ecosystems/habitats are conserved by this method, since it is impossible to have meaningful conservation of a species *in situ* outside the ecosystem of which it is an integral component. It should be mentioned, however, that the mere presence of a species in a conserved ecosystem/habitat is *per se* no guarantee of its survival unless the levels of protection needed for

Box 8.2 Steps involved in top-down approach to conservation of biodiversity (from Ganeshaiah *et al.*, 2001)

1. Compile the spatial data on layers of the ecosystem
   a. Physical parameters such as temperature, rainfall, humidity, altitude
   b. Biological parameters such as distribution of biological diversity, endemism, threatened species, vegetation type etc.
2. Identify the landscape types

Divide the target area into grids and assign the data layers from step 1. Using multivariate tools cluster the grids to arrive at the distinct landscape types. An hierarchical system of crunching the clusters can be adopted such that the number of landscape elements eventually handled may be decided on the basis of resources available. Stratifying the area by overlaying the physical and biological parameters can also identify the landscape elements.

3. Using GIS tools, map the landscape types or clusters and identify a set of large patches within each landscape type as representatives for conservation.
4. Evaluate the area for genetic and biological diversity
   a. Set up sample plots in all the selected large landscape types and evaluate them for their genetic conservation worth.
   b. Estimate the genetic conservation value: Since the genetic conservation values (GCVal) of a site that can be computed as a function of the population are not known, we suggest that the genetic conservation value of a site be computed as a sum of the genetic diversity (Gdi) weighted with the population size (Pi) of the species (i) over a defined number of species as given below:

$$\text{GCValj} = \sum_{i=1}^{S} Gdi \times pi$$

where S = number of species in the j_{th} plot, and GCValj = the genetic conservation value of j_{th} plot.

   c. Estimate the diversity of plant species of the plots.
5. Identify areas with high conservation worth

Those with the highest species diversity and high GCVal can be mapped and the corresponding areas prioritised for conserving the genetic resources of the area.

Box 8.3 Steps involved in bottom-up approach to conservation of biodiversity (from Ganeshaiah *et al.*, 2001)

1. Identifying the target groups

   These could be sets of species that share common economic use and are taxonomically related.
2. Map the hot spots of species diversity and locate the viable patches:

   Contours of species diversity of the targeted groups could be constructed based on a range of data sets on the occurrence and abundance of the species. The data sources could be forest working plans, flora, forest officer's data sets and herbarial collections (Ganeshaiah and Uma Shaanker, 1998, 1999; Ravi Kanth *et al.* 1999). Typically, information on the spatial distribution of different species of the group chosen is collected in a spatial database system. The study region is divided into grids of suitable size and the density of species in each grid counted. From this density data, contours of different species densities are drawn and hot spots of species richness identified. From within these identified hot spots of species diversity, patches containing the viable populations of a reasonably large set of species need to be located for estimating the genetic diversity of the species.
3. Estimating the genetic diversity and identifying the genetic hot spots for conservation

   The genetic diversity of the chosen species in these patches is estimated for further prioritization. Patches with high genetic diversity and reasonably large populations of species shall be the candidates for conservation and hence shall be identified as the 'genetic hot spots' for conserving *in situ* the genetic resources of the target group. Such genetic hot spots for different groups of plants can be integrated to eventually arrive at a network of sites for conserving the plant genetic resources of the area or region.

this species are adequate. Even in areas under active *in-situ* conservation, proper management strategies are needed to maintain the viable populations of a threatened species requiring conservation.

Both domesticated and wild taxa can be conserved *in-situ* through a network of protected areas throughout the globe (FAO 1991c). Although the problems involved in *in-situ* conservation are enormous, the results obtained thus far are encouraging, in spite of the fact that the overall extent to which it conserves habitats and species, especially in the tropics, is not clearly known. A survey of protected areas worldwide revealed that the *in-situ* method conserves habitats and species to a reasonably good extent. Many protected areas are largely free of human activity.

Protected areas: Introduction

Areas of natural habitats/ecosystems protected under *in-situ* conservation are called **protected areas.** The 1994 IUCN *Guidelines for Protected Area Management Categories* (IUCN 1994a, b) define a protected area as follows: 'An area of land and/or sea specially dedicated to the protection and maintenance of biological diversity, and of natural and associated cultural resources, and managed through legal or other effective means'. Protection of areas is a recent phenomenon dating back only to about a century in most countries, but several centuries in India, Thailand and China. There are many types of protected areas with differing degrees of protection, permanency and purpose. No single category of protected area can satisfy all needs and over the past many years a wide variety of protected area categories have evolved. Some have evolved to fulfil local requirements, while others, e.g. National parks, have a much broader role in conservation.

The confusing range of protected areas can be classified broadly into the following: Scientific reserve/Strict nature reserve, National park, Natural monument/Natural landmark, Managed nature reserve/Wildlife sanctuary, Protected landscapes and Seascapes, Resource reserves, Natural biotic area/Anthropological reserve, Multiple-use management area/ Managed resource area, Biosphere reserves, World heritage sites (natural) etc. (Mackinnon *et al.* 1986). A new category of World Heritage Sites called **Cultural landscapes** has been recognised which acknowledges the 'complex interrelationships between man and nature in the construction, formation and evolution of landscapes'. Each type of protected area is stated to be particularly suited to one or many plant conservation objectives (Table 8.2). It is very difficult to distinguish one type of protected area from another as considerable overlapping exists between them; hence only the most commonly recognised protected areas are discussed here.

Today, there are over 9832 protected areas (IUCN categories I-V), including 1508 National Parks, of approximately 9.25 million km^2 (almost the size of the USA) or about 8.2% of the Earth's land surface. A further 40,000 smaller protected areas cover another 5% of the land area (Duthie 1997; McNeely 1994; UNEP 1995). The goal recommended by IUCN, however, is preservation of a cross-section of all major ecosystems to the extent of 13 million km^2, or about 10-12 % of the Earth's surface.

Biosphere Reserves and National Parks

Introduction

Protection of an ecosystem in the form of a **Biosphere Reserve** dates back to just a century and half. Originally these reserves were created not with the intention of protecting biodiversity, but as tourist attractions and for their aesthetic appeal; they protected the plants and animals living there only incidentally. For example, in North America, the world's first National Park was established in 1847 at Yellowstone, with the initial emphasis on recreation rather than conservation. In the UK, the establishment of Nature Conservancy in 1949 was instrumental in popularising the already existing National parks and Biosphere reserves as means of *in-situ* conservation. Only in the last 50 years have Biosphere reserves and National park systems been exclusively created to retain biodiversity and ecological processes and hold great promise for efficient conservation of habitats/species. The emphasis in such efforts was on conservation of ecosystems and conservation of species as components of ecosystems.

Table 8.2 The most significant types of protected area for conservation of plants (from Given 1984)

Primary conservation objective	*I*	*II*	*III*	*IV*	*V*	*VI*	*VII*	*VIII*	*IX*	*X*
Maintain sample ecosystems in natural state	1	1	1	1	2	3	1	2	1	1
Conserve genetic resources	1	1	1	1	2	3	1	3	1	1
Maintain ecological diversity and environmental regulation	3	1	1	2	2	2	1	2	1	1
Provide education, research, and environmental monitoring	1	2	1	1	2	3	2	2	1	1
Produce timber, forage on a sustained basis	-	-	-	3	2	-	3	1	3	-
Conserve watershed condition	2	1	2	2	2	2	2	2	2	2
Protected sites and objects of cultural heritage	-	1	3	-	1	3	1	3	2	1
Stimulate rational, sustainable use of marginal and rural areas	2	1	2	2	1	3	2	1	2	2

Type of protected area: I- Strict nature/wilderness area (includes two categories: Strict Nature Reserves and Wilderness areas); II- National Park: Protected area managed mainly for ecosystem protection and recreation; III- Natural monument/landmark: Protected area managed mainly for conservation of specific natural features; IV - Habitat/species management Reserve; V- Protected landscape/seascape; VI- Resource Reserve; VII- Anthropological Reserve; VIII- Multiple use area; IX- Biosphere Reserve; X- World heritage site.

I to VI are 1994 IUCN categories (see UNEP 1995).

Ratings: 1 = Primary objective for management of area and resources.

2 = Not necessarily primary but always included as an important objective

3 = Included as an objective and whenever resources and other management objectives permit.

Objectives and goals of reserves

Nature Reserves are now developed across the world for one or more of three primary biological motivations: (i) To preserve large and functioning ecosystems; in such an effort, the resources are expected to provide adequate conditions for the long-term survival of evolving ecosystems. (ii) To preserve biodiversity with a view to conserving the maximum possible number of species and communities. (iii) To protect particular species or groups of species of special interest, which are especially threatened. These three goals and the reserve types developed to fulfil them are not mutually exclusive; one can take care of the one or both of the other two. The more a reserve can take care of all three goals, the more desirable and successful it will be. A fourth motivating goal may be added, namely the perpetuation of plants for continuing and sustainable harvest. In addition to these four goals, other cogent objectives should be added (McNeely *et al.* 1987): (i) to maintain and promote genetic diversity, (ii) to serve the requirements of education and research, (iii) to conserve water and soil, (iv) to manage wildlife, (v) to serve as sites of recreation and tourism, (vi) to protect social and cultural heritage, (vii) to maintain scenic beauty and aesthetic integrity, (viii) to promote options for the future and (ix) to promote integrated development within the reserve and between reserves.

Biosphere reserves will also play an increasingly important role in the conservation and utilisation of wild relatives of crops, although the resource value of such wild relatives available in most reserves of the world is still poorly known. Less than a third of the world's biosphere reserves have been inventoried for their component species and the status of their populations. There is no systematic documentation of wild relatives of useful taxa and land races of cultivated plants in these reserves. Further, only a few of the world's biosphere reserves are in the centres of crop origin (see Chapter 4) or in regions of great species diversity. However, even the scant data available indicate that well over 20% of the reserves have populations of wild relatives of major crops, and with better documentation and inventories this percentage is likely to exceed 50. It may be further stated that less than 10% of the total species and allelic diversity of the major crop gene pools are currently maintained *in situ* in Bioreserves. Of these, only a small proportion is conserved at adequate levels to withstand imminent threats to them as well as to meet the needs for germplasm for breeding programmes.

Land races located in reserve areas are to be conserved in their home environment as a part of dynamic conservation, since once displaced by improved cultivars, their *in-situ* conservation, as a deliberate measure, becomes very difficult. True, any attempt to retain 'primitive farms' that contain land races for a long term in a radically changed social and technological environment would be self-defeating. However, one can propose to establish 'Crop Reservations', areas of 0.56 to 1 ha in size, wherein a land race could be maintained under the management of a local agricultural officer. This is practised in Iran and Turkey even today (Kuckuck, in Bennett 1968, p. 32, 61). These can be allowed to be subjected to environmental changes resulting from agricultural development through the use of fertilisers, pesticides, innovative technologies etc. as well as the genetic changes mediated by natural hybridisation, mutation and natural selection. Further details are given below.

Nature reserves must promote **representativeness** at local, regional, national and international levels. The word 'representative' connotes many different things, including representation in species composition, physical structure of vegetation (such as forest, grassland etc.), parameters of phenomena such as species diversity, and percentage of endemic species and human land-use patterns and intensity. However, it is difficult for any one Nature reserve to be 'representative' in all the foregoing senses. The purpose of identifying representative areas is to have a comprehensive network of protected areas to safeguard the full range of global biodiversity. For example, the guidelines for selecting a tropical forest habitat as a protected area are provided in Box 8.4 (McNeely *et al.* 1987).

World biosphere reserve programmes

The Biosphere Reserve programme is a worldwide one of international scientific co-operation. It combines biodiversity conservation with scientific research, monitoring of environment, training and demonstration in bioresource management, environmental education, and local co-operation and co-ordination. The real purpose of this programme is to promote a harmonious 'marriage' between conservation and sustainable development. The Biosphere programme was essentially the outcome of the efforts of UNESCO; this organisation started the Man and Biosphere (MAB) programme (Gregg 1988). The past role and future directions of this programme were discussed in the International Conference on Biosphere Reserves in Seville, Spain in 1995. The conference adopted 10 key directions to develop MAB's three main functions of conservation, development and logistic support. As of June 1995, there were 324 reserves located in 82 countries covering some 2,115,000 km^2. The most important among them are: Dinghushan Biosphere Reserve of South China, Beni Biosphere Reserve of Bolivia, Manu Biosphere Reserve of Peru, Guatopo Biosphere Reserve of Venezuela, Kwakwani Reserve of Guayana, Luquillo Reserve of Puerto Rico, and Virgin Island Biosphere Reserve of St John.

Design of biosphere reserves

The design of a Nature Reserve is very crucial to its success (Frankel and Soulé 1981; Margules and Nicholls 1988; Margules *et al.* 1982; May 1975; Spellerberg 1991). Much discussion has ensued about reserve designs, due largely to lack of clarity or precision as to the aims of particular reserves and their perceived lifetime. The aims, already mentioned above, are mainly dictated by an amalgam of ecological, economic, social and political compulsions.

Six critical issues determine the success of a reserve which must perforce be seriously considered while designing it, namely: (i) reserve size, (ii) inclusion of spatial and temporal heterogeneity and dynamics, (iii) ideal geographic context, (iv) connection of various reserves (i.e., contiguity) on a regional basis, (v) regard for natural landscape elements and (vi) creation of zones of different uses within the reserves (Given 1996).

(i) **Reserve size**: Ideally, reserves should be large. Larger reserves are much better than smaller ones for the following reasons: They maintain individual species, biodiversity and ecological functions, especially the homeostasis of ecosystems, energy cycle, food chain and trophic levels. Larger reserves will often contain the minimum viable population (MVP) size of constituent species; they will maintain genetic diversity better; they do not have **edge effects**; they will be able to minimise the impact of external variables and to accommodate disturbances much better. In the past, the size of many Nature Reserves and National Parks was determined primarily on aesthetic and ecological grounds rather than genetic considerations of those species especially requiring conservation. A good example is *Eucalyptus globulus* in Victoria and Tasmania, where the stands retained in reserves are genetically less desirable (Eldridge *et al.* 1993).

How large should a nature reserve be? This is a highly debated question. The size of a reserve should be based on computation of the MVP of that/those species which is/are the chief target(s) for conservation in it through population viability analysis (PVA). Almost all computations made to date for reserve size have been based on animals, especially larger carnivores. Claims such as those of McNeely *et al.* (1990) that 'many plants can survive for several centuries in a forest niche scarcely larger than the diameter of its leaf rosette' give a false notion

Box 8.4 Guidelines for selecting a system of tropical forest habitats for protection (from McNeely *et al.*, 1987)

Step 1 : Survey

Step 2 : Establish criteria

A. Ecological criteria
(1) Dependency
(2) Naturalness
(3) Uniqueness
(4) Diversity
(5) Integrity
(6) Representativeness

B. Scientific and educational criteria
(1) Convenience
(2) Monitoring benchmark
(3) Research history
(4) Demonstration
(5) Process relationship
(6) Awareness

C. Social and economic benefit criteria
(1) Economic benefit
(2) Social acceptance
(3) Recreation
(4) Tourism
(5) Landscape
(6) Demonstration

D. Pragmatic criteria
(1) Urgency
(2) Opportunism
(3) Management
(4) Feasibility
(5) Availability

Step 3 : Select areas to be included in the system

Step 4 : Establish the system

Step 5 : Manage the individual protected areas

about the total area requirements of plant species, since such claims do not take into account all aspects in the life cycle of a plant species. In fact, computation of reserve size based on MVP of plant species is the need of the hour.

A critical survey of all world reserves revealed that they vary markedly in size and degree of protection. Sizes range from about one hectare to millions of hectares, i.e., are extremely disparate. A survey of 919 of the world's more significant Nature Reserves showed that 76% were less than 100,000 ha in area and only 3.5% exceeded a million ha. In Australia, for example, there are 3429 reserves covering 6.5% of the continent's total land area. Seventy-four per cent of these reserves are less than 1000 ha in size, while only 2.6% exceed 100,000 ha; the nine largest reserves comprise 37.5% of the protected area (Given 1996).

(ii) **Spatial and temporal heterogeneity and dynamics:** Spatially and temporally heterogeneous areas are better suited for developing a reserve than homogeneous areas, if the goal is to maintain very high biodiversity. The basis for this fact is that nature is dynamic and changes over time and space through biotic and abiotic disturbances and that a spatially heterogeneous reserve better accommodates these disturbances than a homogeneous one by offering every species a diversity of habitat types at any given point of time. If a habitat patch is changed or disturbed or otherwise becomes unsuitable for a given species, other appropriate habitat patches may be available for colonisation in a heterogeneous reserve. Heterogeneity also promotes the occurrence of metapopulations of a species.

(iii) **Ideal geographic context**: Context refers to the nature and location of habitat patches within a reserve as well as in the larger landscape of the reserve. Within the reserve, each habitat patch is spatially located in such a way that there will be movement of organisms from one habitat patch to another. At the total landscape scale, the entire reserve functions in the context of a surrounding, non-reserve area of the landscape, which can have critical consequences, both positive and negative, for the reserve.

(iv) **Connection of different reserves**: Reserves need to be contiguous so that they can maintain the overall integrity of the physical environment and the basic biogeochemical processes of the region in question. Reserves can be made contiguous through **corridors**, which are strips of areas similar to reserves connecting two or more of them. These corridors allow movement of species and thus recolonisation. Corridors are believed to counter the fragmentation effects of reserves, which are too small to keep MVP of many species. Corridors also help in reducing erosion by wind and water, in providing shelter to living organisms and in increasing the aesthetic beauty of the landscape (Hobbs 1992). Corridors should also function as viable habitats similar to reserves (Simberloff *et al.* 1992). However, adequate data are not available to categorically assert that corridors play a significant role in the migration of species across the landscape.

(v) **Natural landscape elements**: Natural landscape elements include such features as valleys, drainage basins, ridges, streams, ecotones, slopes, canyons, habitat peninsulas or any other distinctive feature. The more the diversity of natural landscape features, the more enhanced the value of a reserve. Any modification of the natural elements, such as road laying, highways and railway lines, agricultural fields, timber lands, industries and human dwelling will make the reserve inferior.

(vi) **Creation of zones within a Reserve**: The protected Nature Reserve should have a central core with suitable habitats and landscape elements and a **buffer zone** to absorb edge effects. Edge effects include changes in temperature, relative humidity and light, more exposure to wind, elevated levels of tree mortality, increased leaf fall and decrease in population size of the constituent species. In very small protected areas, the entire area, in reality, becomes an 'edge'. Edge effects are particularly important for threatened taxa (Janzen 1986). The more widely adopted technique for minimising edge effects and maintaining core areas is to establish buffer zones around the reserve. These areas may be defined as areas peripheral to reserves 'which have restrictions placed on their use to give an added layer of protection to the Nature Reserve itself' and to compensate local tribals and villagers for loss of access to core areas of the reserves (Mackinnon *et al.* 1986).

Although different kinds of buffer zones are recongnised (Fig. 8.2), two main types known to serve rather different functions are important. 'Extension buffering' extends habitats of the core area into the buffer zone, allowing much larger populations to survive than may be possible in the core area alone. It is particularly useful for small populations of extremely rare species. 'Socio-buffering' allows activities such as raising of crops or harvesting of bioresources from the reserve in such a way as to discourage use of the protected core area by people native to the reserve area who traditionally use these resources. In other words, these activities are allowed in the buffer zone to compensate for loss of traditional harvesting rights and privileges in core areas (Hanks 1984; Oldfield 1988). Since local people disturb the core zone, buffer zones can provide both biological and social benefits without the imposition of mandatory restrictions (for details, see Oldfield 1988 and Box 8.4.).

The shape of the reserve area should be compact with biologically meaningful boundaries. Areas should contain year-round habitat requirements for animals associated with target plants, especially the essential pollinators/dispersers. Total isolation of one habitat from another should be avoided. Allowance should be given for manipulative management of the reserve.

Biological aspects of Reserve design

Over the last five decades, emphasis has centred on size, shape and other physical aspects of reserve design, largely ignoring the biological aspects. True, biogeographic ideas and theories dominated the physical aspects of reserve design, which are also paramount in assessing the biological aspects of reserve design. The most celebrated dynamic equilibrium theory of island biogeography (MacArthur and Wilson 1963, 1967; Preston 1962) was extended to reserves because of the obvious analogy between true islands and reserves, which are 'islands' of natural vegetation in an otherwise 'unnatural' and hostile landscape (May 1975). Just as for true islands, for reserves also the precise number of species at equilibrium depends mostly on size of the area at their disposal. This is termed the 'area effect'. Equilibrium additionally depends on the distance from the source of immigrants. This is termed the 'distance effect' (Frankel and Soulé 1981). Both effects profoundly influenced application of the island geography theory to the design of nature reserves (Roche and Dourjeanni 1984). Some authors pointed out, however, that often aspects of the theory have been uncritically accepted and applied to reserves with no serious thought given to the differences between true islands and reserves (Spellerberg 1991; Wright and Hubbel 1983). This objection notwithstanding, the island geography theory has had a profound effect on the design of reserves for conservation.

The major problem in reserves meant for conserving a single species or a group of selected taxa is that any effort to manage the reserve too intensely for the selected species will result in the general decline of its total biodiversity. This may even lead to the loss of particular components of biodiversity. If, on the other hand, the reserve does not receive targeted approach to management, the specific species may get lost and the reserve becomes vulnerable to contamination from gene flow and invasion of inferior elements. Hence a judicious combination of strategies to safeguard both targeted species and the entire biodiversity of the reserve should be practised.

Vest Pockets and Garrison Reserves

Lands with recognised economic value for alternate uses usually have smaller or fewer reserves. Such reserves are termed **Vest Pockets** and **Garrison Reserves**. They are very small, sometimes only a few hectares in extent. They can be made to survive for a longer time only with very intensive management care. In many places, these reserves represent the small remnants of populations or natural communities. The difficulties in managing small reserves for conservation are: (i) Lack of high level diversity initially. To achieve this through intense management in the long term is also difficult. (ii) They suffer from greater impact through edge effects. (iii) They are very expensive to maintain because of the intensive management required. However, several smaller reserves in an area can be brought under a network for conservation management. (iv) The most serious problem is that adjacent land use is very different from that

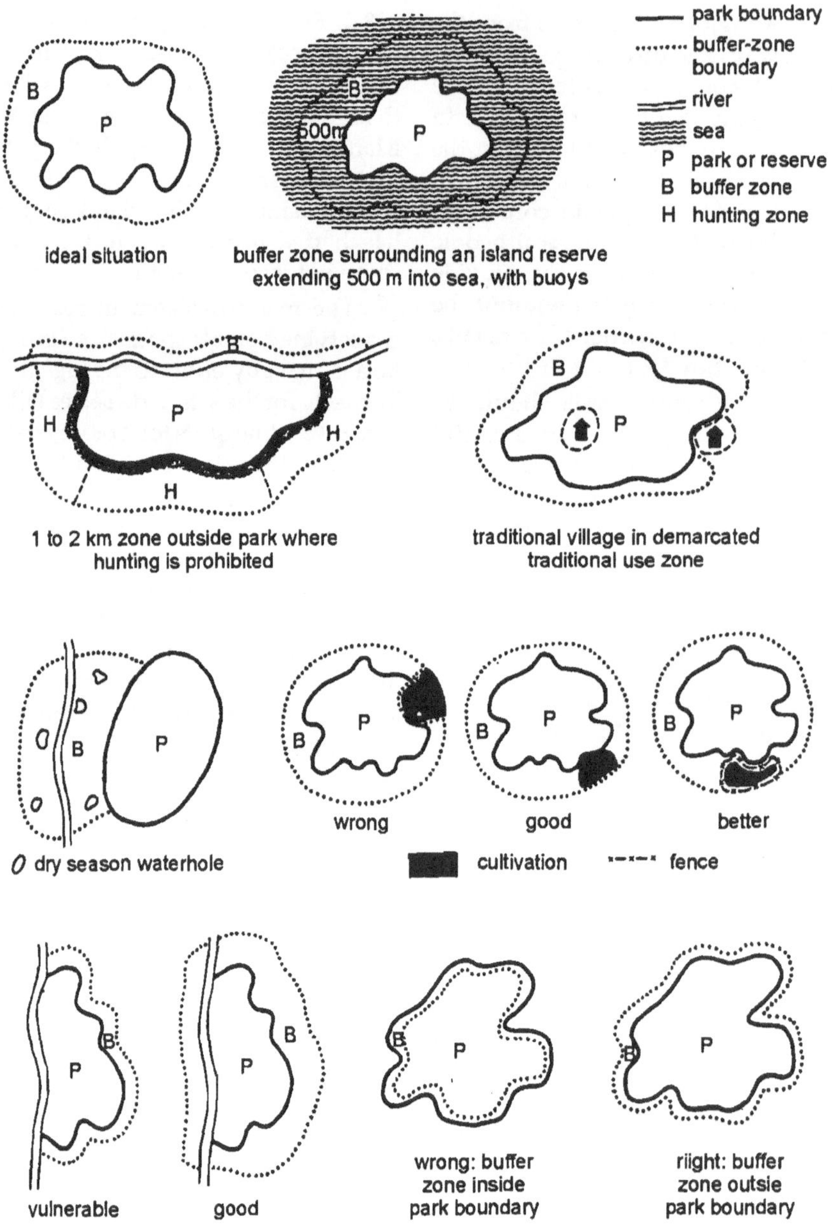

Fig. 8.2 Different kinds of buffer zones for protected areas (from Mackinnon *et al.*, 1986).

in the Reserve. In several countries, small reserves may be the only type available for conservation or only smaller reserves feasible due to compelling cultural, social, political and/ or economic factors.

To what extent have protected areas established and maintained so far served their true function of protecting and conserving biodiversity? Unfortunately, too little data has been gathered to answer this question satisfactorily. A recent study done by Bruner *et al.* (2001) is worth mentioning, however. They assessed the anthropogenic threats to 93 protected areas in 22 tropical countries and found that the majority were not only successful in stopping land clearing and other efforts at ecosystem damage, but even showed significant recovery of natural vegetation.

It is very essential to remember that great biodiversity exists even outside protected areas. This understanding forces us to realise that there is an absolute need to adopt a broad landscape or bioregional approach to conservation and management of biodiversity (Heywood 1997b). It must also be pointed out that to date very little effort has been put into conservation efforts outside protected areas.

On-farm and Home Garden Conservation

On-farm Conservation involves the maintenance of traditional crop varieties or cropping practices and systems by farmers within traditional agricultural systems. The crop varieties maintained 'on-farm' are often known as **land races**. The farmers sow and harvest the crops regularly and every season keep aside a portion of the harvested seeds for resowing. This ongoing practice has safeguarded land races of several crops highly adapted to the local environment; in many instances these crops contain locally adapted alleles that have proved very useful for specific breeding programmes. A detailed analysis of on-farm conservation throughout the globe showed that seed-propagated crops (such as seed and grain crops, vegetables, forages and fodder species), vegetatively propagated crops (such as potato, sweet potato, yam, cassava, taro and *Xanthosoma*) and even wild and semicultivated species were conserved as land races. It is true that land races yield much less than modern cultivars but on-farming conservation ensures the protection and preservation of ancient land races and wild species for posterity. Therefore, the traditional farmer/tribal person growing such land races must not only be helped with subsidies, but the land races traditionally cultivated by him should be continuously protected and monitored. Jana (1993) went further and stated that one should eschew the notion that *in-situ* conservation of land races is only for safeguarding breeding materials; it should be practised for its own sake. However, undertaking such conservation steps poses a challenge in traditional areas 'without a return to or the preservation of traditional cultural systems, which may be unacceptable or impracticable under political systems' (Qualset *et al.* 1997). Most traditional agroecosystems are under the process of modernisation to varying degrees in different parts of the world. A few selected cases of ongoing maintenance of land races by local farmers may be mentioned here: land races of potato in the Tulumayo Paucartombo Valleys of the Central Andes of Peru, of maize in southern Mexico, of wheat in western Turkey and Italy, or of rice in the Chaing Mai Valley of Thailand.

Home Garden Conservation is a smaller-scale conservation effort made on plants grown in home, kitchen or back yard gardens. Most of these taxa are either vegetables/fruits or ornamentals. These often have indigenous germplasm in the form of land races, obsolete cultivars and rare species (Hawkes 1983). For example, Hawkes (1983) listed 45 species of cultivated and 25 of wild medicinal taxa conserved in such gardens in a village in Central Java. In Mexico, the Huastec Indians manage serveral home gardens, which form the place for more than 300 species (Alcorn 1984). Esquirel and Hammer (1992) described the role of such gardens, called *Conucos*, in Cuba; these are relatively larger gardens where the traditional Cuban farmers maintain local cultivars of crops.

For conservation of land races and home garden taxa, incentives must be provided. The provision of subsidies for farmers cultivating land races has already been mentioned. In addition to this, the following other incentives are suggested (Qualset *et al.* 1997). (i) Perverse incentives already being given for conservation of modern crop varieties should be discontinued. (ii) Amenities in rural and tribal areas should be enhanced. (iii) The development of niche markets should be encouraged. (iv) Facilities to expedite commercialisation of the products of land races should be built. (v) Amateur conservationists, farmers and landowners, natural resource groups and scientists should be encouraged to conserve land races. (vi) Government and other institutional interventions should be minimised. (vii) Detailed scientific studies (especially population genetic studies) of land races should be done. (viii) Breeding programmes in land races should be established.

Ex-situ *Conservation*

Ex-situ conservation can be followed for both wild plants and domesticated crops. While the first section under this heading is concerned mainly with domesticated crops, all subsequent sections are applicable to both wild and domesticated taxa.

Germplasm Collections

These refer to assemblages of genotypes or populations often maintained as research materials for plant breeders, crop evolutionists, ecogeographers, taxonomists, phytopathologists etc. Germplasm collections consist of samples of domesticated taxa and their wild relatives, which are maintained either *in vivo* in the form of plants, seeds, tubers and other propagules, or *in vitro* in the form of single cells to parts of whole plants.

Phillipe de Vilmorin at Verrieres near Paris attempted the earliest substantial collection of a crop plant, wheat, in the middle of the 19th century. Other collections, not only in wheat but other crops as well, followed closely in England, Germany, Sweden and Australia. These collections consisted mainly of land races obtained from different scientists, breeders or traditional farmers in different countries. However, the first systematically assembled germplasm collections were the result of several plant exploration expeditions conducted in the 1920s and 1930s by N.I. Vavilov and his colleagues at the USSR Institute of Plant Industry, now known as the Vavilov Institute of Plant Industry (Vavilov 1949-50).

Inspired by Vavilov's initiatives, systematic explorations and collections of germplasms intensified, particularly in the areas identified by Vavilov as centres of genetic diversity. For example, 34 potato collection expeditions were undertaken by Hawkes (see Hawkes 1970) in South America between 1925 and 1967. Research institutions, in particular international agricultural research centres, played an increasingly important role in germplasm collections of many crop taxa. Many individual scientists were responsible for establishing very good germplasm collections of crops such as tomato, safflower, okra, cucurbits etc. As a result, there are now about 6.2 million **accessions** of 80 different crop taxa stored in 1320 gene banks and related facilities in 131 countries (FAO 1996a). Of these, wheat germplasm accessions constitute 14%, rice 7%, maize 4% and other crops 75%.

The institutions involved in germplasm collections are different, are operating in different geographic locations and are working with different purposes in mind. The same is true of individual scientists. Hence, there are large differences in the origin and purpose, size and geographic or taxonomic coverage, and the biological status of the material between the germplasm collections attempted and executed so far. Nonetheless germplasm collections often contain unique and very valuable genetic materials worthy of being identified, maintained and exploited. Germplasm accessions, individually as well as collectively, are not only to be made available to all kinds of users of the world's genetic resources presently, but also to be conserved for the use of future generations.

What should be the size of the plant germplasm collection? No single figure can apply to all collections or to all crops. However, it is recommended that 10-100 is suitable to a collection of cytogenetic stocks of species, 10,000 and above for a world collection of any major crop such as wheat or rice. Collections should represent and cover all possible variations that could be exploited for crop improvement and consequently the larger the collection, the more likely it will include all possible genetic information. Yet the prohibitive cost involved imposes limits on the indiscriminate accession of germplasms.

Redundancy is a major problem in germplasm acquisitions. This can be avoided or minimised by setting up a **core collection** versus a **reserve collection** (Frankel and Brown 1984). The former refers to a minimal set of collections that more than adequately represent the genetic diversity present in the total collection. The latter refers to an accession not selected for the core set but nevertheless conserved. Further details on these collections can be found in Brown (1989).

Lastly it should be emphasised that germplasm resources, although primarily intended for breeding and related purposes, have been/may be used to unravel evolutionary processes and taxonomic affinities, to identify centres of origin or diversity of crops and to discover the distribution patterns of genetic variability or particular features or genes, and to relate these patterns to environmental variables to which the plants are subjected.

Botanic Gardens

Although a satisfactory definition is still wanting, a botanic garden can be broadly defined as a place of collection of living plants grown for educational, recreational, economic, medicinal or scientific purposes. A botanic garden is the most important form of *ex-situ* conservation[2]. The most widely known function of botanic gardens is to assemble and maintain a diversity of plant species in the open or in greenhouses for reference and study. Botanic gardens conduct or facilitate research in diverse aspects of plant science, especially in taxonomy. They have a long history of concern for taxa of economic value, dating back to the medieval gardens of the 14th century of Europe and further back to the 7th century "Nandavanas" or "sacred groves" of India. The first recorded European botanic garden was probably the one established in 1545 at the University of Padua in northern Italy, which even in 1591 was reported to contain around 2000 species (Given 1996). In India, epigraphical inscriptions speak of well-established temple gardens (*Nandavanas* or sacred groves) even by 7th century A.D. (Krishnamurthy 1999; Swamy 1973). From the 17th century onwards botanic gardens have played a significant role in the introduction and exchange of useful plants between Europe, the Americas, Africa and Asia.

It has already been noted that botanic gardens form the single most important institution involved in the *ex-situ* conservation of wild plants. The very first comprehensive attempt to involve world botanic gardens in conservation was made in 1953 at an International Colloquium of the Sub-Committee of Botanic Gardens of the Union of Biological Sciences (IUBS). As a result, in the next year the International Association of Botanic Gardens (IABG) was formed. This was followed by the compilation of an International Botanic Gardens Directory (see Larsen *et al.* 1987). In 1968, the late Ronald Melville used information drawn from world botanic gardens in preparing the International Red Data Book. In the Kuala Lumpur symposium of 1974, the roles and goals of tropical botanic gardens were discussed at

---

[2]Some botanic gardens have large stretches of natural wild areas, for example many gardens of South Africa, the Royal Botanic Gardens of Sydney, Singapore Botanic Gardens and even the Kew Botanic Gardens. Wild plants conserved in these natural stretches of botanic gardens should actually be considered *in-situ* conservation (see Prance 1997).

length. Meanwhile, the Threatened Plants Committee (TPC) entrusted the plan for conservation of threatened taxa to the IUCN Botanic Gardens Conservation Secretariat (BGCS; now BGCI). The BGCS, first through a conference held at the Kew Gardens in 1975 (Simmons *et al.* 1976), emphasised the need for conserving threatened taxa in the world botanic gardens. The first International Botanic Gardens Conservation Congress was held at Las Palmas, Canary Islands in 1985, where planning for conservation of taxa through the botanic garden was further strengthened. This conference examined critically the role of botanic gardens in the context of a world conservation strategy and recommended the establishment by IUCN of BGCS 'to co-ordinate and monitor the plant conservation activities of the botanic gardens of the world' (Jackson 1989). The BGCS was set up by IUCN on the first of January 1987, which subsequently became an independent body in 1990.

The second International Botanic Gardens Conservation Congress was held at Ile de la Réunion in 1989. A drastic change in direction was introduced in this congress; the major emphasis shifted to germplasm conservation in the botanic gardens, with plants of economic value principally targeted for conservation over threatened and endangered species. The BGCS was advised to give priority to monitor and promote research into conservation of useful taxa, including germplasms of major and minor crops (Jackson 1989). However, propagation and reintroduction of rare and endangered species are to be actively encouraged but with lower priority. This in reality meant the continued conservation of around 35,000 endangered species in botanic gardens, in addition to greater effort on 15,000 or so economically important species including a number of medicinal taxa (Heywood 1991a). This, according to Frankel *et al.* (1995), amounted to a 'north-south' distinction between the two contrasting strategies of botanic gardens. The BGCS thus has to lay greater emphasis in the tropics of the Southern Hemisphere on threatened species, which have greater economic value (the 'south strategy'), while in the Northern Hemisphere emphasis is to be on endangered species, irrespective of their economic value (the 'north strategy'). This new Botanic Gardens Conservation Strategy (WWF/IUCN/BGCS 1989) was elaborated further by BGCS. It recommended botanic gardens to participate in *in-situ* conservation, with small reserves of 5 to 50 hectares solely entrusted with the conservation of specific individual species with their genetic diversity. The BGCS now maintains the database on *ex-situ* conservation assembled by the World Conservation Monitoring Centre (WCMC), and records about 60,000 rare and threatened plant species held in about 400 institutions worldwide (Jackson 1991).

There are many kinds of botanic gardens and arboreta (BGCS 1989): (i) traditional, state-supported gardens, with associated herbarium and laboratories (e.g. Berlin, Kew, Bogor, Peradeniya); (ii) municipal or civic supported, sometimes with associated herbarium and laboratories, normally open to the public (e.g. Gothenberg, Glasgow, Nantes); (iii) university gardens with an associated herbarium and laboratories, usually open to the public (e.g. Cambridge, Berkeley, Hamburg, Montpellier); (iv) private, often with some state support, with an associated herbarium and laboratories, invariably open to public (e.g. Missouri Botanical Garden, Fairchild Tropical garden at Miami); (v) private, without state support, usually lacking a herbarium or laboratory (e.g. Les Cedres, Maurimurta); (vi) Government/State arboreta with an associated herbarium and laboratories (e.g. US National Arboretum) or without herbarium (e.g. Westonbirt); (vii) university arboreta with an associated herbarium and laboratories (e.g. Arnold Arboretum); (viii) private arboreta, with or without herbarium or laboratories (e.g. Hilliers, Morton arboretum, Bickelhaupt); (ix) botanical-zoological gardens/parks (e.g. Hongkong, Wilhelma Stuttgart) and (x) agrobotanical gardens (e.g. Godollo, Gatersleben, Castelar).

There are more than 1500 botanic gardens/ arboreta throughout the world (BGCS 1989; UNEP 1995) and this number should have increased further in the past few years. Of these 1500 gardens, about 800 were believed to be very active in plant conservation process. Of these, 582 have a seed bank facility. All these gardens together at the time of survey contained well over 3 million plant taxa, excellent proof of their importance in active plant conservation programmes at that time itself. There is considerable imbalance in the global distribution of botanic gardens (Table 8.3). Europe has 540 gardens, USA 451 (Heywood 1990), Africa 82, South America 66 and India 79 (8 large and 71 smaller gardens; Chakraverty and Mukhopadhyaya 1990; Nayar 1997; accounting for about 8500 taxa). Most tropical countries, which account for more than 90% of the world diversity, have very few botanic gardens. It is gratifying to note, however, that many of the most recently created botanic gardens have been housed in tropical countries. The collections maintained in botanic gardens are very diverse. Orchids, succulents, bromeliads, bulbous taxa and temperate trees are particularly well represented. Tropical trees, epiphytes, carnivorous taxa, palms, cycads etc. are rather poorly represented.

The problems encountered by botanic gardens are innumerable. Many botanic gardens, especially in the tropics, are poorly financed and badly organised and managed. Trained technical persons are wanting. Also, a large percentage of plants growing in such gardens are of low conservation priority. In many cases the genetic diversity maintained in these gardens (also in some well-organised gardens) is inadequate for conservation purposes. The majority of endangered species are tropical while the well-organised gardens are predominantly located in temperate regions; these temperate gardens have taken up much of the burden of preserving tropical species in greenhouses with simulated tropical climate or in the form of seed storage. A simple, most effective and least costly way to overcome this problem according to Raven (1976), is to ensure the perpetuation of endangered species by introducing them successfully into the horticultural trade; botanic gardens can take up this introduction very seriously. Already, endangered plants in the wild such as *Ginkgo biloba*, *Franklinia alatamaha*, *Encephalartos woodii*, *Swietenia humilis*, *Dracaena draco* and others have been perpetuated in this manner. Another way of resolving this problem is through effective co-ordination between botanic gardens situated in

Table 8.3 Geographic distribution of botanic gardens and arboreta by region (from Heywood 1987)

Region	*No. of gardens*	*No. of arboreta*	*Population*
Europe	392	64	500,899,461
Former USSR	121	34	266,674,000
South-west Asia	14	2	134,508,085
South & South-east Asia & Oceania	107	50	1,322,957,070
Australia & New Zealand	47	8	18,125,857
China, Japan, Mongolia, & Taiwan	66	13	1,144,269,000
North Africa	10	—	89,170,000
Tropical Africa	28	2	329,846,340
South Africa	14	1	25,590,000
North America	98	92	252,219,951
Middle America	36	—	93,133,378
South America	51	1	236,979,117
Caribbean	20	—	28,764,205

different parts of the globe. The BGCI is already co-ordinating very well by disseminating information, promoting *ex-situ* conservation of threatened plants by fixing priorities, providing technical guidance and data as well as supporting weak gardens financially and otherwise. BGCI has a worldwide membership of about 320 botanic gardens.

Information is still scant regarding the impact of botanic garden management on the actual conservation status and survival of threatened species. There are several known cases of threatened species surviving exclusively due to botanic gardens (Falk 1987), for example *Sophora toromiro* from Easter Island and *Commidendron rotundifolium* from St. Helena. Some other taxa that have become extinct in the wild but are still maintained in botanic gardens are listed in Table 8.4. As mentioned elsewhere, the mere presence of a species in a botanic garden is no guarantee for its conservation, however. Hurka (1994) argued that one significant way wherein botanic gardens could promote conservation is through education and through their influence on public opinion. Botanic gardens should give greater prominence to local flora. A good example for this type of service turned out in conservation is the Missouri Botanic Garden (St. Louis, USA), which through the establishment of the Centre for Plant Conservation in 1984 in the garden has already collected and maintained all the endangered species of the USA in the living state, supplemented by a seed bank.

Seed Banks

A **seed bank** is a collection of seeds stored in a viable state for posterity. This term is also used in a different context,, i.e., accumulation of ungerminated seeds usually found in the soil (Archibold 1989). Soil is a kind of gene bank to which seeds are added year after year. The number of seeds that accumulate over the years can be prodigious, as seen in a study made in Denmark. One block of top soil—1 m square and 20 cm deep—yielded roughly 135,000 seeds, of which about 50,000 were living and could germinate under the correct conditions (Koopowitz and Kaye 1990); such an accumulation was labelled a 'Natural seed bank'.

The artificial seed bank is a very good device for conservation of plants. Conservation is both safest and cheapest if life processes are reduced to a low level, and seed preservation in seed banks fulfils this requirement admirably. Conservation through seed storage aims to preserve as much as possible the genetic integrity of individuals or populations of a species. Although frequencies of genotypes or alleles may undergo inevitable changes due to storage, gene or allele erosions are minimised, and recombination with alien material is avoided. Preclusion or reduction in natural selection, genetic drift, natural hybridisation and destruction by parasites or loss through human error are also ensured. No other genetic effect has been observed.

A seed bank is one of the most efficient methods of *ex-situ* conservation for sexually reproducing species. Seeds suitable for long-term storage are termed **orthodox, conventional** or **desiccation–tolerant.** Such seeds can be stored for a long time without substantial loss of vitality and without much genetic changes (Roberts 1975, 1989). Roberts (1975) defined orthodox seeds as 'seeds for which the viability period increases in a logarithmic manner as one reduces the storage temperature and the moisture content of the same'. Seeds, which cannot be stored in seed banks under the given conditions, are termed **recalcitrant.** Jackfruit, citrus, avocado, coffee, tea, maple, cinnamon, nutmeg, oak, chestnut, mango, cocoa, coconut and rubber belong to this category. It is difficult to conserve recalcitrant seeds. It is estimated that 20% of the world's total plants produce recalcitrant seeds that do not survive in low temperature and/or dehydrated conditions. They have no natural dormancy and die if not allowed to germinate immediately. It has also been found that the

Table 8.4 Some taxa 'extinct in the wild' but conserved in botanic gardens (data from IUCN/WCMC and Michael Maunder, in Prance 1997)

Sl. No.	*Taxon*	*Native Country*
1	Anthurium leuconeurum	Mexico
2	*Arctostaphylos uvaursi* ssp. *loebreweri*	USA
3	Bromus verticillatus	UK
4	Calandrinia feltonii	Falkland Island
5	Ceratozamia hildae	Central America
6	Commidendrum rotundifolium	St. Helena
7	Cosmos atrosanguineus	Mexico
8	Erica verticillata	South Africa
9	Encephalartos woodii	South Africa
10	Franklinia alatamaha	USA
11	Graptopetalum bellus	Mexico
12	Helichrysum selaginoides	Tasmania
13	Lysimachia minoricensis	Minorea
14	Opuntia lindheimeri	Mexico?
15	Paphiopedilum delelanatii	?
16	Sophora toromiro	Easter Island
17	Tecophilaea cyanocrocus	Chile
18	Trochetiopsis erythroxylon	St. Helena
19	Tulipa sprengeri	Turkey
20	Dombeya acutangula	Rodrigues
21	D. mauritiana	Mauritius
22	*Vernonia shevaroyensis**	S.India

* Extinct in wild, while a single tree is growing in the botanic garden cum orchidarium of the Botanical Survey of India at Yercaud, Salem, Tamil Nadu, S. India

majority of orthodox seeds are either from temperate regions, where dormancy is enforced by cold winters, or from arid regions where dormancy is enforced by hot and dry climate. Rain forest seeds are often recalcitrant.

Among orthodox seeds, much importance is often given to small seeds, which occupy less space. The FAO constituted a panel of experts on plant exploration and introduction, which formulated certain standards and procedures for storage installations meant for long-term seed conservation (FAO 1975). The recommended conditions for seed storage are -18°C or less temperature, use of airtight sealed containers, and seed moisture content of 5-2%. The seed bank needs proper and continuous power supply. Medium-term storage can be effected at 6-1°C (Cromarty *et al.* 1982; Dickie *et al.* 1984; Ellis *et al.* 1985; Hanson 1985; IBPGR 1985). Storage at the temperature of liquid nitrogen has given encouraging and problem-free results. It is also less costly than the conventional seed storage. However, the moisture content needs to be strictly controlled. Orthodox seeds can be kept under storage for very long periods at sub-zero degree centigrade, if previously dried to about 5 to 8% moisture content. Although longevity varies from taxon to taxon, orthodox seeds can be stored for 5 to 25 years at 0 to 5° C (medium-term storage) or for up to 100 years if stored at -10 to -20 ° C (long-term storage). The basis for this increased period of viability with decreased temperature treatment is as follows (Koopowitz and Kaye 1990): For every 5°C drop in temperature, the life of the seed will double. For example, onion seeds with 10% moisture are viable for 16 weeks at 35°C, but will live for 78 years at 0°C. Dropping the temperature further to -15°C would enhance seed longevity to 624 years (Koopowitz and Kaye 1990). Similarly, the life of a seed will double with each 1% decrease in water content. However, a minimum of 4% water content appears to be required by seeds to remain alive. Keeping the foregoing in mind, when both temperature and water content are judiciously reduced, the two factors multiply the life span of a seed. For example, when temperature drops by 5°C and water content decreases by 1%, the seed lives four times longer (Koopowitz and Kaye 1990). The actual reason for failure in storage of recalcitrant seeds is that a decrease in moisture content of such seeds infringes viability; such seeds, at maturity, have a moisture content of 12 to 31% and any reduction renders them non-viable (see Chin and Pritchard 1988).

Long-term seed storage requires conduction of regular regeneration tests in stored seeds to monitor their viability. The frequency of their germination/regeneration depends on their initial viability, rate of loss of viability due to storage, and storage conditions. The minimum requirement for each regeneration test is 400 seeds. Some seed banks recommend 5% of the seeds stored be subjected to regeneration tests. In large seed banks, regeneration tests are done continuously, on the seeds of one species or the other.

Two types of seed storage are normally practised in a seed bank. As per the first, **base collections** are stored under optimum conditions and are not interfered with until reduced viability is noticed through periodic germination tests. When reduced viability is observed, a new seed generation is carried out. In the second type of seed storage, **active collection** of seeds is done, from which subsamples can be taken periodically for experimentation, exchange, evaluation and display. The active collection need not necessarily be stored under optimum seed storage conditions and can be multiplied periodically by growing a new generation of plants and reharvesting.

The most important seed banks in the world are: The United State's National Seed Storage Laboratory (NSSL) (Fort Collins, Colorado, USA), the Vavilov Institute (Leningrad), the Izmir Centre (Turkey), the Royal Botanic Gardens (Kew) (seeds of 4000 species have already been banked and there is a current programme to bank the seeds of the entire British flora of some 1500 species, Prance 1997) and the Iberian Gene Bank (Madrid). In addition, nearly 580 botanic gardens of the world have developed facilities for seed banks of wild taxa, of which at least 150 have low temperature seed storage facilities. While some seed banks specialise in the storage of seeds of taxa in specific geographic areas, others specialise in seeds of a particular taxonomic group. Some banks concentrate on the seeds of forest trees and some on specific crops. As examples of the latter can be mentioned the 60 seed banks developed and promoted by IBPGR; but even these, until very recently, have not attempted to store seeds of wild relatives of crops. The total number of accessions of PGRFA stored in seed banks worldwide is 3, 610, 428 (FAO 1996a).

There are disadvantages in the seed bank technique as well. It cannot be practised for all seeds. It is also criticised because it 'freezes evolution' of plants; seed germplasms held in banks are no longer continuously adapting to changes in the environment, such as exposure to new races of pests or climatic stresses/changes.

'Test-tube' Gene Banks

Many plants either do not produce seeds (i.e., clonal crops) or are not normally reproduced from seeds so as to maintain intact a highly heterozygous genotype. To these categories belong short-lived plants which are propagated from tubers, bulbs, corms, rhizomes, roots etc. as well as long-lived shrubs and trees, which are propagated vegetatively. These taxa are conserved through the maintenance of their vegetative propagules under appropriate conditions. Propagules of short-lived taxa have a storage life of only months rather than years. Potatoes can normally be kept at 4°C until the next spring or a further period of 12 months but not beyond. Most other roots and tubers have a storage life of less than 12 months. Cuttings of shrubs and trees, either unrooted or rooted, can be stored between -2 and +2°C up to a maximum of five years.

Pollen Banks

Pollen storage is very important, as it can make available directly the pollen required for crossing and breeding works, especially in breeding tree taxa. Since it is now possible, at least in a few taxa, to raise whole plants as haploids from pollen grains, pollen banks have assumed additional importance. The major disadvantage of a pollen bank is that only paternal material can be conserved and regenerated.

Adequate techniques for the fairly long-term storage of pollen are not yet available. Cryopreservation is required for long-term storage of pollen (Towill 1985). If the moisture content of pollen is suitably reduced, according to the requirements of a particular species, pollen can be kept at cryogenic temperatures of -180 to -196 °C for periods ranging up to 6-12 years, but in the majority of species mostly well below one year. Although the exact reasons for the loss of pollen viability/fertility in a very short time under apparently good storage conditions are not known, it is believed that the moisture content of pollen plays a very critical role. Therefore, cryopreservation for long-term pollen storage is unlikely to be within reach for general application until the causes of pollen breakdown under storage are well understood. However, freeze- and vacuum-drying have both been successfully applied to storing pollen of a number of taxa even up to 12 years, usually at a storage temperature of +5 to -18 °C.

Field Gene Banks

This is one of the major methods of *ex-situ* maintenance of taxa that must be conserved. A field gene bank is an area of land where collections of growing plants of species needing conservation are assembled; the assemblage contains as many individuals of the target species as possible in order to sustain genetic diversity. These field gene banks make available plant material for breeding, reintroduction, research especially in population genetics, physiology, microbiology, biochemistry, nutrition and processing technology, and for several other purposes.

Field gene banks are particularly useful in the conservation of perennial species and therefore are of greatest importance in forestry. They have also been established in agriculture/plantation sectors providing the required germplasms of such tropical crops as cocoa, coconut, rubber, mango, cassava, coffee, banana, sweet potato and yam. The IBPGR has initially established/promoted 23 field gene banks for crops. Several of these also contain germplasms of wild relatives of domesticated crops.

The characterisation and documentation of germplasm resources in a field gene bank consist of five operational procedures (IBPGR 1991): (i) **Establishing the origin of the plant material**: This is called **Passport data**. Passport data includes **accession data,** if an accession has been received from a breeder, another institution etc., along with details on site of origin, ancestor or pedigree. It will also include **collection data** if the accession is directly derived from a field collection, along with a description of the collection site. (ii) **Characterisation**: This includes recording all highly heritable and easily identifiable characters of the plant, especially from a fully mature plant. These characters, often called **Descriptors**, must be those that are expressed in all environments. (iii) **Preliminary Evaluation**: It includes details on plant development and recording of characters expressed by the growing plant. (iv) **Further Evaluation:** This includes recording all the reactions of plants to physical stresses and to pathogens and predators. (v) **Management data:** This includes handling of the genetic resource, its distribution/exchange, regeneration and maintenance etc.

Over the years 'the global network of base collections' has swelled to about 50 institutions and these institutions have accepted IBPGR's invitation to participate actively in this network. Now, more than 100 species are being conserved in field gene banks throughout the world. The institutions include some of the International Centres of Agricultural Research (IARCs) and some national centres. The total number of accessions of PGRFA conserved in field banks is 526,300 (FAO 1996a)

The advantage of field gene banks is their provision of materials readily accessible for utilisation and evaluation. A major disadvantage is possible reduction in genetic diversity of the material, thereby increasing its susceptibility to pests and diseases. Furthermore, field gene banks involve large areas of land, which limits the genetic range of the material that can be held.

DNA Banks

A DNA bank may be defined as a 'gene library' in which samples of DNA extract are stored. This provides a new option for accession of plant germplasm (Adams *et al.* 1994; Peacock 1989). The DNA samples are of three kinds: (i) total genomic DNA, (ii) DNA libraries and (iii) individual cloned DNA fragments including RFLP probes, mini- and microsatellites etc. Samples of the first type are made with DNA isolated directly from plant tissues. The DNA library preparation requires another step—fragmenting the isolated DNA with a suitable restriction enzyme and packaging the diverse mixture of fragments into a suitable cloning vector. The purpose of a DNA library is to retain each fragment from the original DNA extract so that the whole genomic DNA is represented in the mixture. The third type of DNA samples—individual cloned fragments—are fixed and genetically pure since each vector molecule in the sample is host to the same DNA fragment.

Stored DNA samples ideally serve two contrasting purposes: (i) they are very convenient experimental materials that can be shipped immediately without quarantine problems and put to use immediately for further analysis and manipulation at the molecular level, and (ii) they are ideally suited to a 'time capsule' approach to conservation. In other words, they are frozen genetic resources and potentially the most stable form of preserved germplasm; they do not require recurrent regeneration to retain their future utility indefinitely.

Concomitantly, DNA banks are afflicted with almost all the problems faced by seed banks, but more acutely. Proper, reliable documentation and labelling of DNA samples are very crucial to their use. Both these and the despatch, receipt and use of these samples require a high level of technical skill. Moreover, establishing ownership and control of DNA samples is a complex 'delicate' problem. These factors limit the contribution DNA banks could make to the conservation of biodiversity. Furthermore, the sophistication of 'immobilisation' techniques and the potential power of PCR amplification methodologies notwithstanding, total genomic DNA samples are virtually non-renewable. Thus DNA banks offer no newer solutions at least for the present, to the conservation of endangered species. DNA storage invariably only allows for recovery of single genes but not of whole genomes (Peacock 1989).

In spite of the above problems, an international network of DNA banks (DNA Bank-Net) for the conservation of genomic DNA has been vigorously promoted (Adams 1993). There are three primary data bases that collect and distribute DNA material: GeneBank

(Bethesda, Maryland USA), EMBL (Cambridge,UK) and the DNA Data Bank of Japan (Mishima, Japan). The main aim of such a network is to make biological resources more widely available for human benefits through increased research work and use. However, it should be understood that it is only an adjunct and not an alternative methodology in conservation. So a DNA bank cannot replace a field gene bank despite being less costly than the latter.

In addition to DNA banks, there are RNA and protein banks in some developed countries. The major protein databases are the PIR-International Protein Sequence Database (Multinational) and the SWISS-PROT Protein sequence database at Geneva. A Ribosomal Database exists at the University of Illinois, USA.

In-vitro Conservation Methods

This form of conservation is very important and is followed throughout the world by many institutions. Most of the *in-vitro* methods have several advantages, such as greater safety from viral attack, lower cost and accommodation for larger numbers (Withers 1989). All the methods involve storage of plant germplasms under *in-vitro* conditions. Meristem tips and buds are usually stored; such storage is especially followed for those taxa whose seeds are recalcitrant. Very often the meristems or buds from intact plants or the adventive 'embryos' produced directly from *in-vitro* grown explants or calli are coated with materials such as alginate or neutral gums and stored. Such resultant structures are often called '**synseeds**'. Very rarely calli are stored *in vitro*. Around 1500 wild taxa have been stored *in vitro* in various institutions around the world. The number of accession of PGRFA conserved *in vitro* is estimated to be 37,600 (FAO 1995).

In-vitro storage is labour-intensive and costly since subculturing is necessary at regular intervals to test viability of the stored tissue. The annual average total cost for conservation for one accession is estimated to be $82.92 (see Virchow 1998). The other disadvantage is the risk of somaclonal variations that develop under *in-vitro* conditions.

Ecosystem Restoration

Ecosystem or Habitat restoration[3] (more often treated under a separate discipline called '**Restoration Ecology**') is a very rapidly growing and intellectually exciting aspect of conservation science. It refers to the task of 'fixing' damaged ecosystems. Thus, ecological restoration can be crudely but effectively defined as 'making nature' (Jackson 1992). The Society of Ecological Restoration (SER) defines it as 'the process of intentionally altering a site to establish a defined, indigenous, historical ecosystem'. The goal of this process is to emulate the structure, function, diversity and dynamics of the specified ecosystem (SER 1991, p. 4). The growing interest in this area is already reflected in the founding of the SER in 1987 and more recently in the launching of its new journal *Restoration Ecology*.. An important edited book entitled *Restoration of Endangered Species* (Bowles and Whelan 1996) has also been published.

Habitat restoration efforts have to be delicately balanced after due consideration of the economic and political demands of humanity on the one hand, and the biological requirements of the species/habitats targeted for restoration, on the other. This leads to the situation wherein conceptual issues important to habitat restoration range from the sociopolitical (including the management of restoration efforts), to the biological such as population biology, landscape ecology etc. (Bowles and

---

[3]Some confusion exists between the term 'restoration' and the alternative (?) terms such as 'rehabilitation', 'reclamation', 're-creation' and 'ecological recovery'. However, the term 'restoration' is preferred because of wider usage (see Meffe and Carroll, 1994 for more details on this issue).

Whelan 1996). Brown (1996) drew the attention of restorationists to the fact that they should not ignore the situation wherein society could suffer missed economic opportunities due to land being dedicated to restoration instead of to other direct uses. In many countries, restoration is attempted at several smaller sites for many reasons including financial, logistic, bureaucratic and biological.

Reintroduction is one type of restoration in which a species is intentionally introduced into its native habitat and range from which it disappeared or became extirpated for some reason or the other (Cairns 1986). Reintroduction is also called **translocation** (Given 1996; Templeton 1996). Reintroduction, as a rule, is done from materials of a taxon assembled and cultivated *ex-situ*; often the same sources are depended on for occasional replenishment. During reintroduction care must be taken to see that the plant becomes firmly established in the original habitat and that a critical minimum viable population is established for proper reproduction.

Plant reintroduction is a high-risk and high-cost strategy of ecological restoration (Maunder 1992). With both annuals and perennials, the success rate of reintroduction reaches the highest with materials derived from the transplant site and thus the microhabitat adaptation becomes a distinct advantage. In certain instances, however, reintroduction has involved the translocation of genotypes across geographic ranges, resulting in the introduction of 'incorrect' genotypes where they do not belong. Reintroduction demands expertise in management skills and enough resources for operations ranging from collection of plant materials, raising seedlings, choice and preparation of transplant sites, planting, protection and further safeguarding of transplant populations.

Reintroduction calls for knowledge and information about the taxon's key genetic, demographic and ecological traits that affect its vulnerability to stochastic extinction events. In addition, information on the causes of, or circumstances contributing to, the decline of the original population size is needed for any attempt to counter adverse ecological factors such as the presence of predators, parasites, competitors, nutritional imbalance or deficiencies, or the decline or absence of beneficial co-occurring taxa (such as pollinators and fruit/seed dispersers) or adequate water supply etc. (Lande 1988; Menges 1991). Details on the autecology of the target species are also very essential.

A recent survey indicated that about 250 projects in over 25 countries involving about 35 plant families have been undertaken thus far to introduce species into their native habitats. However, very little has been done in developing countries, including India, in the tropical belt. The most important reintroductions include *Stephanomeria malheurensis* (Asteraceae) in Oregon, USA from where it became extinct, *Pediocactus knowltonii* (Cactaceae), a highly endemic taxon in New Mexico, *Sophora fernandeziana* (Leguminosae) in the Chilean island of Juan Fernandez, *Ruizia cordata* (Sterculiaceae), a highly endangered shrub in the Indian Ocean island of Réunion, *Gentiana nivalis* (Gentianaceae) in Scotland, and *Trochetiopsis melanoxylon* (Sterculiaceae) in the island of St. Helena.

Ecosystem restoration and species reintroduction are not, however, without problems. There is scope for making innumerable mistakes, even devastating ones that could lead to ultimate loss of the target species. In the coming years, unfortunately, ecosystem restoration will be more difficult to execute since the rate of ecosystem/species loss is increasing at an alarming rate.

In-situ or Ex-situ Conservation?

It has repeatedly been argued both in favour and against *in-situ* and *ex-situ* conservation efforts. However, after a detailed and critical analysis of these arguments, one necessarily concludes that the safest and perhaps also the most effective conservation strategy is that which combines these two complementary methods. *Ex-situ* conservation is to a large extent subsidiary and complementary to *in-situ* conservation. The latter has the potential for long-term preservation of ecosystems, species and populations under conditions of continuing adaptations. Tree species are preferably conserved *in-situ* since they require considerable space to conserve the required critical minimum viable population, which varies from 500 to 5000 individual trees (Prance 1997). *In-situ* conservation also protects the associated animals and microbes (such as pollinators, dispersers, rhizobia and mycorrhizae), thereby enabling free energy flow. The *in-situ* approach is essential in places whose flora has not been adequately inventoried. *Ex-situ* conservation affords the freedom to select a particular population, species or ecosystem for priority conservation on geographic or ecological grounds, for educational reasons, or because of the mere fact that the system in question is endangered/threatened in its natural environment. In fact Article 9 of the Convention on Biological Diversity states that *ex-situ* activities should be undertaken as far as possible and as appropriate, and predominantly for the purpose of complementing *in-situ* measures.

Ex-situ Conservation of Microbes

Both efforts and information with respect to *ex-situ* conservation of microbes are lacking. However, a wide range of techniques are available for the preservation of microbial taxa and their strains. Of these, lyophilisation through freeze-drying and cryopreservation in liquid nitrogen are the most efficacious for long-term storage. It should be mentioned, however, that as yet not all micro-organisms are amenable to preservation by these methods. In fact, because of this problem the world network of microbial collections contains only a tiny fraction of the microbial species present in the environment. For example, estimates place at least 36 major divisions in the domain Bacteria, but only four are even moderately well represented in culture collections. For many of these divisions not even one representative species is present in culture collections. It is therefore suggested that for non-culturable taxa, DNA banks should be established (Fuerst and Hugenholtz 2000). The development of programmable coolers is enabling protocols to be devised for the cryopreservation of even those microbes previously considered recalcitrant. Even where species cannot be grown in pure culture, host tissue containing these microbial species or samples of other substrates such as soil can be preserved by cryopreservation. The readers are advised to refer to the works of Kirsop and Doyle (1991) and the World Federation for Culture Collections, WFCC (1990) in which guidelines for the establishment and operation of such collections are provided. At present the World Data Centre for Micro-organisms (WDCM), established by WFCC and sponsored by UNEP, UNESCO and RIKEN holds a database on about 800,000 (786,328 to be exact) micro-organism strains held by about 500 (482 to be exact) collections from about 60 countries. These include fungi (45%), bacteria (43%), viruses (2%) and others (plasmids, plant cells, algae etc.).

Special mention must be made here of the Budapest Treaty and the Microbial Type Culture Collections (MTCCs). The Treaty came into force early in 1977 and was amended in late 1980. At present 44 states and three intergovernmental industrial property organisations are signatories of the Budapest Treaty. The signatories include India. As per this Treaty, certain organisations involved in microbial culture collections are recognised as 'International Depository Authorities' (IDAs). Anybody/any organisation desirous of securing a patent involving a micro-organism is mandatorily required to deposit the microbe in any one of the IDAs mentioned in Table 8.5.

Social Approaches to Conservation

Loss as well as conservation of diversity is an issue of great social concern, since biodiversity has very great intrinsic value in a sociocultural system (Shiva *et al.* 1991). However, the intimate relationship between society and biodiversity has not yet been fully realised by many people in this mechanised world. This is evident from the OTA report (OTA 1987, p.127) which states: 'Social and political processes influencing how biological diversity is perceived and valued are the least well understood and, in the long run, the most important factors affecting success of on-site diversity maintenance'. In fact, sociologists, anthropologists and historians have to develop

Table 8.5. Microbial culture collections currently holding IDA status (from Sekar and Kandavel 2002)

S.No.	*Country*	*Depository Institution*
1.	Australia	Australian Government Analytical Laboratories (AGAL)
2.	Belgium	Belgian Co-ordinated Collections of Micro-organisms (BCCM™)
3.	Bulgaria	National Bank for Industrial Micro-organisms and Cell Cultures (NBIMCC)
4.	Canada	National Microbiology Laboratory, Health Canada (NMLHC)
5.	China	China Centre for Type Culture Collection (CCTCC)
		China General Microbiological Culture Collection Centre (CGMCC)
6.	Czech Republic	Czech Collection of Micro-organisms (CCM)
7.	Germany	Deutsche Sammlung von Mikooganismen und Zellkulturen GmbH (DSMZ)
8.	Spain	Coleccion Espanola de Cultivos Tipo (CECT)
9.	France	Collection nationale de cultures de micro-organismes (CNCM)
10.	United Kingdom	Culture Collection of Algae and Protozoa (CCAP)
		European Collection of Cell Cultures (ECACC)
		CABI Bioscience, UK Centre (IMI)
		National Collection of Type Cultures (NCTC)
		National Collection of Yeast Cultures (NCYC)
		National Collections of Industrial, Food and Marine Bacteria (NCIMB)
11.	Hungary	National Collection of Agricultural and Industrial Micro-organisms (NCAIM)
12.	Italy	Advanced Biotechnology Centre (ABC)
		Collection of Industrial Yeasts (DBVPG)
13.	Japan	International Patent Organism Depository (IPOD)
14.	Republic of Korea	Korean Cell Line Research Foundation (KCLRF)
		Korean Collection for Type Cultures (KCTC)
		Korean Culture Centre of Micro-organisms (KCCM)
15.	Latvia	Microbial Strain Collection of Latvia (MSCL)
16.	Netherlands	Centraalbureau voor Schimmelcultures (CBS)
17.	Russian Federation	National Research Centre of Antibiotics (NRCA)
		Russian Collection of Micro-organisms (VKM)
		Russian National Collection of Industrial Micro-organisms (VKPM), GNII Genetika
18.	Slovakia	Culture Collection of Yeasts (CCY)
19.	United States of America	Agricultural Research Service Culture Collection (NRRL)
		American Type Culture Collection (ATCC)
		IAFB Collection of Industrial Micro-organisms
20.	Poland	Polish Collection of Micro-organisms (PCM)

very important descriptions of social factors affecting biodiversity maintenance at specific sites, although some effort has already been made towards this goal. It is also important to remember to educate people that science alone cannot protect biodiversity. Society and social and cultural values must be called in as well in the conservation and protection of nature and its bioresources. This section highlights how society practised biodiversity conservation in the past by attaching ethical, aesthetic and religious values to it. The role of NGOs in persuading and pressurising society to fight against attempts by the State to degrade and deplete bioresources on the pretext of developmental activities and welfare measures to people is also emphasised.

Sacred Groves

The present technocratic and scientifically oriented society mistakenly considers that religion is not interested in protecting and managing biodiversity. The truth, however, is that religious values very often help to protect biodiversity. The practice of protection of patches of forests with temples in their vicinity has long been in vogue in India and a few other parts of the world. In some instances, forest patches or gardens with local floristic elements (often called ***Nandavanas***) have been specially created near established temples and declared sacred to ensure their protection and conservation. Such **sacred groves** and gardens dedicated to the worship of the Presiding Deity of each temple are mentioned in ancient Greek, Latin American and Indian literary works as well as in epigraphical records and copper plates of these countries. Data also come from folk traditions, history and traditional knowledge passed on through several generations. There are reports of sacred groves in the Near East, Europe, North and Sub-Saharan Africa, India, S.E. Asia, Oceania, China, Japan, Siberia and the Americas (Hughes and Chandran 1998; Hughes and Swan 1986). There are several references to sacred groves in the **Old Testament** also. Touching plants (and animals associated with them) in these sacred groves and gardens was forbidden to all except the temple priest, and his too restricted to offerings to the Presiding temple Deity and curing the ailments of local people (the temple priest was invariably the village doctor). The groves were considered the property of Gods, who may be male or female, and represented in the temple as a slab of stone, a hero stone, *sati* stone or trident. Since sanctity was ascribed to the plants of such groves and since spiritual beings were believed to reside in such places, ordinary human activities were voluntarily precluded. These activities included tree-felling, gathering of wood/fuel and plants and leaves, hunting, fishing, grazing by domestic animals, ploughing, planting and harvesting, and dwelling. This ensured the conservation and preservation of the local vegetation for posterity. It should not be thought that such groves and gardens and the idea behind their establishment passed away with the ancient world. On the contrary, several sacred groves and temple gardens persist in many parts of India and a few other countries even today, in spite of the fact that adjacent forests and vegetation have been totally or partially lost. For example, *Ginkgo biloba*, the living fossil, is presently growing in one of the largest 'seminatural' populations at Tian Mu Shan, near the Kaishan Buddhist temple in China (del Tredici *et al.* 1992).

In India, several thousands of sacred groves and temple gardens of various sizes (from clumps of trees to several hundred hectares) and diverse floristic composition have been reported in all vegetation types ranging from evergreen forests to desert/arid vegetation. About 4215 sacred groves covering an area of 39,063 hectares are estimated to be distributed in India (Malhotra 1998). Groves are reported in areas from the highest mountain peaks to sea level (Kushalappa and Bhagwat 2001). The largest sacred grove, Mawflong, occurs in Meghalaya near Shillong. Sacred groves are known by several names in India depending on the place, *Kan* in Karnataka and part of Maharashtra, *Kavus* in Kerala, *Deorai* in Madhya Pradesh, *Devarakadu* in northern Karnataka and Goa, *Orans* in Rajasthan, *Mawflong* in Assam and Meghalaya, *Koilkadu* in Tamil Nadu etc. Floristically such groves even today contain taxa lost/endangered in adjacent regions. Several endemics are today reported only from such groves, for example *Dipterocarpus indicus, Myristica fatua, Pinanga dicksonii, Manilkara hexandra* etc. Responsibility for protecting groves and enforcing rules were assumed by the local community since the grove was an integral part of the village society. Local autonomy was given to these groves, although epigraphical records indicate that the committee in charge of temples and associated groves included a representative of the state/king of that region. The priest of the temple was often the only person authorised to enter the grove and that too only for collecting flowers/fruits to offer to the presiding deity or herbals to cure diseases of the local people (Krishnamurthy 1999; Swamy 1973). Sacred groves, wherever still present, should be preserved and restored for several reasons, including their value as historical evidence for the relationship of human beings to nature.

Sthalavrikshas

Sthalavrikshas is a Sanskrit word meaning temple trees. In India every temple, whether Hindu, Jain or Buddhist, had and still has a specific tree taxon cultivated very near to the *sanctum sanctorum*, the place where the Presiding Deity/Idol is erected. That particular tree is the temple tree for that temple. The idea behind this is that the people of the village where the temple is located should be very much concerned to protect trees in general and express their society's sentiment by selecting a representative local tree taxon as the temple tree. This is yet another form of people's regard for nature, nature worship and an expression of societal concern for the conservation of plants. Ancient Tamils, for example, thought that Gods resided in trees and by worshipping trees they worshipped God. They further thought that different Gods resided in different types of trees and hence they planted the specific temple tree in temples meant for specific Gods. As many as 65 temple trees are recorded in Tamil Nadu alone, each in specific temples.

People's movements for Biodiversity conservation

People throughout the world are feeling uneasy about modern developmental schemes, which have resulted in the substantial loss of biodiversity, thereby denying them a just, equitable and sustainable resource base. Expression of such resentment has been spontaneous in many places and has resulted in resistance movements to retain control over national resource bases and to save them from organised ecocides. The concern of the local society in all such cases has been to prevent the total destruction of the renewable resources on which their lives and livelihood depend. Although there are several such resistance movements, only the most important are described here.

Chipko Movement

'Chipko' means 'to hug' or to embrace in affection. The chipko movement is a success story in environmental conservation and credit for it goes to the women of Reni village of the Alaknanda catchment area in the Uttarkhand region of the Himalayas of India, who sought to protect the local forests called *Myka* ('mother's house'). It was prompted by an order issued in the 1960s by the Government of India to cut trees in the Himalayan foothills on a large scale. This led to severe floods in 1970 which swept away 6 bridges, 16 footbridges and over 600 houses, in addition to destroying crops over hundreds of hectares, affecting no less than 100 villages in the area. The ***Chipko Andolan*** (Chipko revolution) was born when the women of the village sought to save the forests in this extremely sensitive region of the Alaknanda catchment by marching to them shouting slogans and hugging the trees when contractors were ready to cut them down. As a result of this pressure, the contractors had to abandon their operations (Bhat 1987).

The same story was repeated in 1975 when 200 women in Gopeshwar, Chamoli district, waged a war against the cutting of oak trees by the district administration. Women of Bhyundar village in the lower reaches of the famous Valley of Flowers also went on the warpath with their 'Chipko'. In 1980, women of Dungri-Paitoli villages in the Chamoli region of Uttarkhand also resorted to the Chipko agitation to save their forests. The action was repeated in the 1980s in Bached village of Gopeshwar by 200 women, who prevented the felling of 1600 trees. The local women argued that even dead trees should not be removed, as the lumbering practices would invite soil erosion.

The mother organisation of the Chipko movement was the Dasholi Gram Swarajya Mandal (DGSM) led by Mr.Chandi Prasad Bhat and Mr.Sunderlal Bahuguna. Chipko was only one of the *Andolans*. The DGSM educated the villages about the importance of forests and their conservation through lectures, ecodevelopment camps, tree plantation programmes, construction of simple water distribution schemes, and setting up microhydel projects with locally available technologies.

The Chipko movement became the spearhead for communities elsewhere. In the state of Karnataka a similar people's movement was organised, the 'Appiko movement' to save forests.

Chico River Dam and Tribal Campaign

The Cordillera highlands of the northern Philippines witnessed a successful struggle for self-determination by the tribals of the area who traditionally tapped the forests for their livelihood. The government planned construction of four dams along the Chico River, two each in the Kalinga-Apayao and the Mountain Province. The Bontok and Kalinga tribals resisted this step as the project would displace an estimated 100,000 persons and submerge 2000 ha under rice cultivation and 2500 ha under coffee and fruit plantations. After a great struggle by the tribals and the Cordillera People's Bedong Alliance (CBA), comprising 130 organisations, the Chico River dam project was shelved.

Participatory Forest Management

A young forest officer, Ajith Kumar Banerjee, initiated PFM in Arabari, state of West Bengal, India in 1972. It subsequently expanded to other states of India and became a very significant turning point in the history of forest conservation not only in India, but other Asian countries (for details, see Swaminathan 2000). The essential feature of PFM is that the State and Society become partners in the management of forest resources. The State continues to own the resources but the benefits are shared, especially access to non-timber forest products for the community. This leads to the community developing an economic stake in the preservation of forests, which ensures its conservation and sustainable exploitation. Thus, forest degradation can be reversed through PFM, which has been observed in several forest tracts of India over the last 25 years.

Others

Several other instances are known worldwide wherein societal or people awareness has aided the conservation of precious areas rich in biodiversity. Examples include prevention of construction of a dam across Nam Choan on the Kwae Yai River in Thailand and in the Narmada River in India, and protection of Sertani lake in Indonesia from exploitation for a hydroelectric project.

Role of Universities and Other Educational Institutions in Biodiversity Conservation

Needless to say, biodiversity and its conservation have been largely conceived and guided by the academic community in universities and other institutions of learning (Temple 1995). Academic interest in the science of biodiversity and conservation is strongly motivated by the high level of scientific rigour, multi- and interdisciplinary approaches, and innovative measures requisite for tackling biodiversity problems, which only an academic community possesses. Governmental agencies and private groups interested in biodiversity may not have the expertise necessary for solving all the problems; their target is usually limited to specific categories of biodiversity for study and/ or conservation. Although biodiversity science has emerged with contributions from useful theories and hypotheses (such as the Island Geography Theory), new theoretical contributions have been few in recent years. These can only be 'brainstormed' by academics in institutions of higher learning. Moreover, a complex web of problems remains untackled in the biodiversity crisis (see Chapter 1); this web likewise can only be unravelled by innovative, mission-oriented and crisis-driven academicians.

Another need of the hour is to improve the liaison between academicians, States, NGOs and other conservation practitioners. All the concerned parties should/must work together in resolving biodiversity issues. Further, academicians and academic institutions ought to play a major role in imparting biodiversity awareness through environmental education of the public that is both general and, where necessary, more specific. The general goals of environmental education (including biodiversity awareness) are quite simple: (i) Foster clear awareness of, and concern about, economic, social, political, and ecological interdependence in urban and rural areas. (ii) Provide every person with opportunities to acquire the knowledge, values, attitudes, commitments and skills needed to protect and improve the environment and its resources. (iii) Create new patterns of behaviour of individuals, groups and society as a whole towards the environment (International Environmental Education Program 1985). Education must be seen as a two-way process between environmental managers and professionals on the one hand, and society and people on the other.

Within this broad-brush approach to environmental education, facts about biodiversity and its conservation must be related to people's lives as well as to nature's needs. Often popular support for conservation of charismatic species or useful species is relatively high, but it is difficult to convince people of the importance of 'odd' taxa or unrealised genetic resources.

Biodiversity Awareness Programmes

The interrelated components of biodiversity education are awareness, real-life situations, conservation and sustainable development. An awareness programme is an essential precursor to formal learning. Rare plants or flagship species will generate interest; plants that appeal to people may create awareness. Awareness can also arise as a result of an environmental crisis. NGOs, environmental groups etc. can create awareness. Awareness programmes must be built upon and, where possible, consolidated by formal education. There is an enormous variation in background and fundamental perceptions of students, both within and between cultures. This suggests the need for a **template approach** (Pitt 1987) to education on biodiversity, in which different relevant examples are inserted into a basic text for wider use. The template approach allows the preparation of textbooks and supplementary resources with wide general application. Specific examples and emphasis can be inserted for particular regions or levels of education. This allows students to relate to those things of biodiversity with which they are familiar.

The division of education into elementary, secondary and tertiary segments and into discrete subjects does not always fit well with the requirements of teaching biodiversity. The solution may be to recognise levels of complexity in education rather than formal, age-related grades. This would allow teachers to introduce aspects of biodiversity at appropriate levels of complexity (MacFie 1987). A balance between classroom based and outdoor (or real-life) teaching is essential and the latter can be achieved through visits to natural habitats, botanic gardens and industries using plants.

At the tertiary level, biodiversity studies tend to be strongly oriented towards the subject *per se*. New careers in environmental education, ethics, conservation, economics, law etc have come in. Information about Natural History Societies and biodiversity organisations and their activities could inspire students. Biodiversity clubs, formed in some countries, should be organised in all countries. Films, tapes, wildlife shows, flower shows, slide-tape programmes, video programmes or even Endangered Species Fairs can be arranged. In addition, tourism, posters, wall-charts, booklets, leaflets, low-cost books, trekking etc. may also help to propagate awareness. The Anglade Institute jointly run by St. Joseph's College at Tiruchirappalli and Jesuit College at Kodaikanal in India is doing yeomen service in this context.

Biodiversity Education Resources

Some countries cannot afford to launch and maintain their own programme for biodiversity education. Help is available from international environmental agencies, which fund the following aspects/programmes.

1. Educational aid to organise courses and to prepare study materials involving and relating to local people.
2. Workshops to train people in biodiversity awareness and education, who in turn train and educate others.
3. Preparation of educational materials in the local language for field biodiversity awareness programmes.
4. Courses on Reserve management, Sustainable utilisation, Herbarium & Museum management etc.

For instance, the International Environmental Education Programme (IEEP) handles all of the above. UNESCO and UNEP jointly founded IEEP. Among its achievements are the following: A series of international and regional meetings, regular free communication of information in five languages to more than 13,000 industries and institutions, including 900 Environmental Education (EE) institutions and 300 EE projects, undertaking of pilot and experimental EE projects, with involvement in over 13 countries of more than 260,000 pupils and 10,000 teachers, educators and administrators. There is also an International Centre for Conservation Education in England, which provides a global focus for practical conservation education activities in developing countries.

Media

Media can play an outstanding role in biodiversity education and conservation (Bellamy 1979). Books, journals, newspapers, bulletins, radio and television form the important media for creating biodiversity awareness. Three separate television channels—National Geographic, Animal Planet and Discovery— are doing great service towards this goal. David Bellamy's 'Botanic Man' serials and programme in Thames television have already popularised plant science and conservation throughout the western countries. There are also special magazines such as *Pied Crow, Outreach* (supported by WWF/IUCN), *Down to Earth*, The Hindu Supplements on Environment etc. for this purpose.

Sustainable Development

Even today, the term 'sustainable development' remains enigmatic. Moreover, it continues to be treated, regrettably, more as a socio-economic and political concept, than a scientific or technological one. What is sustainable in a socioeconomic perspective may not really be so in an ecological sense and vice versa. What is sustainable in one place or at one time may not be so in another place or at another time, i.e., temporal and spatial constraints play a role. The net result is that the concept of sustainable development is becoming increasingly 'complex and amorphous'. Today the number of definitions of sustainable development is thought to exceed 100 (Khoshoo 1998). Defining sustainable development in exact terms acceptable to all has proven very difficult. One of the best ways to understand sustainability is to know that 'the rate of harvest from a renewable system must never exceed the rate of annual increment' of that system. If the harvest respects that limit and if no major environmental disturbance occurs, the system can go on forever. Sustainable biodiversity development consists of three important but overlapping steps: *save it, know it,* and *use it* (Janzen and Gámez 1997). These three components have penetrated the current concept of biodiversity conservation, so much so that conservation is now inseparably intertwined with sustainable development.

Sustainable development became an issue on the environmental agenda in the second half of the 1980s, especially after the publication of the book *Our Common Future* by the World Commission on Environment and Development (WCED). Now sustainable development has become a composite discipline including science,

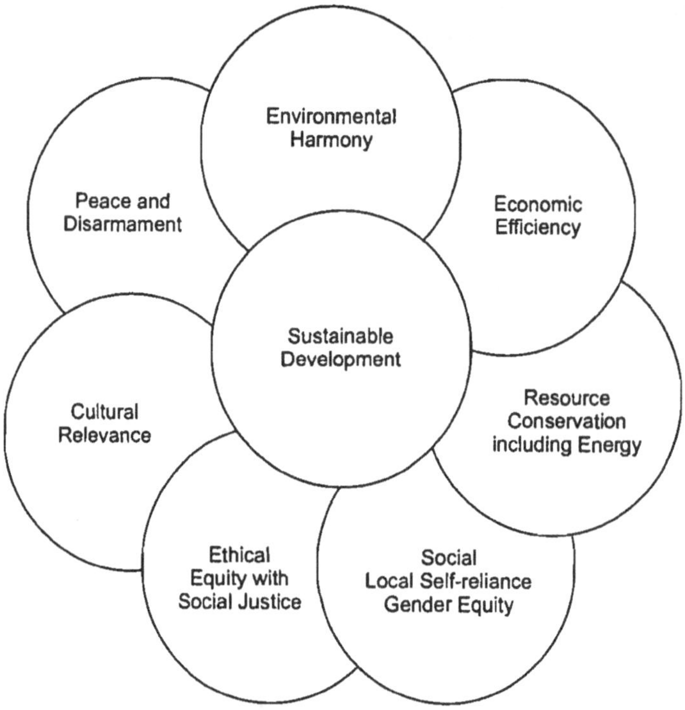

Fig. 8.3 Major dimensions of Sustainable Development.

technology, sociology, economics, ethics, trade, law and politics, incorporating several dimensions (Fig. 8.3). Five years after the publication of this book and after the Rio summit in 1992 the term 'Sustainable Development' had become very popular throughout the world. Most people were made to believe that sustainable development is a panacea for all the environmental ills and problems facing the earth, which led to considerable euphoria. A special UN Commission on Sustainable Development was formed, which had to report through ECOSOC to the General Assembly of UNO. In the last decade, consequently, an unusually large number of books and papers have appeared on this topic, with the gradual realisation that sustainable development is not in actuality a panacea for all environmental ills.

Originally the WCED described sustainable development as development that ensures meeting 'the needs of the present without compromising the ability of future generations to meet their own needs' (WCED 1987). This implies that we should impose voluntary limits on the use of bioresources now for the benefits of our children and grandchildren. However, western countries have been rather indifferent to the concept of sustainable development, holding it would be 'a drag on their development' (for more information, see Chapter 9).

9

MANAGEMENT OF PLANT BIODIVERSITY

Introduction

There is an ever-growing demand for bioresources by humankind. It is fully realised now that these growing demands can no longer be met by tapping the still unexploited bioresources or by trade-offs between goods and services (Ayensu *et al.* 1999). Any nation can increase food supply by converting forestlands to agriculture but in doing so it curtails the supply of goods and services rendered by forests which are of equal or greater importance. The projected climatic changes may also exacerbate the act of balancing supply and demand of bioresources. Sectorial approaches to management of bioresources made sense when trade-offs among goods and services were modest or unimportant. They are not sufficient today 'when ecosystem management must meet conflicting goals and take into account the interlinkages among environmental probables'. Therefore, we are now forced to look for an integrated approach to biodiversity management (Ayensu *et al.* 1999).

It is worth noting that the governments of the world's various nations had already made an important, but unnoticed, commitment to nature (including biodiversity) conservation and management through the World Charter for Nature, adopted by the General Assembly of the United Nations in 1982 (McNeely *et al.* 1990). This Charter recognises that humanity is part of nature, that every form of life is unique and warrants respect, and that continued benefits from nature depend upon the management and maintenance of essential ecological processes. In other words, the goal of biodiversity management is to strike the optimal balance between biodiversity conservation, 'advancing human sustainable living' and benefit sharing. Successful management depends on two things: (i) The social, political, economic and cultural contexts within which management objectives are pursued should be properly understood by policy-makers and managers, and (ii) Proper tools and methods should be selected to attain the aforesaid objectives. An integrated, predictive and adaptive approach to biodiversity management requires three basic types of information: (i) reliable site-specific baseline information on all aspects of biodiversity; (ii) knowledge on how the value generation (in terms of goods and services) in specific ecosystems will respond to changing environments and (iii) integrated regional models that incorporate the biophysical, economic and technological changes (Ayensu *et al.* 1999). The scientific community must immediately take steps to mobilise all its knowledge in a manner that can increase awareness, provide information, capacity building at local, regional and national levels and informed policy changes that will better manage the Earth's biodiversity. Biodiversity management can be brought about effectively through committed organisations at the national and international levels (both governmental and non-governmental) which frame policies and methodologies for execution. They also collect/collate vital data, store them and distribute them to the needy. In addition, multilateral and national treatises, conventions and legal systems help in the effective management of biodiversity. Biodiversity information and knowledge are made available to any one interested through well-organised databases. In fact, Article 7d of CBD points out the requirement to 'maintain and organise by any mechanism, data derived from identification and monitoring activities'. Similarly, Article 17 of CBD is concerned with information exchange. Nowadays biodiversity and bioinformatics have grown to be made for each other. Biodiversity management therefore requires skills in interdisciplinary areas such as Biology, Economics, Anthropology, Engineering, Forestry, Agriculture, Oceanography, Sociology, Management Science, Geography, Geology, Computerisation etc. This chapter provides details on biodiversity management.

Organisations Associated with Biodiversity Management

Two categories of organisations are necessary for efficient management of biodiversity. The first, already stated in the preceding paragraph, is primarily concerned with framing policies and methodologies for execution besides serving as sources of data. The second is involved in financing biodiversity-related projects, workshops and other activities. Some organisations, such as IUCN, UNESCO and UNEP, are involved in both activities.

Organisations Primarily Involved in Framing Policies and Methodologies for Execution

IUCN

IUCN stands for International Union for Conservation of Nature and Natural Resources. It is a federative membership organisation, composed essentially of governments or governmental agencies, and scientific, professional and conservation organisations. Its foundation was laid in 1948 at the International Conference held at Fontaineblen, sponsored jointly by UNESCO and the Government of France. The present name IUCN was given to this organisation in 1956. IUCN is also now designated as the World Conservation Union.

IUCN initiates and promotes scientifically based conservation measures in co-operation with the UN and other intergovernmental agencies, and with its sister organisation, the WWF (Worldwide Fund for Nature and Natural Resources). It functions and operates through several commissions and committees. The standing Commissions of IUCN include those on Ecology, National Parks and Protected areas, and Environmental policy. IUCN, through its Commission on National Parks and Protected Areas, has developed a system of classification of different types of protected areas, based on several management objectives. This system has 10 different types of protected areas, of which the World Heritage Sites and Biosphere Reserves have international standing and recognition (for more details, see Chapter 8). IUCN'S Commission on Environmental Policy is largely responsible for initiating negotiations way back in 1978, when the initial drafts for the Biodiversity Convention were prepared; in 1988 the IUCN circulated a comprehensive draft among the participating countries at the UNEP Governing Council. This was followed by the Rio Convention (for more details, see **page 154**). IUCN'S Species Survival Commission (SSC), established in the mid- 1960s, has now approximately 3500 members in 135 countries. It aims at promoting action to arrest biodiversity loss and to restore threatened species to safe population levels. The SSC is divided into 95 Specialist Groups, which cover different taxonomic groups, geographic areas or subject areas.

IUCN has published among others, the following documents/books: (i) *World Conservation Strategy* (1980) in collaboration with WWF and

UNEP, (ii) *Caring for the Earth—A Strategy for Sustainable Living* (1989) in collaboration with WWF and UNEP and (iii) Red Data Books (RDBs). RDBs are prepared by SSC in collaboration with WCMC. The first RDB on plants, published by Dr. Ronald Melville in 1970, was updated in 1973 as the IUCN Plant RDB. RDBs provide information on the current status and conservation requirements of the threatened plant species of the world. For each threatened species details are provided on distribution, population status, habitat, ecology and potential conservation measures. Red Data volumes are divided on taxonomic or geographical bases. The most important books relating to plants are *IUCN Plant Red Data Book* by Lucas and Synge (1978), *The Plant Red Data Book of Rodrigues* by Strahm (1989), *Draft IUCN Red List Categories* (1993), *IUCN Red List Categories* (IUCN 1995) and *IUCN Red List of Threatened Species 2000* compiled by C.Hilton-Taylor, (iv) IUCN Bulletins, (v) IUCN's first edition of the *United Nations World List of National Parks and Equivalent Preserves* (IUCN 1992) and subsequent lists (IUCN 1994a,b), (vi) *Global Biodiversity Strategy: Guidelines for Action to Save, Study, and Use Earth's Biotic Wealth Sustainably and Equitably* (WRI/IUCN/UNEP, Washington DC 1992).

IUCN convenes General Assembly meetings every third year. Its actions are reported in IUCN Bulletins. IUCN has also organised/supported several conferences/workshops on biodiversity. It has about 60 nations and 128 governmental agencies as members. Gland, in Switzerland, is the headquarters of IUCN.

UNEP

UNEP stands for United Nations Environment Programme. It is a UN agency engaged in the co-ordination of intergovernmental measures for monitoring and protecting the environment, achieving sustainable development and resolving biodiversity issues. It has the following threefold mandate: (i) to create awareness on global environmental problems, (ii) to build consensus on actions addressed towards these problems and (iii) to promote and support such action programmes. UNEP was formed subsequent to the UN Human Environment Conference at Stockholm in 1972, initially with an Executive Body (the Governing Council) of 50 members. A voluntary UNEP fund was established to finance the projects of the UNEP. For co-ordinating all activities of UNEP, an Environmental Co-ordination Board was created.

UNEP has taken up, among others, the following major activities (Olembo 1996): (i) biodiversity country studies, (ii) global biodiversity assessment (GBA) projects and (iii) biodiversity data management (BDM) projects for capacity building in developing countries and improved in-country networking of biodiversity information. This effort is made to empower the countries to effectively manage their biodiversity data and information and to underpin the sustainable use of their genetic resources. The 'National Biodiversity Country Studies, Strategies and Action Plans' were taken up to underpin and reinforce in-country biodiversity and assessment and planning processes, to identify national priorities for action, and to provide a baseline for monitoring effectiveness of national actions, policies and programmes on conservation and sustainable use of genetic resources. The UNEP Biodiversity Country Studies Projects, initiated in 1991, were funded by Global Environmental Facility (GEF). These are implemented in developing countries in co-operation with donor countries and UNDP. By about 1995, 19 studies had been completed and several more were in progress.

The GBA was commissioned and published after the CBD was signed. The Scientific and Technical Advisory Panel (STAP) of the GEF recommended the need for a very comprehensive review of current knowledge of biodiversity to the UNEP. It was to meet this challenge that UNEP commissioned the GBA in 1993 and publication readied in 1995 November during the second conference of the parties to the CBD. The GBA was written by thirteen teams of experts involving 300 authors from over 50 countries; in addition, several hundred scientists from more than 80 countries, covering different disciplines of biology, economics, sociology etc. peer-reviewed different parts of the text. The assessment was divided into the following sections: Introduction; Characterization of biodiversity; Magnitude and distribution of biodiversity; Generation, maintenance and loss of biodiversity; Biodiversity and ecosystem functions—basic principles; Biodiversity and ecosystem functions—biome analysis; Inventorying and monitoring of biodiversity; The resource basis for biodiversity assessment; Data and information management and communication; Biotechnology; Human influences on biodiversity; Economic values of biodiversity; Measures for conservation of biodiversity and sustainable use of its components (Heywood 1997a; UNEP 1995). For political reasons and the fact that the GBA was not commissioned by governments but by a UN agency, assessment of country-level diversity was deliberately omitted, as were policy recommendations. Information contained in this assessment can be used for a wide variety of purposes, including conservation planning, monitoring and sustainable use of biodiversity; it can also be used by governments for assessing the state of environment in territories under their purview (see details in UNEP 1995).

UNEP operates the Earth Watch Programme and funds publication of the magazine *Earthscan*. It has held/supported several conferences, workshops and meetings. In India, it has been involved in a number of projects including the Tree Planting Programme and the Ganga Action Plan.

UNEP has produced several seminal publications/documents in addition to *Global Biodiversity Assessment* (published by Cambridge University Press, 1995) mentioned above. These have brought into focus the pertinent issues of biodiversity conservation and its sustainable use. Other important publications to date are: (i) *World Conservation Strategy*, (ii) *Global Marine Biological Diversity: A Strategy for Building Conservation into Decision Making*. Centre for Marine Conservation, in association with IUCN/WWF/UNEP and The World Bank. E.A.Norse (ed.). Island Press, Washington DC (1993). (iii) *Caring for the Earth—A Strategy for Sustainable Living*, in association with IUCN/UNEP/WWF, Gland, Switzerland (1991). (iv) *The World Charter for Nature*. (v) *Global Biodiversity Strategy*, in association with WRI and IUCN, Washington DC (1992), which provides guidelines for action to save, study and use the Earth's biodiversity sustainably and equitably. (vi) Endangered Species of the World under the title *Blue Book*. (vii) *Global Biodiversity: Status of the Earth's Living Resources*, WCMC, Chapman & Hall, London (1992). (viii) *The Global Biogeochemical Cycles*. (ix) *World's Lakes Inventory*. (x) *The Norway/UNEP Expert Conference on Biodiversity, Proceedings*. O.T. Sandlund and P.J. Schei (eds.) Directorate for Nature Management/Norwegian Institute for Nature Research, Trondheim (1993). (xi) *From Genes to Ecosystems: A Research Agenda of Biodiversity*. O.T. Solbrig (ed.). IUBS/SCOPE/UNESCO, Paris (1991).

UNEP is headquartered in Nairobi, Kenya.

UNESCO

UNESCO stands for United Nations Educational, Scientific and Cultural Organisation. It was established in 1945 as a UN Agency. In 1948 it funded for the establishment of the IUCN. UNESCO has also assisted in the creation and operation of networks such as the MIRCENS (Microbial Research Centres), Biosphere Reserves and other protected areas, and marine research stations. MIRCENs are the outcome of a joint effort by UNESCO, UNEP and the International Cell Research Organisation. A worldwide network of microbial culture collections was established and by 1992 there were 16 such collections, which are now known as MIRCENs (Table 9.1). Activities of MIRCENs typically include collection, maintenance, testing and distribution of microbes, and training of personnel. Though each MIRCEN works on its mandatory priorities, they all work together in strengthening the network.

The International Man and Biosphere Programme (MAB) was initiated by UNESCO in the early 1960s, took its final shape in 1968, and was actually launched in 1971. One hundred and ten countries co-ordinate in the MAB. Several UNESCO-MAB documents have already been prepared, wherein the objectives of the network of Biosphere Reserves, the characteristics which the Reserve must display and the action plans of these Reserves are detailed. Subsequently several MAP networks were established. These include the EuroMAB, USMAB, MAB-CYTED (Ibero-American Programme), CBRN-MAB (Chinese network) and MAB-GEF (Central European network).

UNESCO has also identified World Heritage Sites and listed them. A World Heritage Fund was created and is being managed by UNESCO's World Heritage Committee; the annual budget for this fund is 2 million US dollars.

UNESCO, with the help of IUCN, was instrumental in the preparation of the draft for the Rio summit and convention. It, along with the International Union of Biological Sciences, launched in 1991 the famous co-operative scientific programme on biodiversity called 'DIVERSITAS' for studying the origin, maintenance, loss etc. of biodiversity. UNESCO in co-operation with IUBS and SCOPE has published the following book: *From Genes to Ecosystems: A Research Agenda for Biodiversity.* O.T. Solbrig, 1991. The headquarters for UNESCO is in Paris.

Table 9.1 MIRCENs recognised by UNESCO

1. Ain Shams University, Faculty of Agriculture, Shobra-Khaima, Cairo, Arab Republic of Egypt
2. Applied Research Division, Central American Research Institute for Industry (ICAITI), Ave, La Reforma 4-47 Zone 10, Apdo Postal 1552, Guatemala
3. Cell Culture and Nitrogen-Fixation Laboratory, Room 116, Building 011-A, Barc-West, Beltsville, Maryland 20705, USA
4. Centre National de Recherches Agronomiques, d'Institut Sénégalais de Recherches Agricoles, B.P. 51, Bambay, Senegal
5. Department of Bacteriology, Karolinska Institutet, Fack, S-10401 Stockholm, Sweden
6. Departments of Soil Sciences and Botany, University of Nairobi, PO Box 30197, Nairobi, Kenya
7. Fermentation Technology MIRCEN, ICME, University of Osaka Suita-shi 656, Osaka, Japan
8. Fermentation, Food and Waste Recycling MIRCEN, Thailand Institute of Scientific and Technological Research, 196 Phahonyothin Road, Bangkok 9, Thailand
9. Institute for Biotechnological Studies, Research and Development Centre, University of Kent, Canterbury CT27TD, UK
10. IPAGRO, Postal 776,90000 Porto Alegra, Rio Grande do Sul, Brazil
11. Marine Biotechnology MIRCEN, Department of Microbiology, University of Maryland, College Park Campus, Maryland 207742, USA
12. Mycology MIRCEN, International Mycological Institute, Ferry Lane, Kew, Surrey TW9 3AF, UK
13. NifTAL, Project, College of Tropical Agriculture and Human Resources, University of Hawaii, PO Box "0", Paia, Hawaii 96779, USA
14. Planta Piloto de Procesos Industriales. Microbiologicos (PROIMI), Avenida Belgranoy Pasaje Caseros, 4000 S.M. de Tucuman, Argentina
15. University of Waterloo, Ontario, Canada N2LK 3G1, and University of Guelph, Guelph, Ontario N1G 2W1, Canada
16. World Data Centre on Collections of Microorganisms, RIKEN, 2-1 Hirosawa, Wako, Saitama 351-01, Japan

WWF

WWF stands for Worldwide Fund for Nature and Natural Resources. It was established in 1961 and is headquartered in Gland, Switzerland. WWF International has several affiliated national units. The Indian unit was established in 1969 at the time of the XII General Assembly of the IUCN, held in New Delhi. The WWF International is controlled by a Board of International Trustees, while the national units are managed by separate national teams. For example, the Indian unit has a Board of 8 trustees, with its headquarters in Mumbai. It has a network of 18 State and Divisional Units. WWF has initiated several specific conservation programmes in more than 24 countries, with importance given to endangered fauna and flora.

The logo of WWF is the panda, as designed by Gerald Watterson.

ICSU

ICSU stands for International Council of Scientific Unions. It represents 80 of the world's scientific academies or research councils; 50 nations are involved through national representation. Many of these nations are signatories to the Biodiversity convention. ICSU is composed of 20 scientific unions including eight in biology, which have direct interest in biodiversity. These eight unions are: International Unions of (i) Biological Science (IUBS), (ii) Biochemistry and Molecular Biology (IUBMB), (iii) Pure and Applied Biophysics (IUPAB), (iv) Immunology Societies (IUIS), (V) Microbiological Societies (IUMS), (vi) Nutritional Sciences (IUNS), (vii) Pharmacology (IUPHAR) and (viii) Physiological Sciences (IUPS).

The ICSU also supports 19 bodies that are interdisciplinary. Many of these are concerned with biodiversity. These include SCOPE (Scientific Committee on Problems of the Environment), IGBP (International Geosphere-Biosphere Programme), SCOR (Scientific Committee on Oceanic Research), COGENE (Committee on Genetic Experimentation) and CASAFA (Committee on Application of Science to Agriculture, Forestry and Aquaculture). The most important programme with explicit responsibilities in biodiversity undertaken by ICSU is DIVERSITAS jointly managed by IUBS, SCOPE and UNESCO.

The DIVERSITAS programme was launched in 1991. Its purpose is to combine activities in research inventory work and surveillance with training and awareness-raising. It is required by ICSU as one of its most important action programmes in environmental matters and has already obtained backing from numerous National Committees. DIVERSITAS has four main themes, which include inventorying and monitoring of overall biodiversity spatially and temporally and the study of genetic diversity for the conservation of wild relatives of cultivated plants. DIVERSITAS covers all levels of biodiversity, from genes to ecosystems, incorporating both marine and terrestrial environments. Special attention is shown to the rather neglected aspects of microbial biodiversity and aquatic biodiversity. It should be emphasised that DIVERSITAS appreciates the revival of taxonomy as the most relevant area and currently gives it the greatest priority. This programme also strives to remove the current dichotomy between the biodiversity agenda and priorities of developed and developing nations. DIVERSITAS promotes the use of UNESCO'S MAB network as the focal sites for inventory programmes. Several countries such as Brazil, China, France, Japan and the USA have started their own DIVERSITAS activities, but co-ordination between these has not yet been effected.

BioNET-International was initiated by the intergovernmental CAB-International in 1993 with the idea of helping to generate biosystematic self-reliance in the developing countries (Jones and Cook 1993) through the mobilisation and pooling of already existing biosystematic resources as well as through the transfer of skills, knowledge, scientific expertise and other resources provided by expert institutions of developed countries. BioNET has planned a series of interlinked, subregional Technical Co-operation Networks in developing countries, known as 'Locally Organised and Operated Partnerships' or LOOPs, supported by BIOCON and a central technical secretariat. Initially BioNET had a particular interest in microrganisms and some animal groups but subsequently included other groups depending on local needs. Four priorities have been identified for the initial activity of LOOPs: Information and communication services, personnel training, rehabilitation of collections, and development and application of new resources. LOOPs have been organised in Europe (EuroLOOP) (includes 60 institutions from 22 countries) and the Caribbeans (CARINET). Four other LOOPs in East Africa (EAFRINET), South-east Asia (ASEANET), South Pacific (PACINET) and South Africa (SAFRINET) contemplated earlier (Jones and Cook 1993) have now been realised.

FAO

FAO stands for the Food and Agriculture Organisation. It is a specialised agency of UNO established as early as 1945. FAO has several member nations. Although initially formed to take care of food and agriculture, FAO is now the chief organisation for protection and conservation of germplasms of all food plants and animals. FAO has its head office in Rome, Italy.

FAO programmes traditionally support developmental efforts in fisheries, forestry and agriculture. FAO's Global System for Conservation of Plant Genetic Resources was set up in 1983 to co-ordinate activities related to genetic resources concerning food and agricultural production (Glowka *et al.* 1994). It included two major developments: (i) The International Undertaking on Plant Genetic Resources, which aims to ensure exploration, collection, conservation, evaluation and utilisation of plant genetic resources of present and future importance. (ii) The Commission on Plant Genetic Resources (CPGR), an international forum of more than 125 countries, which promotes and implements activities of international undertaking. Later FAO policies and actions tended to have a larger impact on biodiversity. For example, after recognising the rapid loss and degradation of biodiversity throughout the world, in 1989 FAO added Biotechnology and the Tropical Forestry Action Plan (TFAP) as priority areas. Numerous activities subsequently undertaken or co-ordinated by FAO are designed to mitigate biodiversity loss, especially in tropical forests.

TFAP is a programme of FAO intended to halt the destruction of tropical forests and to promote their sustainable development. The programme was initiated in 1985 and focuses on five issues: fuelwood and agroforestry, land use and watersheds, forestry management for industrial uses, tropical forest ecosystem conservation, and strengthening institutions for research, training and education. Initially, development of national level TFAP for every country with tropical forests (about 70 countries) was required and the FAO carries out all action plans mentioned above through national TFAPs.

FAO has assisted the UNESCO in the Man and Biosphere Programme and has co-operated with IUCN in the 'Caring for the Earth' strategy. Together with UNDP and the World Bank, FAO sponsors the Consultative Group on International Agricultural Research (CGIAR), which supports more than 6 International Agricultural Research Centres (IARCs).

The idea of establishing Genetic Resources Centres (GRCs) emerged from the very vast and still unparalleled collections assembled by N.I. Vavilov, his colleagues and the USSR Institute of Plant Industry. In the decade 1920 to 1930, this institute became a highly developed and successful GRC. The GRC idea was further molded at a conference convened by FAO in 1961 (Whyte and Julen 1963) and subsequently another conference organised by FAO and International Biological Programme (IBP) in 1967 (Frankel and Bennett 1970). Interestingly, the term 'Genetic Resource' was first introduced in the latter conference. The decade 1965 to 1974 saw the emergence of what has been called the 'genetic resources movement', the initiatives for which came largely from the FAO panel of experts. The Consultative Group on International Agricultural Research (CGIAR) was founded in 1971 jointly by the World Bank, FAO and UNDP. FAO experts also played a prominent role in a meeting convened at Beltsville, USA in 1972 by FAO and the CGIAR, which made recommendations for a global network of GRCs and for their co-ordination. These efforts ultimately led to the establishment of 16 IARCs as well as 227 seed banks from 1971 onwards; these IARCs consist of a consortium of donor countries and Foundation and Developmental Banks sponsored by the World Bank, UNDP and FAO. At present there are more than 16 IARCs, most of which are specifically responsible for germplasm conservation of specific crops and actively collect the requisite material on a worldwide basis. For example, the IRRI (International Rice Research Institute) based in Manila, Philippines is concerned with the collection and maintenance solely of rice germplasm (both cultivated and wild) from every place throughout the world in which this plant is known.

Another organisation was established under IARC in 1974, namely IBPGR (International Bureau of Plant Genetic Resources), now known as the International Plant Genetic Resources Centre (IPGRI). Its head office is in Rome, Italy with affiliated NBPGRs (National Bureau of Plant Genetic Resources) in several countries. The IPGRI plays a major role in the collection, documentation and conservation of 'useful plants' and their wild relatives. It has developed a standardised description on a crop-by-crop basis by printing gene bank catalogues and directories of germplasm collectors, and by holding germplasms of crop material. It also advises and encourages repositories of plant material to use standard passport data, to systematise holdings and to discard redundant or 'junk' samples, i.e., poorly documented ones. With the assistance of IPGRI, 90% or more of the known land races of crop taxa and over 510,000 accessions are now held by IARCs and seed banks.

CAB International

CAB International is an intergovernmental organisation. It provides very important research data together with scientific and developmental services for agriculture, forestry and related fields throughout the world. It also houses the world's largest bibliographic database in the form of CAB abstracts on research and scientific development. CAB International comprises four constituent International Institutions—Entomology, Mycology, Biological Control and Parasitology.

WCMC

These initials stand for the World Conservation Monitoring Centre. It was founded in 1988. It is a joint undertaking of the IUCN, UNEP and WWF. It serves as a primary source of information on the conservation status of many species of plants (and animals) and monitors conservation and sustainable development. It maintains a database of published literature, unpublished reports, government reports, and references to conservation organisations, contacts and correspondents throughout the world. The detailed biological information available with WCMC covers the distribution, ecology and status of 52,000 plant species; data include population size, potential or actual threats and occurrence in cultivation. Also listed are key sites of high biodiversity and the locations and importance of about 16,000 protected areas worldwide. Mapped digitised data are also available.

ISBI

Driven by concerns of human effect on bioresources and with a view to integrating ecological sciences with resource management and development, the Ecological Society of America proposed the Sustainable Biosphere Initiative (SBI) with a list of ecological agenda. These agenda were further expanded in a workshop in Mexico in which leading ecologists from the world over recommended the establishment of a co-operative programme as a global venture, the International Sustainable Biosphere Initiative (ISBI). The central goal of the initiative is to 'facilitate the acquisition, dissemination, and utilisation of ecological knowledge to ensure the sustainability of the biosphere' (Ramakrishnan 1992).

The concept of sustainability implies the current use of ecological systems with compromising the needs or options of future generations, which obviously involves trade-offs. One of the fundamental objectives of the ISBI is to achieve a better understanding of these trade-offs. Highly interactive participatory research/activity is visualised wherein ecologists join hands with planners and administrators, resource managers, specialists in several disciplines and society at large (see Chapter 8, **Fig. 8.3**). Three important components are visualised by ISBI for research priority: (i) diversity and sustainability, (ii) sustainability in changing biosphere and (iii) human dimensions of sustainability (Fig. 9.1).

Fig. 9.1 Research priorities in three facets of sustainability identified by ISBI.

Organisations Involved in Financing Biodiversity Management

It is often said, 'Conservation without financial resources is just conversation'. Allocation of adequate finance is essential for proper and effective conservation. There should be a suitable match between conservation needs and financial allocation at all levels. This section deals with organisations/ institutions involved in funding biodiversity initiatives.

GEF

GEF stands for Global Environmental Facility. At the annual IMF-World Bank Development Committee meeting of 1989, France suggested the creation of environmental protection activities that would provide benefits to the global human community. By November 1990 an agreement was reached between 25 countries on this suggestion and GEF was created. It was decided that the World Bank, UNDP and UNEP would co-operate in administering the GEF. GEF was required to develop a mechanism for not only establishing a new multilateral fund but also for distributing concessionary finance on loan or grant for the purpose of protecting the 'global commons'. By March 1991, 21 countries had committed approximately US $1-4 billion to the GEF fund over a three-year pilot stage. GEF would accept proposals for funding in four areas: (i) protecting the ozone layer, (ii) limiting greenhouse gas emission, (iii) protecting biodiversity and (iv) protecting international waters.

The main mandate of GEF with reference to biodiversity is to preserve specific areas of the world which contribute Goods and Services to humankind, such as harvestable materials for medicines or industrial products, genetic resources for food products and the regulation of climate and rainfall patterns. The GNP of the nation receiving grants from GEF should be less than US $4000 and the nation should also have UNDP programmes.

The GEF provided US $300 million for biodiversity during 1991- 1994 (McNeely 1996). The most important biodiversity projects funded by the GEF during that period include: Congo Tropical Forest Preservation, Bwindi Forest Conservation in Uganda, Trust Fund for Environment Conservation of Bhutan Wildlife, Protected Areas Management in Laos, Wetland Protection in Elkala National Park of Algeria, Forest Biodiversity of Poland and Conservation of Biodiversity in 20 Protected areas of Mexico. In addition, GEF funded a number of technical assistance projects in East Africa, Vietnam, South Pacific, Choco regions of Colombia, Guyana and the Amazon. The World Bank manages the GEF. After 1994 the GEF financed several programmes on biodiversity in various parts of the world.

WHF [C]

WHF stands for World Heritage Fund. This is one of the best known International Trust Funds for Biodiversity-related activities. WHF was established pursuant to the World Heritage Convention (WHC) in 1972. WHF raises funds by a combination of voluntary and compulsory contributions from contracting parties to the tune of 1% of their contributions to the regular budget of UNESCO every two years. The WHF committee is composed of 21 members elected by the contracting parties (117 parties as of 1993). Funding is given by WHF to protect cultural landscapes and World Heritage Sites (WHS) established by UNESCO. The annual budget of WHF is around US $2 million. Assistance is given for studies, provision of experts, training, supply of equipment etc., but assistance covers only part of the budget and the states are expected to contribute a substantial share for the project or programme.

Biodiversity Legislation and Conventions

Introduction

Primitive man, even from the hunter-gatherer stage, was highly dependent on the various elements of biodiversity; he had developed an unwritten code for the sustainable use of biodiversity. Such codes exist even today in several tribal pockets throughout the world. During the evolution of human society and civilisation, these unwritten codes were replaced by legislation. Only a few decades ago, however, environmental law emerged as a distinct branch of law in order to regulate the activities of man towards the biotic and abiotic components of the environment.

Effective legal protection for biodiversity can be provided only after a synthesis of social structure, policy and regulation has been effected. Society must first realise that the fuel for this process is an understanding of the biotic community and our place in this community; we must realise that we are only fellow members of the Earth's biotic community. Biodiversity laws differ from other laws in that they, instead of governing relationships between persons or between persons and society, strive to protect biodiversity, the destruction of which affects all of humanity. However, the implementation and enforcement problems of these laws are almost insurmountable. In several countries, a large number of environmental laws and treatises are really 'paper tigers' because governments lack the will or ability to enforce them. The most noteworthy example is the lack of enforcement of a legal ban on clearing tropical forests. In other words, there is no use of legislating biodiversity protection until our society is 'structured to make adherence to such protections possible' (Given 1996).

Throughout the world, biodiversity laws started as specialised sub-branches of agriculture and/or forestry laws; initially they dealt specifically with regulation of the exploitation of wild species and the establishment of protected areas. Only slowly did they evolve as specialised biodiversity laws extended into laws pertaining to planning and land-use legislation. From the initial regulatory and punitive status, biodiversity laws are now increasingly developing to provide a framework for the establishment of procedures and Institutions destined to facilitate and encourage biodiversity conservation and management programmes, to make biodiversity into a public service and to promote better public awareness of biodiversity (Given 1996).

International Biodiversity Laws

International laws have two dimensions: public and private. With reference to the first dimension, international laws govern the activities and relationships between nations, although the principle of State sovereignty dominates them. Consequently, nations are not strictly bound by such international laws; their consent is very essential while approving or enacting such laws. Consent is often obtained through the signing of a treaty relating to these laws by duly empowered persons officiated by the concerned country and the treaty subsequently ratified by an act of parliament or equivalent body of that particular country. With reference to the private dimension, it can be mentioned that private law operates within the context of the public law, and controls the activities and relationships of individuals and non-governmental organisations.

It is evident from the foregoing that treaties have become the backbone of international laws. Treaties are contracts providing for benefits to both the contracting nations and, therefore, if one nation fails to comply with its treaty commitments, the other can retaliate by refusing to discharge its own obligations. Very recent developments in International Laws have resulted in treaties laying down general rules that the contracting nations should absolutely commit themselves to, failing which they self-defeat themselves. But although binding upon the contracting countries, these treaties are very difficult to enforce (de Klemm 1990). This inherent drawback can be nullified to some extent, however, through the establishment of appropriate institutions, such as Conferences of the Nations and Secretariat. These would continually review implementation of the contents of the treaties, encourage and promote co-operation between the nations, and provide a forum wherein cases of non-compliance are discussed and solutions reached. Other suitable steps towards implementation of the treaties without problems are: (i) underscore the obligation of contracting nations to provide periodic reports on the actions taken by each nation in implementing the contents of the treaties; (ii) empower the Conference or Secretariat to adopt specific recommendations relating to the treaty and (iii) allow admission of non-governmental organisations as observers at meetings of the Conference (Lyster 1985). The sources of international laws can be four: (i) international conventions, (ii) international custom (these first two as a result of treaties mentioned above), (iii) general principles of law and (iv) judiciary decisions and teachings of the most highly qualified publicists such as the International Court of Justice Statute 1948 (UNEP 1995).

The most important international conventions/treaties are mentioned in Table 9.2.

Convention on Biological Diversity

Also called the **Rio** or **Earth Summit**, the Convention on Biological Diversity (CBD) is a major landmark in biodiversity management, regulation and utilisation. This convention was the result of very intense political interest in biodiversity and several years of intense biodiplomacy. Preparations for CBD were initiated by UNEP in 1987 with the formation of

Table 9.2 Most Important International and Regional Conventions on Biodiversity

Sl.No.	*Convention*	*Year*	*Place*	*Importance*
1.	International plant protection convention	1951	Rome	Global
2.	Convention on the High Seas	1958	Geneva	Global
3.	International convention for protection of new varieties of plants (UPOV)	1983	Geneva	Global
4.	Convention on wetlands of international importance, especially as waterfowl habitats (also called the Ramsar convention)	1971	Iran	Global
5.	UNESCO convention for protection of the world's cultural and natural heritage	1972	Paris	Global
6.	Convention on international trade in endangered species (of wild fauna and flora) (CITES)	1973	Washington DC.	Global/ Regional
7.	International Tropical Timber Agreement (ITTA)	1983	Geneva	Global
8.	Convention on nature protection and wildlife preservation in the Western Hemisphere	1940	Washington	Regional
9.	African convention on conservation of nature and natural resources	1968	Algiers	Regional
10.	Convention on conservation of European wildlife and natural habitats	1979	Berne	Regional
11.	ASEAN agreement on conservation of nature and natural resources	1985	Kuala Lumpur	Regional
12.	Protocol concerning protected areas and wildlife fauna and flora in the eastern African regions	1985	Nairobi	Regional
13.	Protocol concerning specially protected areas and wildlife in the wider Caribbean region	1990	Kingston	Regional

an Ad hoc Working Group of Experts. This Group of Experts met in 1988, followed by a meeting in 1991 of the Intergovernmental Negotiating Committee for a CBD. The agreed text for CBD was adopted by 101 countries in Nairobi in May 1992. The CBD was launched in June 1992, along with establishment of the Global Environmental Facility (GEF). The Convention came into force on 29th December 1993. This was followed by the first meeting of the contracting countries in the Bahamas in Nov-Dec. 1994. By February 1995, 168 countries had signed this Convention, while by April 1995, 188 countries had ratified its tenets.

The CBD was aimed at reaching a consensus among all contracting countries since there was (and still is) acute dissention between countries over biodiversity utilisation and conservation. Developed countries felt (and many still feel) that conservation of biodiversity, wherever it may be located in the world, is a common concern of humankind, while developing countries tended to show a strong 'country driven approach' with reference to the use of biodiversity for their overall economic development (Khoshoo 1996).

The preamble to the CBD states that the contracting parties (the Nations) are aware of the general lack of information and knowledge regarding biodiversity and of the urgent need to develop scientific, technical and institutional capacities to provide the basic understanding requisite for planning and implementing appropriate measures.

The CBD has agreed to the proposal of 42 Articles and two Annexures. Article 1 describes the objectives as follows: 'to conserve the maximum possible biological diversity for the benefit of present and future generations and for its intrinsic value'. Article 3 emphasises the fundamental principle that the conservation of biodiversity is a common concern for all people and that the States should have the responsibility to ensure that their bioresources are developed in a sustainable way. Articles 7 and 8 emphasise *in-situ* and *ex-situ* conservation; they further emphasise that countries benefiting the most from biodiversity exploitation should also contribute the most to its conservation. Articles 14-17 deal with access to genetic resources and transfer of technology. A conference of parties to the CBD is held periodically. Since their first meeting, the parties have met four more times, the latest in Nairobi (15-26 May 2000).

Trade-Related Intellectual Property Rights

The other international treatises relating to biodiversity are the International Copyright Act of 1886 and the Berne Convention of 1875 (consolidated through several later treatises and Acts). The first provides nations with legal mechanisms for protecting, under their own laws, each other's copyrighted works and materials. The Berne Convention for the protection of literary and artistic works provides legal protection and control of biodiversity information. The World Intellectual Property Organisation (WIPO) is responsible for administering the Berne Convention and other agreements regarding Intellectual Property Rights (IPRs). TRIPs refers to Trade-related Intellectual Property rights and is a term used to describe that branch of law which protects the use of thoughts, ideas and information of commercial value.

As already stated, the law of IPR is meant to protect the legitimate rights of those who produce original work and serves as an incentive. The primary formal mechanisms within the purview of this law are **Copyright**, **Patent**, and **Trademark** besides protection of undisclosed information, industrial designs, geographical indications, layout designs of integrated circuits, and anticompetitive practices in contractual licences (OECD 1996; Sekar and Kandavel 2002). IPR is granted for a stipulated period (usually 20 years) and when this period expires, the patented information is available in public domain. Copyright work should be original and should be the result of a creative process, although national laws differ slightly in the strict enforcement and interpretation of the copyright provisions. This creates problems, especially for biodiversity databases. Patents are traditionally used to protect property rights related to devices or machines. However, at present patents are given to software, specific information, ideas and even to biological organisms. The breaking down of distinction between software and information has blurred the strict adherence to patent and copyright provisions originally contemplated. Wide-area networks have thus increased the 'fluidity and accessibility of data by orders of magnitude' (UNEP 1995). Application and enforcement of copyright and patent principles have been increasingly challenged in view of the aforesaid problems. The public in many countries of the world is increasingly seeking freedom of access to information and public databases. How much of such information is accessible and how many are subject to privacy is a teething problem in many countries.

TRIPs agreement came into effect on 1st January 1995 (see Website http://www.wto.org/wto/intellec/intellec.htm for the text of TRIPs). It also paved the way for the World Trade Organisation (WTO). Between 1996 and at the latest in 2005 all WTO member countries must apply the provisions of TRIPs agreement. Countries which do not yet meet the TRIPs standards have been given a period of five years to change their laws; this period is extendable for another five years for the least developed countries. India, as a member of WTO, is obliged to change her patent laws latest by 31st December 2004 to bring them in line with the requirements of TRIPs. Approximately 70 developing countries have signed the TRIPs agreement. To date TRIPs is the most comprehensive multilateral agreement on Intellectual Property Rights (IPR). Since India is a signatory to the General Agreement on Trade and Tariffs (GATT) and a member of WTO, it is obliged to meet all the articles of TRIPs. TRIPs also outlines features on minimum standards, procedures and remedies for enforcement, and dispute settlement. It may be noted that each country has the option to frame its patent laws within the broad framework defined in the WTO agreement. In this era of biological sciences, resources of biodiversity are increasingly used both in conventional and in modern ways, the latter often through biotechnological methods. Biotechnology uses living organisms or their parts to extract, make or modify products of great commercial value, and also to develop individual cells of higher organisms and of microbes for specific uses. IPR for biotechnology is presently 'in a state of flux'. Two main systems of IPR currently obtain for biotechnology: rights in plant varieties and patents, both of which provide exclusive and time-limited rights of exploitation. More details on Rights in Plant Varieties are provided on page 160, while problems relating to Patents, especially in connection with indigenous knowledge systems, are described on page 187 .

CITES

CITES stands for Convention on International Trade in Endangered Species of Wild Fauna and Flora. CITES, known in its early days as the 'Washington Convention', was the result of extensive lobbying by IUCN for international trade controls in species that are prioritised for conservation. The necessity for controlling international trade in endangered species, although recognised as early as 1911, and provisions to that effect incorporated in several international conventions on the conservation of wildlife, could not be realised in practice for a long time (Lyster 1985) —not until CITES came into force (Favre 1989). CITES was first arranged in 1973 in Washington DC. The trade agreements reached in this convention were enforced in 1975. By about 1994, the number of countries which agreed to enforce the recommendations of CITES, had increased from an initial 10 to 125. By 2000, the number of signatories had reached 140. This list includes India.

CITES is a major global biodiversity protection statute and is intended to be followed internationally. Its goal is 'to regulate the complex wildlife trade by controlling species-specific trade levels on the basis of biological criteria' (Trexler and Kosloff 1991). It includes all species 'threatened with extinction, which are or may be affected by trade'. Thus, CITES focuses primarily on the species category. Species and extinction are the important criteria, but species and geographical criteria are not satisfactorily covered. Rather, the statute of CITES is primarily based on listed endangered species and on laws promulgated in individual countries. This leads to a situation wherein a particular species listed in CITES may be legally obtained and traded from one country, but not from another.

CITES Appendix I lists 675 critically threatened and endangered species. Commercial trade in Appendix I species is totally prohibited. Any non-commercial transaction involving these species, say for instance for scientific research, is subjected to the issue of import and export permits. However, before issuing such permits, scientific authorities between the two countries must certify that the transaction will not be detrimental to the survival of the species in question.

Appendix II lists species (and products derived therefrom), whole genera and even entire families of plants that are not as seriously threatened as those listed in Appendix I, but may become threatened if trade in them is not strictly regulated. Export of specimens listed in Appendix II requires an export permit, which may only be issued on the approval of the scientific authority of the concerned country. This authority must state in the approval certificate that the trade in that particular specimen will not be detrimental to the survival of the concerned species. A total of 21,000 plant species and 3700 animal species are listed in Appendix II. The plant families included are

Cataceae, Cyatheaceae, Cycadaceae and Orchidaceae. Virtually all countries do trade in Appendix II species but Indonesia, Philippines, Thailand, Argentina, Bolivia, Guyana, Cameroon, Tanzania and Mali have the heaviest exports. The most important importing countries are the USA, European countries, Japan, Singapore, Hong Kong, China and Taiwan.

Appendix III lists those species which are rare in one country but fairly common in others. CITES provides the option for countries to restrict trade in these species and allow export only on specific export permits. Import of these taxa into any country requires presentation of a certificate of origin as proof that the species in question does not originate from a country for which this species is listed as rare. Among the 50,000 species or so included in this Appendix, the only plant taxa are five species of trees from Nepal.

All the legal provisions of CITES apply not only to whole live or dead specimens of listed species, but also to their readily recognisable parts and derivatives; however, a small number of exceptions are listed.

CITES arranges conferences for the signatory parties once every two years. It has a permanent secretariat, for which support is provided by UNEP. It is based in Switzerland. Most countries have now ratified CITES. Turkey, Mexico and New Zealand are some of the notable countries that have not yet become a party to CITES.

International trade in wild species and its products is now worth more than 4-5 billion US dollars per year. In fact, it is the second largest illegitimate business after narcotics. In a typical year, it is estimated that at least 38 million plants are imported into the USA alone, many without valid permits and CITES approval. However, there are several instances in which CITES has been highly successful in controlling the illegitimate trade and extinction of specific taxa. The role of CITES, therefore, is to provide the nations of the world a legal framework for combating the illegal trade in endangered taxa.

There have been, expectedly, many problems in the proper and strict implementation of CITES. Some have been removed through increased co-operation between trading nations, recommendations in national and international conferences, and effective action by the CITES secretariat. Other problems continue unresolved. The most important among the remaining problems are: (i) better enforcement of CITES' legislation' (ii) avoidance of overemphasis on animals to the neglect of plants; (iii) preparation and provision to enforcement officers of good identification manuals for the plants listed in CITES; a customs official may not know the difference between, for example, *Cattleya skinneri* (listed in Appendix I) and *C. forbesii* (listed in Appendix II). Once illustrations are available, suspicious plants could be referred to knowledgeable authorities; and (iv) strengthening import and export control machinery at all levels. The almost global acceptance of CITES, very encouraging indeed, should not make us complacent. There are very clear loopholes in the biodiversity legislation of many countries that permit them to evade the spirit of CITES. Several incidences are also known wherein CITES norms were not strictly implemented by customs officials. The CITES document itself may threaten species. There are no real problems with either Appendix I or III of CITES, but the restrictions applied to Appendix II may be harmful to the long-term survival of endangered species. Reports from Mexico and some parts of Latin America state that when clearance is given for development of agricultural fields, native plants are left to rot in heaps at the edge of fields due to trade restrictions. Thus, CITES covers international trade but does nothing to protect taxa in the home country. Restriction of trade at the international level with no protection at home vitiates the spirit of conservation; species can still be destroyed. CITES unintentionally impedes the rescue of many tropical orchid species as well. Since they cannot be traded with other countries due to the blanket protection provided by Appendix III of CITES, they are doomed to extinction when the tropical trees on which they grow are cut down.

The major function of CITES may in reality turn out not to be restriction in trade, but rather in alerting lay governments to international concern for endangered species. Awareness in a number of countries, still indifferent to the fate of their biodiversity, is a need of the hour.

Another organisation, called TRAFFIC (Trade Record Analysis of Flora and Fauna in Commerce) monitors international trade of species. The Indian unit was established in 1991 in the WWF-India headquarters in New Delhi. A few years ago, TRAFFIC conducted an experiment at major international airports of countries that are signatories to CITES to monitor how closely the provisos of CITES were adhered to. The experiment involved carrying a cactus plant through customs in several countries. The plant was either declared or prominently displayed at the customs counter. Only the US and the (former) Soviet Union confiscated the cactus. Confiscation was based, however, on the absence of a proper phytosanitary or health permit and NOT the fact that all cacti are on Appendix II of CITES. Not one of the customs officers encountered during this experiment understood or even seemed aware of CITES, in spite of the fact that some officers did have a copy of the CITES document near-by.

Ramsar Convention

Also called the Convention on Wetlands of International Importance, the Ramsar Convention is an international treaty drawn up in 1971 at Ramsar in Iran. The convention came into force in December 1975. This convention expects its contracting nations to promote the wise use of wetlands situated in their territory; it also requires the contracting parties to designate certain wetlands for inclusion on a list of wetlands of international importance. Such wetlands of international importance are called **Ramsar Sites**. The contracting nations should mandatorily make national wetland inventories; they should also establish nature reserves on wetlands and provide adequately for their maintenance. It is also expected that the nations should train personnel to manage and research wetlands. The contracting parties are further obligated to maintain the ecological character of the listed wetlands. In other words, the convention provides for the international co-operation for wetland conservation. The major parameter for a wetland to qualify for inclusion in the international list is the presence in it of rare, vulnerable, endangered or endemic plants/ animals. As of 1994, 81 States were contracting parties to the Ramsar Convention and 654 Ramsar wetlands have been designated covering an area of more than 43 million hectares (Navid 1994).

A 'Wetlands Conservation Fund' was established to assist countries in implementing the objectives of the Ramsar Convention. Funding is provided only to those countries that have contracted for wetland conservation activities relating to any one of the following fields: improvement of management of Ramsar Sites, designation of new Ramsar Sites, promoting wise use of wetlands, training personnel in wetland management and organising promotional activities such as seminars, workshops, educational programmes etc.

International Undertaking on Plant Genetic Resources and Farmers' Rights

The International Undertaking on Plant Genetic Resources (IUPGR) was adopted in 1983 to strengthen the rights of suppliers of genetic resources and to counterbalance the increasing protection of technology emanating out of genetic resources (Virchow 1998). This undertaking ruled that genetic resources should not be freely accessible under the slogan that 'plant genetic resources are a heritage of mankind'. IUPGR is a non-binding agreement and hence initially many countries paid it scant heed. From 1987 the

IUPGR has added some Annexures including Farmers' Rights. As of May 1997, 111 countries have adhered to the tenets of the IUPGR; notable exceptions are Brazil, Canada, China, Japan, Malaysia and the USA.

IUPGR assures the conservation, use and availability of PGRFA by 'providing a framework recognising the past, present and future contributions of farmers to the maintenance, improvement and provision of PGRFA' (Plant Genetic Resources for Food and Agriculture), a programme now known as **Farmers' Rights**. It provides for internalisation of benefits (i.e. sharing of benefits amongst farmers themselves without the benefit being usurped by others) through some kind of joint property rights (rights jointly held by the farmers). The Farmers' Rights infact envisage farmers themselves to hold the responsibility of conserving land races (Hardon *et al.* 1994). Implementation of Farmers' Rights is hampered, however, by the lack of international funding. Farmer's Rights are discussed further in Chapter 11.

UPOV Convention and Rights in Plant Varieties

Germany and the USA were among the only few countries before 1960 that gave international protection rights to plant varieties. But in 1961 due to pressure from plant-breeding industries, an International Union for the Protection of New Varieties of plants called 'Union pour la Protection des Ontentions Vegetales' (UPOV) in French, was formed in Geneva and the UPOV convention signed immediately. UPOV was established with a view to regulating international trade of protected varieties and ensuring that the member States acknowledge the efforts and achievements of breeders of new plant varieties by granting an exclusive property right (Hardon *et al.* 1994). Initially there were 10 contracting parties but subsequently several others joined the Convention. The UPOV Convention was revised in 1972, 1978 and 1991. The USA endorsed the Convention after amendments were made in 1972. Now there are 32 member states, of which 26 are in the process of implementing UPOV at the national level (FAO 1997).

The UPOV Convention requires that each country adopt within eight years of becoming a member, national legislation ensuring protection to at least 24 genera or species, in accordance with the provisos of the Convention. A plant variety is protectable under the UPOV system if it is distinct, uniform and stable (DUS) (see Chapter 4)) and satisfies a novelty requirement. DUS criteria are checked by the national authority responsible and usually by growing the variety over at least two seasons. Duration of protection depends on the national legislation and on the species to which the variety belongs, but generally for 20-30 years. In addition to the incentive to farmers in the form of exclusive rights, the UPOV system provides 'Breeder's Exemption', i.e., the protected varieties are freely available for further research and development of new varieties. UPOV also confers on the holder the right to **'sell the reproductive material'** (e.g. seeds, cuttings, whole plant) of the protected variety, but not **'consumption material'** (e.g. fruit)(see Table 9.3 for more details). For further discussion on UPOV in relation to Farmers' Rights see Chapter 11.

ITTA and ITTO

ITTA stands for International Tropical Timber Agreement. It came into force on 1st April 1985. The contracting parties entered into the agreement 'recognising the importance of, and need for, proper and effective conservation and development of timber forests' and concomitantly ensuring optimum utilisation of such forests in a sustainable manner and maintenance of ecological balance. Tropical timber reforestation and afforestation are encouraged at national and international levels and more than 66 projects and 35 pilot studies on these aspects have been approved and supported. ITTO (International Tropical Timber

Table 9.3 Comparison of main provisos of PBR under the UPOV Convention and Patent Law (from Swaminathan 1997)

Proviso	*UPOV 1978 Act*	*UPOV 1991 Act*	*Patent Law*
Protection coverage requirements	Plant varieties of nationally defined species ❖ Distinctiveness ❖ Uniformity ❖ Stability	Plant varieties of all genera and species ❖ Novelty ❖ Distinctiveness ❖ Uniformity ❖ Stability	Inventions ❖ Novelty ❖ Inventiveness ❖ Non-obviousness ❖ Industrial application and usefulness
Protection term	Minimum 15 years	Minimum 20 years	17-20 years (OECD)
Protection scope	Commercial use of reproduction material of the variety	Commercial use of all material of the variety	Commercial use of protected matter
Breeder's exemption	Yes	Not of essentially derived varieties	No
Farmer's privilege	In practice, yes	Up to national law	No
Prohibition of double protection	No species eligible for protection can be patented	—	—

Organisation) has published guidelines for sustainable forest management of tropical forests and conservation of biodiversity in such forests (ITTO 1990, 1992, 1993, 1998a, b).

Problems Related to the Legal Status of Plants

The legal status of plants invariably compounds the problem of strict enforcement of biodiversity laws. It is very different from the legal status of animals, and is relatively more ambiguous. Wild animals are usually characterised by law as *res nullius* (Given 1996). This concept, borrowed from Roman law, emphasises that animals, especially wild ones, cannot be subject to ownership even by the land on which they occur, unless they have been legally obtained. In some countries, more recently enacted laws have replaced this Roman concept of the legality of animals, by characterising wild animals as public property. Whether such a change would effect better conservation of animals remains to be seen. Plants, in contrast, are rooted to a place and whoever owns the place is the owner of the plants as well; the owner is therefore free to exploit his plants in whatever way he or she wants. This, however, does not apply to certain specific taxa of plants in some countries. In India for example, sandalwood tree is the property of the government wherever it grows.

Public or private ownership of wild plants, perhaps with the exception of trees and a few other economically useful taxa, generally does not suffice *per se* to discourage their collection or destruction by third parties. People continue to collect wild species of plants even from private lands and this freedom, in fact, is embodied in legislation or even in constitutions in countries such as Norway, Switzerland and Bavaria. Thus it becomes practically very difficult to enforce biodiversity laws with reference to wild plants (Given 1996).

The other problem concerning the legal status of plants is the group ownership of particular areas such as pastures, tank and river bunds etc. In many parts of the world, especially in Western Europe and Japan, pastures, for example, are the common properties of a group of persons, usually a village group (Bromley 1986; Runge 1986). In India and some South-east Asian countries large areas of land are under the ownership of temples or under government and quasi-governmental control; the latter is true of 'wastelands'. Such ownership is traditionally passed on through several generations and is often based on custom and not on written law. Therefore, it becomes the right and duty of every member of the village group to safeguard the common property; and calls for joint decision-making on any issues arising out of these properties (Oakeson 1986). The problem with common ownership is that attention is focused on the conservation and sustainable utilisation of only a particular bioresource in the property and that the preservation of non-target species is not considered; in fact the latter are often destroyed with impunity (Given 1996).

In conclusion it may be stated that wild plants, when occurring in a private property, can be destroyed by their very owners, and when occurring as open access resources can be collected or even destroyed by anyone almost any where. Therefore, any legal protection to wild plants should be viewed against these backgrounds (Given 1996).

Plant Collection and Trade Controls

Uncontrolled mass collection of wild plants, especially of slow-growing or rare categories, can contribute significantly to their depletion or even extinction in the wild. The extinction of *Tricholepidea adamsii*, an endemic mistletoe of New Zealand (Given 1996), and of *Tecophilaea cyanocrocus*, an endemic Liliaceae member of Central Chile (Chilean Forest Service 1989), are documented examples of species lost due to overcollection.

Uncontrolled collection of wild plants was mainly due to the development of trade, particularly international trade, in many species for the extraction of drugs /medicine, for ornamental or for educational/research purposes. A documented example for the first is *Rauwolfia serpentina*, a plant of very great medicinal potential, now almost extinct from natural localities in India, and for the second may be cited many orchids, cacti and lilies. A very important category of uncontrolled collection is known from many South-east Asian countries such as India. In many universities/colleges it was (and is even now in some places) mandatory for every student to submit herbarium sheets of wild plants representative of as many taxonomic groups he was studying as possible. On average, a post-graduate student would submit 300 different species of plants, and there were about 400 post-graduate students specialising in botany in southern Indian universities alone every year. This was almost the sole reason for depletion of plant wealth in many parts of the Himalayas and the Eastern and Western Ghats in southern India. Fortunately, this practice has been discontinued in some universities and other universities have restricted the number of herbarium sheets to be submitted by each student to a maximum of 50.

Laws have been enacted in several parts of the world to enforce partial or total collection controls. Total control consists of an absolute prohibition to collect whole plants or parts thereof of those taxa enlisted in the threatened category. Exceptions are generally made for scientific research, subject to granting of special permits. Partial controls prohibit the uprooting or digging up of subterranean parts or restrict the collection of aerial parts to a small number of twigs (often to less than 20 pieces). In many countries, lists of taxa that can be collected this way have been prepared. In several countries partial or total collection controls apply to all land, whether public or private, and to all persons including landowners. In certain

European countries, this restriction does not apply to landowners, provided the plant to be collected is not traded. In Australia, Switzerland and South Africa, strict restrictions are imposed to collect any species of plant, whether useful or not; collection is allowed in a specified quantity using specified collection methods only after obtaining special permits (Given 1996).

Information on trade controls in fully or partially protected plant species was provided above in the section on CITES. For non-protected plant taxa, however, trade controls are very essential, supplemented by controls applied to all other links in the trade chain (Given 1996), such as licensing of traders, nurseries etc. Buying or selling of plants from persons other than licensed sellers or growers should be prohibited. Such trade controls exist, for example, in parts of Australia and South Africa. Ideally, laws should be uniformly enacted at a national level since interprovincial or interregional trade within a nation may not pose problems. As a good example the Lacey Act of the USA may be cited. Uniform National level legislation cannot be enacted in countries such as Australia for constitutional reasons (see Good and Leigh 1986).

National Legislation

Much legislation has been enacted in different nations of the world to strive to protect important habitats and biodiversity elements. Although beset with considerable political and practical difficulties, and remaining almost always deliberately ambiguous and largely ineffective (Given 1996), such legislation does help in conservation efforts. As an example may be mentioned the UK Wildlife and Countryside Act. Furthermore, very strict and effective legislation does exist in some countries, for instance the Irish Nature Conservation Act, the Swiss Canton of Zurich, Finland's law on land-use, the French Nature Protection Act of 1976 and Denmark's land-use laws.

Biodiversity Information: Management and Communication

Introduction

Readers have to understand that there is an urgent need for resonance between the needs of biodiversity science and scientists on the one hand, and databases on the other. Biodiversity workers live throughout the world and are all interdependent. They need information, and not just local, but regional and international as well. Moreover, biodiversity science depends critically on high-level concepts on biomes, ecosystems, floras and faunas, hot spots, genetic resources, alien taxa etc. (Bisby 2000). Hence a strong effort for the collection, documentation, management and distribution of biodiversity information is needed so that effective decisions on managing bioresources can be made. Such information is also required for enacting national and international legislation and laws.

The use of biodiversity information is triggered by three principle categories of motivations: public policy (involves compliance with laws, rules, legislation, regulations and/or treaties), private sector (needed to advance commercial interests relating to breeding, ecotourism, bioprospecting involving biotechnology etc.) and public interest and cultural motivations (to advance the conservation and sustainable management of bioresources). The important aspects involved here are **data collection** from the real world, **storage of data, analysis of organised and integrated data** (if necessary mathematical modelling as well) so as to obtain useful and pertinent **information**, derivation of **knowledge** from such information through further analysis, interpretation and understanding and finally the attainment of **wisdom** (here, taking wise, proper and efficient biodiversity management initiatives and actions) through the intelligent use of knowledge. The author is reminded here of the famous poem 'Choruses from the Rock' by T.S. Elliot.

Where is the Life we have lost in living?
Where is the wisdom we have lost in Knowledge?
Where is the knowledge we have lost in information?

To these lines of T.S Elliot the following might be added:

Where is the information we have lost in data?
Where is the data we have lost in databases?

Libraries

These are the main sources of information provided through collections of both published and unpublished literature and facilitating its exchange. Most libraries are regional and at best national. However, there are several international libraries. The most important among them are those located at the Asian Institute of Technology, CGIARs (which have separate centres for forestry, tropical agriculture, maize and wheat improvement, potato and plant genetic resources), Institute Français de Recherche Scientifique pour le Dévéloppémént en coopération, International Centre for Integrated Mountain Development, International Centre for Living Aquatic Resources Management, International Development Research Council, International Waterfowl and Wetlands Research Bureau, National Library of Agriculture (USA), The Natural History Museum (UK), Royal Botanic Gardens Kew and Edinburgh (UK), Smithsonian Institution (USA and Panama), UNO, and IUCN—The World Conservation Union. Most of these libraries are thematic, but all provide vital information on biodiversity.

Bibliographies

Information relating to biodiversity is found in the literature pertaining to biological sciences, forestry, agriculture, wildlife and conservation biology as well as in the literature on economics, social sciences and even on legislation and law. These can be searched using the key words 'biodiversity' and 'biological diversity'. In addition, biodiversity information can be found in the so-called 'grey literature' (i.e., reports from NGOs, consultative groups and govt. departments). Most of this grey literature is unpublished and extremely difficult to trace.

Periodicals

Several periodicals contain articles on biodiversity. The names of the most important appear in **Ulrich's International Periodicals Directory**. In addition, several newsletters are available. The most important periodicals are listed below.

- Biodiversity Letters
- Biological Conservation
- Conservation Biology
- Ecology
- Oikos
- Oecologia
- Journal of Biogeography
- Journal of Ecology
- AMBIO
- Annual Review of Ecology and Systematics
- Ecography
- Trends in Ecology and Evolution
- Biodiversity and Conservation
- Biotropica
- Biodiversity
- Threatened Plants Newsletter
- Journal of Intellectual Property Rights
- BDM Updates

Databases

Data refers to 'observations, measurements or facts referenced to some kind of accepted standard, which are subsequently integrated, processed, interpreted or otherwise manipulated to produce information' while Information is the 'knowledge (product) derived from the analysis and interpretation of data' (Busby 1997). Data should be stored, managed and readily made available for integration with other data so that information can be generated from data easily and used as and when required for whatever purpose.

All the nations of the world have now more than adequately realised their wider regional and global responsibilities regarding their biodiversity wealth and conservation as well as the pressing need to manage the biodiversity information and data generated thus far. In addition, individuals, local communities, industries, NGOs and other institutions have also realised that to make proper decisions and manage biodiversity, they need to develop databases and their own information system frameworks. Details about biodiversity databases are given here.

The absolute need for effective organisation, management and use of data and information on biodiversity is already reflected in many international agreements and legislation such as the CBD, CITES etc. For example, as already stated, Article 7d of CBD indicates the requirement to 'maintain and organise, by any mechanism, data ....' and Article 17 of CBD is concerned with the exchange of information. The CBD, however, has not laid down any operational framework for achieving information exchange. In response to the requirement of data management and information exchange on biodiversity, the UNEP and WCMC together designed and submitted to GEF the project proposal on "Biodiversity Data Management (BDM) capacitation in developing countries and networking biodiversity information". The BDM is a UNEP/GEF project funded by GEF to the tune of US $4 million. This project was commenced in June 1994. The Bahamas, Egypt, Poland, Chile, Ghana, Thailand, China, Kenya, Costa Rica and Papua New Guinea were the ten countries that participated in the first phase of this project. The overall objective of BDM is to enhance the capacity building of developing countries in biodiversity data management relating to the implementation of CBD. A subproject agreement was drawn up with WCMC to prepare a set of BDM support materials. A Guide to Information Management was prepared in addition to the Guidelines for National Institutional Survey. The Electronic Resource Inventory of UNEP provides a wide range of information and reference directories on software, hardware methodologies, standards, common practices, data sources, key organisations and exemplary projects related to biodiversity management (Duff 1997).

There are many biodiversity-related initiatives at the national level that are closely linked to the BDM project. These include Biodiversity Country Studies, National Biodiversity Strategies and Action Plans (NBSAP), National Environmental Action Plans (NEAP), National Conservation Strategies (NCS), National Sustainable Development Strategies, National Tropical Forest Action Plans (TFAP), National Forestry/Wildlife Master Plans, Protected Area Systems Plans etc. The BDM Newsletter 'BDM UPDATE' provides complete information on relevant issues and events.

Taxonomic Databases Working Groups for Plant Sciences SA2000 and other Taxonomic Databases

Taxonomic Databases Working Group (TDWG) was established by the untiring efforts of Dr. V.H. Heywood, a noted plant taxonomist, in 1985. TDWG seeks to create worldwide mechanisms for data exchange between botanical databases through the agreement of data models, data structures and standards, and data exchange mechanisms. TDWG has already published a series of standards for areas such as author names and abbreviations, geographical areas, and an international transfer format for botanic garden plant records (Plant Taxonomic Database Standards No.1, Hunt Institute for Botanical

Documentation, Pittsburgh) (ITF 1987). At the IUBS General Assembly in 1994, it was agreed to extend the activities of TDWG to groups other than angiosperms and some progress has been made. The ILDIS (International Legume Database and Information Service) has established a botanical diversity database for the 17,000 known legume species (Zarucchi *et al.* 1993). It is a co-operative project involving more than 20 research groups from five continents; the information system generated is available internationally. The readers are advised to refer to the following also: F.A. Bisby R.M. Polhill, J.L. Zarucchi, B.R. Adams and S. Hollis. 1994. **Legumeline (ILDIS Phase I Database).** Bath Information and Data Services, Bath., and *Designs for Global Plant Species Information System.* Oxford University Press, Oxford.. F.A. Bisby, G.F. Russell and R.J.Pankhurst (eds.).

Systematics Agenda 2000: Charting the Biosphere (SA 2000) is a programme not only aimed at discovery and research but also at documenting and synthesising knowledge about global species diversity within the next two to three decades. It is jointly compiled by the American Society of Plant Taxonomists, Society of Systematic Biologists and the Willi Hennig Society, in co-operation with the Association for Systematic Collections (SA 2000 1994a,b) and is supported by the US National Science Foundation (NSF). Twenty-seven Standing Committees involving over 300 scientists representing a broad array of institutions and specialities were/are involved in the SA 2000 studies. Details are summarised in Box 9.1.

The benefits expected from this ambitious programme are as follows: (i) attainment of more knowledge about an increased number of useful plants; (ii) an improved database, which will help in conservation and bioresource management; (iii) provision of knowledge to help in selection of new and improved crops of food and medicinal value; and (iv) to provide baseline data for monitoring global climate and ecosystem changes, rates of species loss etc. SA 2000 also aims at capacity building in developing countries to enable them to acquire/build collection-based infrastructures such as herbaria, seed/pollen/microbial/gene banks etc. SA 2000 also recognises the need for national research centres manned by professional systematists.

The NSF of US created in 1995 'Partnerships for Enhancing Expertise in Taxonomy' (PEET) (web: nhm.ukans.edu/peet). PEET initiated a new field of Biodiversity Informatics. A database known as 'Species Analyst' (habanero.nhm.ukans.edu.) was first established with information initially available with Kansas Natural History Museum collections. Subsequently collections available elsewhere were included to provide data on 12 million specimens. There is a proposal to increase it to 38 million specimens. 'Species Analyst' also forwards geographic information about a given species in each collection to the San Diego supercomputer centre. There, a programme called **GARP** developed by David Stockwell, maps that information and based on the environmental data available for those sites, predicts the species' 'environmental' niche and its overall distribution.

Other Databases on Biodiversity

Hundreds of databases on biodiversity have now been created throughout the world and space limitations preclude listing all of them here. The most important are mentioned below and, wherever needed, a brief description given.

(i) IOPI World Plant Checklist (Bisby *et al.* 1993; Burnett, 1993)

(ii) BIMS (Biodiversity Information Management System). A relational database for monitoring the conservation status of species, wildlife habitats and protected areas) (Mackinnon 1994).

(iii) BRAHMS (Botanical Research and Herbarium Management System). A database on botanical collection system.

Box 9.1. The missions and goals of Systematics Agenda 2000 (from SA 2000. 1994a, b).

Mission 1: To discover, describe and inventory global species diversity

•

- to survey marine, terrestrial and freshwater ecosystems to achieve a comprehensive knowledge of global species diversity;
- to determine the geographic and temporal distributions of these species;
- to discover, describe and inventory species living in threatened and endangered ecosystems;
- to target groups critical for maintaining the integrity and function of the world's ecosystems, for improving human health and for increasing the world's food supply, and
- to target the least-known groups of organisms.

Mission 2: To analyse and synthesise the information derived from this global discovery effort into predictive classification systems that reflect the history of life

- to determine the phylogenetic relationships among the major groups of organisms, thus providing a conceptual framework for basic and applied biology;
- to discover the phylogenetic relationship of groups of species that are critical for applied biology, targeting species that are important for human health and food production, as well as for conservation of the world's ecosystems;
- to discover the phylogenetic relationships of groups of species that are of critical importance to the basic biological sciences, such as those having broad relevance for experimental science and those critical for maintaining the integrity and function of ecosystems, and
- to develop more powerful techniques and methods for systematic data analysis.

Mission 3: To organise the information derived from this global programme into an efficiently retrievable form to best meet the needs of science and society

•

- to develop systematic, biogeographic and ecological databases of species information based on species housed in the world's natural history collections;
- to integrate data from specimens housed in systematic collections with information contained in GIS databases, thus providing a means to monitor past and present effects of global change on species distributions and extinction;
- to develop linkages among databases for the efficient retrieval of all available information about species and the places in which they occur;
- to develop and implement an information system that can be accessed efficiently by a broad international user community;
- to develop data dictionaries of taxonomic names, geographic localities and other information basic to all systematic databases;
- to develop data products, including guides, keys, electronic floras and faunas and monographic works, and
- to develop mechanisms for maintaining and updating databases and information networks including continuing hardware and software support.

(iv) ENVIS (Environmental Information System, India).

(v) DIALOG. An on-line information source and the longest commercial system available; it contains more than 450 databases containing more than 350 million articles from over 2500 journals; it also contains details on more than 15 million patents and on more than 10 million chemical substances.

(vi) Abstracts of Tropical Agriculture (ORBIT)

(vii) Agricola (DIMDI, Data-star/Dialog)

(viii) AGRIS International (Data-star/Dialog, DIMDI, ESA-IRS)

(ix) Aquatic Sciences and Fisheries (Data-star/ Dialog, DIMDI)

(x) Biological and Agricultural Index (BRS)

(xi) BIOSIS Previews (Data-star/Dialog, DIMDI)

(xii) CAB Abstracts (Data-star/Dialog, DIMDI)
(xiii) GEOBASE (Data-star/Dialog, DIMDI)
(xiv) Life Sciences Collection (Data-star/Dialog, STN International)
(xv) Microbial Information Network Europe (DIMDI, Deutsche Sammlung von Mikroorganismen und Zellkulturen, Gmbtt) (also abbreviated as MINE)
(xvi) Oceanic Abstracts (Data-star/Dialog, STN, ESA-IRS)
(xvii) Remote sensing on-line retrieval systems (ORBIT, ESA-IRS)
(xviii) SciSearch (Data-star/Dialog, DIMDI)
(xix) UBIB UNESCO Bibliography (ECHO)
(xx) SEPASAL (Survey of Economic Plants for Arid and Semiarid Lands) Database by Royal Botanic Gardens, Kew, Surrey.
(xxi) The Chapman and Hall Chemical Database by Chapman & Hall Publishers
(xxii) NAPRALERT (Natural Products Alert)
(xxiii) MEDFLOR (for ethnobiology)
(xxiv) Neogene Marine Biota of Tropical America. Logs marine fossils from the last 25 million years (porites.geology.uiowa.edu/index.htm).
(xxv) Evolution of Terrestrial Ecosystems (ETE). Database developed by the Smithsonian National Museum of Natural History and Hohn Damuth of the University of California, Santa Barbara. It covers both animal and plant terrestrial fossils and includes data on age, species lists, body sizes and diet for nearly 4000 localities, largely from the African Late Coenozoic (web: etedata.si.edu.).
(xxvi) The Palaeobiology database at the National Centre for Ecological Analysis and Synthesis, Santa Barbara (CA) on fossils spanning all time periods on all organisms (www.nceas.ucsb.edu/public.pmpd).
(xxvii) Bacteriology Insight Orienting System (BIOS) (www.-sp2000ao.nies.go.jp/bios/index.html)

Of the 7500 databases available worldwide (as of 1995), 75% are related to biological sciences, of which more than 60% are bibliographic or directory type and the rest are numeric, textual, image or multimedia. Most of these databases, with few exceptions, are available in print, CD-ROM, tape and on-line mode. Though the majority address the global need of users, those developed in the USA, UK, Canada, Australia and European countries satisfy local user needs too.

Distribution of Biodiversity Information

It has now become possible to build comprehensive and integrated biodiversity information systems on networks even though a huge amount of work has yet to be done. Solutions to key issues such as primary attribute data, standards, metadata, custodianship, data management tools, networks etc. are already well developed (Canhos *et al.* 1997; Olivieri *et al.* 1995). Information is becoming more digital and networked. Network publication has become the order of the day. Examples: the Bioline publications (http://www.bdt.org.br/bioline/), 'Tree of Life' (http://phylogeny.arizona.edu/tree/phylogeny.html) and 'Phylogeny of Life (http://ucmpl.berkeley.edu/alllife/threedomains.html). Addresses for electronic journals and newsletters on biodiversity available in the Internet can be found at Electronic Journals VL (http://www.edoc.com/ejournal/) and New Jour websites (http://gort.ucsd. edu/newjour/). Readers are also advised to refer to the websites of 'science' and 'nature' magazines which provide links to a large number of relevant websites on biodiversity: for science magazine (www.sciencemag.org/feature/data/biodiversity2000.shl.) and for nature magazine (http://www.nature.com/nlink/w405/n6783/full/405207ao_fs.html).

The network tools and applications available include Tools using computer networks such as Electronic mail (e-mail), LISTSERV (extension e-mail) and USENET, as well as more popular Network Information Retrieval (NIR) tools such as Telnet, File Transfer Protocol (FTP) (a powerful data exchange tool), Wide Area Information Server (WAIS) (helps in retrieving information by searching indexes of databases), GOPHER (retrieves information through graphic interface), Veronica (acronym for Very Easy Rodent-Oriented Network Index to Computer (Archives), and www (worldwide web).

Digital Libraries have also come into prominence, where every effort is being made to open the way to weaving electronic journals and scientific libraries into a single interconnected database. The Digital Library Initiative (DLI) is a project partly sponsored by the US National Science Foundation (NSF), NASA and the Advanced Research Projects Agency (ARPA) (http://walrus.stanford.edu/ diglib/pub/nsf.announce.html.) (further details are given later).

The best-known computer network is the Internet, a network of networks. The Regional Map to Internet Connectivity (http://info.isoc.org:80/images/mapv14.gif) is regularly updated by the Internet Society, and the International E-mail Accessibility page (http://www.ee.ic.uk/misc/country-codes.html). This has enabled organisations throughout the world to discover new opportunities provided by the information infrastructure and to nurture their development through the establishment of discussion lists, on-line databases, metadatabases, Virtual Libraries (VL), Special Interest Networks (SINs) and search engines (Canhos *et al.* 1997).

Metadatabases

It is well known that data and information on biodiversity are rapidly increasing. Several databases are now available throughout the world. Therefore, a central goal of biodiversity informatics is to develop systems that permit interoperability and knowledge synthesis across a wide array of local/regional systems (Bisby 2000). This has forced the need to use **metadata,** i.e., data about data. Metadatabases hold data about data and are analogous to library catalogues. They facilitate the discovery of existing datasets in institutions and provide additional information about contents, quality and features of the datasets. Good examples of metadatabases are the National Spatial Data Infrastructure (NSDI) provided by the Federal Geographic Data Committee (http://fgdc.er.usgs.gov/), UNEP's GRID Database (http://www.inpe.br/grid/home) and CIESIN (http://www.ciesin.org). GRID (Global Resource Information Database) is a system of co-operating centres within UNEP dedicated to making environmental information more readily accessible to environmentalists and decision-makers and fosters the use of Geographical Information Systems (GIS) and Satellite image processing as tools for environmental analysis. Other metadatabases include the European Environment Agency of the European Union (for environmental data), IUCN Environmental Law Centre (for environmental and biodiversity legal aspects), NASA Global Change Master Directory (for data on global changes), Global Environmental Network Information Exchange (GENIE), Global Land Information System (GLIS) (for data pertaining to the Earth's land surface), Global Environmental Information Exchange Network or INFOTERRA (to facilitate exchange of information on environment), IPGRI (Directories of Germplasm Collections), HEMDisk of UNEP (information on environment monitoring agencies), World Federation for Culture Collections (WFCC) (data on microbes),

ICSU's World Data centres, fossil records on worldwide web (http://sunrae.uel.ac.uk/palaeo/pfr2/pfr.html), plant viruses (http://life.anu.edu. au/./viruses/virus.html), nucleic acid sequences (fly.bio.indiana.edu /11/genebank-sequences) etc.

The most recent and complete metadatabase effort is the Global Biodiversity Information Facility (GBIF) (Bisby 2000; Edwards *et al.* 2000). This provides for interoperability by attempting to draw together basic biodiversity accession records from dispersed sites such as ERIN and Taxaserver (both from Australia), ENHSIN (European Natural History Specimen Information Network) (www.nhm.ac.uk/science/rco/enhsin), ITIS (Integrated Taxonomic Information System) (www.itis.usda.gov), URMO (UNESCO-IOC Register of Marine Organisms) (www2.eti.uva.nl/database/urmo/default.html), IOPI (International Organisation for Plant Information—a global plant checklist) (http://bgbm3.bgbm.fu.berlin.de/iopi/gpc), IPNI (International Plant Names Index) (www.ipni.org/), CONABIO (Comision Nacional para el conocimiento y Uso dela Biodiveridada of Mexico) (www.conabio.gob.mx/), INBio (Instituto Nacional de biodiversidad) (www.inbio.ac.cr/), ABIF (Australian Biodiversity Information Facility) (www.anbg.gov.au/abrs/abif.htm), DIVERSITAS of UNESCO, OBIS (Ocean Biogeographic Information system), 'Tree of Life' (http://phylogeny.arizona.edu/tree/phylogeny.html) and 'Tree Base' (http://herbaria.harvard.edu/treebase/).

Species 2000 is a global metadatabase programme to compile a 'Catalogue of Life' using distributed networking on the Internet. The main aim of this effort is to create a uniform and validated index to the world's known species of plants, animals, fungi and microbes, one for each group of organisms, in the name of Global Species Databases (GSDs). Species 2000 was established by IUBS, ICSU, CODATA (Committee on Data for Science and Technology) and IUMS. It is endorsed by UNEP. It is planning to work closely with GBIF. Species 2000 will involve URMO, ITIS, IOPI, NAPRALERT, NCBI (National Centre for Biotechnology Information, US), GRIN (Genetic Resources Information Network, USDA), SINGER (System-wide Information Network for Genetic Resources) (www.singer.cgiar.org), Flora Base (Plants of Western Australia) (http://florabase.Calm.wa.gov.au), ERMS (European Register of Marine Species) (http://erms.biol.soton.ac.uk), etc.

Virtual Libraries

Virtual libraries (VL) are organised sets of links to items (documents, software, images, databases) on the networks, enabling users to find information that exists elsewhere on the network from one central ('virtual') location. The Worldwide Web Consortium (http://www.w3.org/pub/WWW/) holds the www Virtual Library (http://www.w3.org/hypertext/DataSources/bySubject/Overview.html). This is a distributed subject catalogue and includes a list of virtual libraries on a number of subjects, including those related to biodiversity. The other important virtual libraries are 'Biosciences' located at Harvard University, Cambridge, Mass., USA (http://golgi.harvard.edu/biopages.html.), and 'Forestry' (http://www.metla.fi/info/vlib/Forestry.html). The section on Biodiversity, Ecology and the Environment is maintained by Bryant and Thornhill at the University of California, Irvine, CA, USA (Burley *et al.* 1997), while the www virtual library for fungi is: http://www.keil.ukans.edu/~fungi/.

Special Interest Networks

Special Interest Networks (SINs) are organised as a group of people and/or institutions who collaborate to provide information on a specific subject and consist of a series of participating 'nodes' that contribute to the network's functions

(Green and Croft 1994). More specifically, the nodes are committed to the provision of public access to the unique information available at their sites, accept and store relevant information, provide transparent and organised links to other nodes on the network and co-ordinate their activity with other nodes (Canhos *et al.* 1997). SINs are the modern equivalent of learned societies and publish journals, newsletters, data sets and softwares; they also do library services. As examples of SIN may be mentioned the European Molecular Biology Network (gopher felix.embl-heidelberg.de) and the Biodiversity Information Network (BIN21, http://www.bdt.org.br/bin21/bin21.html). The former is a special interest network that serves researchers in Europe's molecular biology and biotechnology fields; the latter was established to facilitate access to all levels of biodiversity information, from molecular to the biosphere (Canhos *et al.* 1994; 1997). BIN21 is developing affiliations with, and network links to, organisations already carrying out related activities (Canhos D.A.L. *et al.* 1994, Canhos, V.P. *et al.* 1997).

Search Engines are tools for searching and reporting and are fundamental for finding and retrieving information on relevant sites through the use of powerful automated searching tools (Canhos *et al.* 1997). The results are large databases. Examples of search engines in the Internet include Alta Vista (http://www.altavista.digital.com/), Infoseek (http://guide.infoseek.com/), Lycos (http://www.lycos.com/), Yahoo (http://www.yahoo.com/), Net Search (http://home.mcom.com/home/internet-search.html) and the c/net (http://www.search.com/). The Internet Services List by Scott Yanoff is a good overview list of internet resources and is regularly updated (http://www.uwm.edu/ mirror/inet.services.html). The InterNIC Directory and Database Services (http://www. internic.net/ds/dspg01.html) is also a good source for general internet resources.

As a result of CBD and also to comply with the obligations of the Convention, individual countries have established/are starting to establish several specialised networks at country, regional and international levels. ERIN, the Environmental Resources Information Network from Australian Nature Conservation Agency, Canberra (http://kaos.erin. gov.au/erin.html) is one of the exceptions in that it was established even prior to the Rio Convention. The others are from Costa Rica's INBio, the National Biodiversity institute (http://www.inbio.ac.cr/) and the USA (http://straylight.tamu.edu/ bene/bene.html). The INFOTERRA of the UNEP consists of 170 national nodal points co-ordinated from Nairobi. In 1992, the idea to establish global computer networks for biodiversity data took the shape of an initiative known as BIN21 (Biodiversity Information Network, Agenda 21) (http://www.bdt.org.br/bin21.html). Several new national developments emerged as BIN21 nodes: FINBIN, the Finnish Biodiversity Information Network (http://www.csc.fi/biodiv/intro.html); CBIN, the Canadian Biodiversity Information Network (http://www.doc.ca/ecs/biodiv/biodiv.html); BIN-Br, the Biodiversity Information Network, Brazil (http://www.bdt.org.br/ index/binbr/) etc. Unlike other initiatives, BIN-Br is co-ordinated by an NGO, the Andre' Tosello Foundation for Tropical Research and its tropical database. The other databases include the Long-term Ecological Network (LTEN) based in the University of Washington and the projects in Capacity Building for Biodiversity Information Management begun by World Conservation Monitoring Centre (WCMC) based in UK, and the BioNET-International proposed by CAB International to pool the global resources in biosystematics. With support from UNDP under CAPACITY 21, a large number of Sustainable Development Networks are in the planning stage. Recently several special interest groups and networks (SIGN) dedicated to biodiversity conservation have sprung up all over the globe. The most important are: http://conbio.bio.uci.edu/orchid/ (orchids), http://www.labs.agilent.com./bot/ep_home (carnivorous plant database with photos), http://www/issg.org/database/welcome (global invasive species database), www.limnology.org (limnology), http://www.mobot. org/MOBOT/tropicos/moss/, http://www:nybg.org/bsci/hcol/bryo/ (bryophytes), http://ice.ucdavis.edu.US_National_Park_Service/; http://www.conservationorg/science/cptc/consprio/!run_me!.htm (Conservation International—regional conservation analysis project), http://www.inbio.ac.er/ATBI> (All Taxon Biodiversity Inventory), http://www.eti.uva.nl/Database/WBD.html (World Biodiversity Database), http://www.nhm.ac.uk/info/links (Links to Natural History), http://esa.sdsc.edu/biodiv2.htm (Fact sheet from Ecological Society of America), http://esa.sdsc.edu/issues4.pdf. (official publication on biodiversity by Ecological Society of America), http://www.abi.org/ (Association for Biodiversity Information), http://www.nature.nps.gov/im. (National Park Service environmental monitoring protocols and lists), http://www/epa.gov/ceisweb1/ceishome/atlas/bioindicators. (Bioindicators), http://www.uspto.gov/wed/offices/ac/ido/oeip/taf/top97cos.htm (website for obtaining details on leading Patentees in the USPTO), http://www.nybg.org/bsci/hcol/fung/ (Catalogue of Fungi).

Biodiversity Application Softwares

Some of the select Biodiversity Application Softwares are:

ALICE Biodiversity Database System (includes programmes for biologists to design and build their own checklist or biodiversity database) (ALICE 1990)

BG-BASE 4.0 (designed for managing biological information in taxonomy, distribution, conservation and collections management) (O'Neal and Walter 1989)

RAMAS/space (spatially structured population models for Conservation Biology, version 1.3, Applied Biomathematics, New York)

Biodiversity Data Bank 1.0 (desktop tool for storing, analysing and mapping biodiversity data) (Reynolds 1993)

Biological and Conservation Data (BCD) System (facilitates collection, distribution and exchange of information pertinent to preservation of biodiversity) (TNC 1992)

Expert Centre for Taxonomic Identification (provides for worldwide biodiversity Information) (Schalk 1992)

RECORDER 3.2 (provides species inventory for a particular location) (English Nature 1993)

WORLDMAP 2.4 (graphic tool for assessment of priority areas for conserving biodiversity, hot spots, endemism etc.) (www.nhm.ac.uk/science/projects/worldmap)

Williams, P.H. 1994. WORLDMAP priority areas for biodiversity. Using version 3.18, Privately distributed, London, UK

Williams, P.H. 1996. WORLDMAP4 Software and help document 4.1. Privately distributed, London, UK

Faith D.P. and Walker P.A. 1993. DIVERSITY: a software package for sampling phylogenetic and environmental diversity. Reference and user's guide. VI.O. CSIRO Division of Wildlife and Ecology, Canberra

LITCHI software for taxonomy (especially on synonomy) (http://litchi.biol.soton.ac.uk)

CD-ROMs and Diskettes

The distribution of data/information on CD-ROMs and Diskettes has emerged as a major form of data exchange in recent years because of the enormous storage capacity of a single CD-ROM diskette. Since only a one-time expenditure is involved, CD-ROM has emerged, in fact, as an alternative to on-line access and in some cases has totally replaced it. However, it has to be updated frequently. Several CD-ROM and diskette products related to biodiversity are now available. The most important among them are as follows:

(i) African Development Indicators (Diskette) (World Bank)
(ii) AgECONCD (CD-ROM) (CAB International)
(iii) Agricola (CD-ROM) (published by Silver Platter)
(iv) Agris (CD-ROM) (supplied by Silver Platter)
(v) AGRISEARCH (CD-ROM) (supplied by Silver Platter)
(vi) AGROSTAT -PC (Diskette) (FAO)
(vii) Antartica-digital database (CD-ROM)
(viii) Aquatic Sciences and Fisheries Abstracts (ASFA) (CD-ROM) (published by Silver Platter)
(ix) Biological and Agriculture Index (CD-ROM) (published by H.W.Wilson)
(x) Biological Abstracts/RRM (CD-ROM) (supplied by Silver Platter)
(xi) CABCD (CAB Abstract) (CD-ROM) (supplied by Silver Platter)
(xii) Compact International Agriculture Research Library Basic Retrospective Set 1962-1986 (CD-ROM) (CGIAR)
(xiii) Directory of Country Environmental Studies (Diskette) (WRC)
(xiv) Earth Submit (CD-ROM) (IDRC)
(xv) Endangered and Threatened Species (CD-ROM) (Quanta Press)
(xvi) Families of Flowering Plants (CD-ROM) (CSIRO, Australia)
(xvii) GEOBASE (CD-ROM) (Elsevier Publs., Amsterdam)
(xviii) HEMDisc (Diskette) (UNEP)
(xix) Index Kewensis (CD-ROM) (Oxford Univ. Press)
(xx) Life Sciences Collection (CD-ROM) (supplied by Silver Platter)
(xxi) Natural Resources Metabase (CD-ROM) (published by NISC)
(xxii) Ocenographic and Marine Resources (CD-ROM) (published by NISC)
(xxiii) Plant Gene CD (CD-ROM) (CAB International)
(xxiv) PROSPECT (Programmed Retrieval of Species by the Property and End-use Classification of their Timbers) (Diskette) (published by Oxford Forestry Institute)
(xxv) SESAME (CD-ROM) (CIRAD)
(xxvi) TREECD (CD-ROM) (CAB International)
(xxvii) World Resources Data Base (Diskette) (WRI)

Thesauri on Biodiversity and Environment

1. CAB International Thesaurus for Agriculture and Environment
2. The Consortium for International Earth Science Information Network on-line catalogue system
3. EEA, Trilingual Thesaurus for the Environment
4. INFOTERRA Thesaurus of Environmental Terms
5. Japan Information Centre for Science and Technology Thesaurus
6. Glossary of Technical Biodiversity Terms: http://www.wri.org/biodiv/gbs-glos.html.

Directories of Biodiversity Data Sources

A range of directories, catalogues and indexes are now available, which are helpful in locating sources of information on biodiversity. The most important among them are:

(i) Guarino L., Roa V.R. and Reid R. (eds.) 1995. **Collecting Plant Genetic Diversity: Technical Guidelines**. CAB International, Wallingford, UK
(ii) IPGRI. **Directories of Germplasms Collections Series**. IPGRI, Rome

(iii) Sugawara H., Ma J., Miyazaki S., Shimura J. and Takishima Y. 1993. **World Directory of Collection of Cultures of Microorganisms** (4th edn.). WFCC World Data Centre on Microorganisms, Saitama, Japan

(iv) IUCN and IWRB. Directories of wetlands for various parts of the world

(v) IUCN and WCMC. Directories of Protected areas for various parts of the world

(vi) IUCN 1992. **Protected Areas of the World**. IUCN, Gland, Switzerland

(vii) IUCN and WCMC. **Tropical Forest Atlas Series**. Macmillan Publ. Co., London

(viii) WCMC and Royal Botanic Gardens 1990. **World Plant Conservation Bibliography**. Cambridge Univ. Press

(ix) WRI, IIED, and IUCN. 1992, 1993. **Directory of Country Environmental Studies**. WRI, Washington DC

(x) IUCN Species Survival Commission (1994). IUCN Red List categories, IUCN, Gland, Switzerland

(xi) HMSO 1994. Biodiversity: the UK action plan. HMSO, London

(xii) Hilton-Taylor, C. 1996. Red Data List of South African Plants. National Botanical Institute, Pretoria, South Africa

(xiii) IUCN Red List of Threatened species 2000, compiled by C.Hilton-Taylor (www.iucn.org/redlist/2000/index.html,2000).

Catalogues and Indexes of Plant and Microbial Taxa

A. Algae (including Cyanobacteria)

1. Dawson E.Y. 1962. **New Taxa of Benthic Green, Brown and Red Algae published since De Toni**. Beaudette Foundation, Santa Cruz, California.
2. De Toni J.B. 1889-1924. **Sylloge Algarum** (6 vols.). Pavia.
3. Drouet F and Daily W.A. 1956. **Revision of the coccoid Myxophyceae**. Botancal Studies from Butler Univ. **12**: 1-218.
4. Vanlanginghan S.L. 1967. **Catalogue of the Fossil and Recent Genera and Species of Diatoms and their Synonyms**. J.Cramer, Vaduz, Liechtenstein.

B. Bacteria

1. Skerman V.D.B., McGowan V. and Sneath P.H.A. 1989. **Approved Lists of Bacterial Names** (rev. edn.). American Society for Microbiology, Washington DC.

C. Bryophytes

1. Bonner C.E.B. 1962 onwards. **Index Hepaticarum.** J. Cramer, Weinheim, W. Germany.
2. Van der Wijk R, Margadant W.D and Florscütz P.A. 1959-69. **Index Muscorum.** Utrecht, Netherlands.

D. Ferns

1. Christensen C. 1906-65. **Index Filicum** 5 Vols. Hagerup, Copenhagen, Denmark.
2. Jarret, F.M. (ed.) 1985. **Index Filicum Supplementum quintum pro annis** 1961-75, Clarendon Press, Oxford, UK.

E. Flowering Plants

1. **Index Kewensis** 1985 onwards (2 vols., 18 supplements). Clarendon Press, Oxford, UK.
2. **Kew Index** 1986 onwards [Annual]. Clarendon Press, Oxford, UK.

F. Viruses

1. **Virus Identification Data Exchange** (VIDE), CAB International, Wallingford, UK.

G. Fungi

1. Deighton F.C. 1969. **A Supplement to Petrak's List 1920-1939.** CAB International, Wallingford, UK.
2. Hawksworth D.L. 1972. **Lichens 1961-1969.** CAB International, Wallingford, UK.
3. **Index of Fungi 1940 onwards** [Twice yearly]. CAB International, Wallingford, UK.

4. Lamb I.M. 1963. **Index nominum lichenum inter anos 1932 et 1960 divulgatorum**. Ronald Press, New York, NY.
5. Petrak, F. 1930-44. **Verzeichnis der neuen Arten, Varietäten, Formen, Namen und Wichtigsten Synonyme.** Just's Botanischer Jahrbücker 48(3), 49(2), 56(2), 57(2), 58(1), 60(1), 63(2).
6. Petrak, F. 1950. **Index of Fungi 1936-1939.** Commonwealth Mycological Institute, Kew, Surrey.
7. Saccardo, P.A. 1882-1931, 1972. **Sylloge Fungorum** (26 vols.). Saccardo, Padua.
8. Zahlbruckner A. 1921-40. **Catalogus lichenum Universalis** (10 vols.). Bornträger, Leipzig.

The information provided above clearly shows that the bioinformatics revolution is now fully capable of enabling biodiversity researchers to communicate efficiently with each other, thus providing a 'springboard and common language for progress'. It was seriously felt a decade ago that 'Biodiversity components alone are not in threat, but also the accumulated knowledge about each species'. But now, most if not all knowledge has found place in the various types of information systems and databases detailed above.

10

BIODIVERSITY AND BIOTECHNOLOGY

Introduction

Biotechnology is fast emerging globally as a very dominant economic sector and is expected to contribute up to 50% of the world economy in the near future. CBD defines **biotechnology** as 'any technological application that uses biological systems, living organisms or derivatives thereof, to make or modify products or processes for specific use'. Thus, the chief and critical raw material for biotechnology is biodiversity. Many modern technologies, including recombinant DNA technologies, genetic engineering, transgenics, and cell and tissue culture protocols, are employed to obtain newer and newer benefits for human beings. The relationship between biotechnology and biodiversity is multidirectional (UNEP 1995): (i) Biotechnology provides very powerful tools for critical assessment of biodiversity, especially genetic diversity (see details in Chapter 2), and consequently the identification of potential bioresources themselves. (ii) It gives newer methods and guidelines for conservation of biodiversity (see details in Chapter 8). (iii) It enhances the wise and efficient utilisation of bioresources, both as a genetic resource for production and in the remediation of altered/degraded ecosystems. In other words, the increasing application of biotechnology and molecular tools to biodiversity has greatly enhanced the value and availability of bioresources, bioprocesses, and end and by-products for humankind. This is particularly true in reference to: (i) increased availability of food, feed and other renewable raw materials; (ii) improved human health and hygiene; (iii) greater protection of the environment; and (iv) enhancement of biosafety and environment-friendly technologies.

However, since at present only developed countries have the necessary expertise and money to venture into biotechnological exploitation of biodiversity, poor developing countries are disadvantaged in spite of their biodiversity-richness (Khoshoo 1996). This too obviously puts these latter countries at a disadvantage, especially given the advent of CBD and the signing of WTO agreements. Further details on these aspects are discussed in Chapter 11.

Biotechnology and its Role in Assessment of Biodiversity and Bioresources

Biotechnological tools enable critical analysis of the diversity of organic compounds present on the Earth. The development of very sensitive and highly specific bioassays to detect even picogram quantities of potentially useful organic molecules and compounds and automated screening technology allow screening thousands of plant and microbial samples at a very quick pace and efficiently select those with value in pharmaceutical, agrochemical, food, cosmetic and other industries (Komen 1991). We also now possess the expertise to analyse the diversity available in the DNA molecule that controls the production of novel compounds. DNA sequences of several organisms are being unravelled and specialised methodologies have been developed to analyse and identify specific gene products. The genetic diversity and its potential have been studied in several species to date (for details, see Amaral 2001; Lakshmikumaran *et al.* 2001; Ratnam 2001). We are now better equipped to delineate one species from another and accurately estimate the diversity within and between species (for details, see Chapter 2). Recent data have indicated that genetic variations do not always correlate with phenotypic or chemotypic variations and hence biodiversity should be best assessed only on genetic grounds and not exclusively on phenotypic characteristics. Taxonomic systems based on DNA and RNA sequence data or on the basis of variation in gene products should hereafter form the main basis for biodiversity characterisation and assessment.

Biotechnological tools have enabled identification of desirable and elite genes locked up in wild relatives of domesticated macrobes and microbes. They have further made possible specific transfer of such genes to domesticated taxa to produce transgenics that have increased performance, productivity and resistance to abiotic and biotic stresses of all kinds. For example, medium chain fatty acids such as lauric acid form the major source of dietary and industrial oils in the tropics. The production of this fatty acid in seeds is controlled by a specific protein called *acyl carrier protein -BTE*. Cloning the DNA of the gene coding for this protein from *Umbellularia californica* and its introduction into other oil-seed plants can substantially increase the level of laurate (Voelker *et al.* 1992). Other notable examples of useful genes identified and mapped using biotechnology are: Genes responsible for the production of useful variants in corn identified through the use of quantitative trait loci (QTL) and evolutionary history (Doebley *et al.* 1995), hitherto undiscovered genes from wild relatives of potato and tomato (Tanksley and McGough 1997), and several new useful genes in rice and corn and their wild relatives (McCouch 1998).

Biotechnology and its Role in Biodiversity Conservation

Biotechnology can also play an important role in culturing and conserving biodiversity. In *ex-situ* situations, it provides newer opportunities. A new class of non-living collections of living organisms in the form of DNA libraries (of naked DNA and genomic and cDNA as well as isolated chromosomes) and sequence databases has been made possible (for details, see Chapter 8). These profoundly increase the option values of genetic resource collections since the latter are the bases of agricultural development. Such *ex-situ* collections are invariably desirable from organisms of threatened categories, especially wild relatives of domesticated plants/microbes and land races. Such collections can be attempted after genetically characterising the wild relatives/land races through RFLP, RAPD (and its modifications) and Allozyme techniques and ascertaining their option values for possible future use in improving domesticated taxa. Although DNA banks are not substitutes for other conservation methods, they have the distinct advantage of requiring minimum space. The other biotechnological methodologies exploited for *in-situ* conservation of plants include rapid multiplication in a very short time using micropropagation, cryopreservation of vegetative or sexual propagules, synseed technology etc. (for details, see Chapter 8). Under *in-situ* situation, application of biotechnology provides data critical for the best management solutions for conservation of the target species. These data allow assessment of optimal or minimal population sizes for maintaining diversity, so that augmentation can be done to increase the population size through transfer from wild populations or through intense breeding programmes.

Biotechnology and its Role in Utilisation of Biodiversity

In an age when population is exponentially increasing and biodiversity being depleted due to man-made environmental degradation, biotechnology should come to the rescue of humankind by providing greater and efficient means of utilising the available biodiversity. This is possible only if biotechnologists come forward to work in harmony with taxonomists and ecologists to rapidly assemble the basic information required for proper evaluation of the genetic resource potential of biodiversity. Once a country attains the capacity to manage its genetic resources, this will automatically enable it to produce novel products from its own biodiversity.

The following are some of the areas of application of biotechnology in biodiversity utilisation:

(i) Transgenic organisms can be used as sources of proteins and peptides (structural and storage proteins, hormones such as insulin, antibodies and all categories of enzymes), lipids and fatty acids (edible and industrial oils, essential oils), carbohydrates (cellulose, agar, pectins and glues, chitin, many monosaccharides) and secondary metabolites (diverse metabolites that find use as pharmaceuticals, food additives, colorants, flavours and aromas, condiments, biopesticides etc.).

(ii) Organisms can be targeted to specific end uses through application of genetic engineering, breeding and *in-vitro* culture systems, such as enhancing agronomic performance in yield, disease resistance, stress tolerance etc. A major objective of modern agriculture is to improve yield both quantitatively and qualitatively. Two factors that greatly affect yield are diseases and abiotic stresses. These two can be overcome through biotechnological means either by modifying the endogenous genes or by introduction of new genes. The former can be done by induced mutagenesis, while the latter easier method can be effected by utilizing genes already existing in the gene pools of various categories (i.e., primary to quaternary) mentioned in Chapter 2, and introducing the same via 'marker assisted' breeding or genetic engineering. More and more useful genes are being discovered from microbial sources. Some examples are: the heat stable *Taq* polymerase from the bacterium *Thermus aquaticus*, the endotoxin *Bt* gene from *Bacillus thuringiensis* for insect resistance, the *basta* herbicide resistance gene from *Streptomyces hygroscopicus* and a ribonuclease gene for male sterility from *Bacillus amyloliquefaciens*.

(iii) We can improve, maintain or rehabilitate our environment effectively through:

(A) Identification of soil microbes, through DNA fingerprinting, that are elite with reference to nitrogen fixation (bacteria, cyanobacteria), phosphate solubilisation (bacteria, mycorrhizae), mineral leaching (bacteria), pollutant degradation etc. Further engineering and transferring of the key microbial genes involved in the enrichment or cleaning up of soil or water can be subsequently done. Biodiversity is the key resource for the rehabilitation and remediation of degraded and contaminated ecosystems. Biotechnology can become helpful in rehabilitation, for example:

(a) Genetic characterisation of endo-, ecto-, orchid- and ericoid-mycorrhizal fungi, and rhizobia and other nitrogen fixers which may be important as inoculants for reforestation/afforestation of degraded lands or lands subjected to long history of intensive cropping.

(b) Environmental clean-up can be done by suitable microbial systems because bioremediation can be carried out *in situ*, eliminating or greatly reducing costs, and the microbes multiply at the expense of the pollutants, thereby naturally enhancing the remediation rates. However, the following problems are encountered that greatly affect the remediation processes: (i) lack of an adequately known organism capable of thriving on the pollutant and utilising it as an energy source, (ii) limiting environmental features, and (iii) absence of appropriate microbes at the particular pollutant site. The limiting environmental features may include soil toxicity, extreme acidity/alkalinity, presence of anaerobic conditions, pressure of other highly competitive microbes etc. These problems notwithstanding, microbes have been identified which hold promise for effecting a variety of bioremediation processes. We have anaerobes that can grow on toluene, ethylbenzene, xylene and other BTEX class of pollutants (Lovely *et al.* 1994). There are also microbes that successfully grow in and degrade environments rich in polychlorinated biphenyls (PCBs) (Erickson and Mondello 1993; Furukawa *et al.* 1993). We also have microbes that have a capacity for chlororespiration using the dechlorination of a chlorinated substrate as the electron acceptor for energy generation and growth (i.e., organisms that can grow on tetrachloroethane, chlorinated benzoate and chlorinated phenol) (Cole *et al.* 1994; McCarty 1993; Mohn and Tiedje 1992) etc.

Bioremediation cannot be completely carried out by inoculating/ introducing a single microbe; it requires either a consortium of microbes or the construction, through genetic engineering, of organisms with an amalgam of desirable traits. For example, an organism that grows on the explosive material trinitrotoluene (TNT) was recently constructed; it has the capacity to remove the nitrogroups from TNT. To this was added a plasmid that has the ability to grow on toluene (Duque *et al.* 1993).

(B) Identification and use of plants and microbes that are naturally capable of accumulating high doses of heavy metals such as lead, cadmium, zinc, mercury, chromium etc.

(C) Production of biosurfactants (Finnerty 1994), bioplastics and other biodegradable products from microbes and further improving the latter's efficiency by suitable modification of the biosynthetic pathways involved in the production of such substances.

(D) Enhancement of the efficiency of microbes in industrial processes such as microbially enhanced secondary recovery of oils from reservoirs, bioleaching of metals from low grade ores, production of industrial enzymes useful in food and other industries, and enhancement of microbial fermentation for the production of antibiotics, alcohols, organic acids etc.

(iv) Overcoming interspecific and intergeneric barriers in plants through *in-vitro* cultures, *in-vitro* fertilisation, parasexual hybridisation, embryo rescue etc., so that superior hybrids can be produced. *In-vitro* culture can also increase diversity through somaclonal and gametoclonal variations or through *in-vitro* mutagenesis.

(v) Cell and tissue culture technology could be used for the industrial-level production of pharmaceuticals (alkaloids, steroids, flavonoids, terpenoids, enzymes etc.), food additives (carotenoids, anthocyanins, betanins, phenolics), perfumes (rose, jasmine, lavender, sandalwood oil, agarwood oil, *agil* etc.) biopesticides (pyrethrum), polysaccharides, proteins, lipids etc. There is scope for not only increasing the efficiency of production of these substances *in vitro*, but also for manipulation of biogenetic pathways of primary and secondary metabolism. The latter has been especially exploited either for the production of an intermediary compound in a metabolic pathway (for example, vanillin in the lignin biosynthetic pathway) or for biotransformation (i.e., converting one intermediary product of a metabolic pathway into another more useful product) through stereospecific reactions such as hydroxylation, oxidation, hydrolysis, isomerisation and dehydration. As examples of the latter may be cited the production of betamethyldigoxin and a number of monoterpenes. More than 100 major industries worldwide are working on tissue/cell culture, with a great array of target molecules.

(vi) Rapid and efficient clonal propagation of useful plants has been achieved in a number of places, using cell/tissue culture technology.

Adverse Impacts of Biotechnology on Biodiversity

The CBD in Article 8(g) calls for measures to 'regulate, manage or control the risks associated with the use and release of living modified organisms resulting from biotechnology, which are likely to have adverse environmental impacts that could affect the conservation and sustainable use of biological diversity, while also taking into account the risks to human health'. The CBD further advises, in Article 19.3, its signatory countries to safely handle and use any genetically modified organism (GMO).

Interest in research on the potential effects of the release of genetically modified organisms on biodiversity is accelerating. Introduction of GMOs into natural and homeostatic ecosystems is likely to cause the loss of species and habitat diversity; at least there is a strong theoretical possibility for this. Adverse biological effects on non-target populations and ecological and evolutionary disruption may be either the **direct** result of the introduced transgene(s) or alternatively the **indirect** result of changing socioeconomic conditions related to the application of recombinant technologies.

Direct Impacts

Several direct non-target effects on beneficial and native organisms by GMOs have been reported. An example is the transgenic Bt cotton plant (raised to resist the cotton bollworm), which affects a wide array of non-target insects such as butterflies, moths and beetles. Some GM crops have been shown to affect soil ecosystems by decreasing the rate of decomposition of organic wastes, affecting carbon and nitrogen levels and decreasing the diversity of soil microbial populations (Wolfenberger and Phifer 2000). Another possible direct impact of GMOs raised for conferring viral resistance is the likely emergence of new viruses with new biological characteristics through recombination and heteroencapsidation (Wolfenberger and Phifer 2000).

GMOs can cause harm to biodiversity due to the properties of the transgenes *per se* or their transfer and expression in non-target organisms. Adverse impacts on biodiversity through the introduction of GMOs may also result from disturbance of the dynamic population equilibria of ecosystems. Enhanced ability of GMOs to invade natural habitats of native species may lead to reduction of the population size of native taxa and in extreme cases to the total elimination of the populations of certain species, especially threatened ones (Edge 1994). The most important factors augmenting the hazard potential of GMOs are their increased fitness, adaptation to a wide variety of environmental conditions, instability of transgenes, resistance/avoidance to predation and favourable econiche (Strauss 1991).

The highly invasive nature of transgenic crops is considered a perceivable risk and many examples support this probability. The transgene reportedly enhances the invasive property of a species (Wolfenberger and Phifer 2000). The transgenic salt-tolerant rice is known to escape from its confinement and to extensively invade saline areas. Many engineered crops also turn into weeds in agricultural and non-agricultural fields (Tomiuk and Loescheke 1993). The transgenic crop may become uncontrollable in subsequent cropping by producing more persistent 'volunteer' plants or the transgene may be transferred to its sexually compatible wild species/relatives, resulting in the production of more persistent weed populations. Gene flow has been found to occur between rice and perennial wild rice, maize and teosinte, sugar-beet and wild beet, alfalfa and its wild relative etc. (Dale 1994). The potential and established ecological risks arising from the introduction of GMOs have been very well documented (Stewart-Tull and Sussman 1992). In light of this, the effects of introducing transgenic plants with conferred disease resistance, pest resistance, herbicidal/pesticide/ fungicide tolerance, environmental stress tolerance etc. would have to be very seriously considered as part of a risk assessment and biosafety measure before field trials and commercialisation of transgenic crops are attempted.

Risk assessment consequent to the release of a GMO is not easy and involves several sequential steps: (i) *a priori* identification of the potential/ documented hazards, (ii) estimation of their relative importance, and (iii) computation of the probability of adverse impacts (UNEP 1995). The impact probability and its quantification are hardly feasible in the case of biodiversity. Mathematical modelling approaches may help but are only approximations to real situations. Hence risk assessments are often done individually for every GMO released taking into consideration data obtainable from characterisation of the host organism, methods of transgenesis, characterisation of the GMO, intended use/effect of the GMO, ecological relevance of the intended effect of the release etc. Based on the results of risk assessment, GMOs are classified as low, intermediate or high risk organisms.

Certain important measures have been suggested for minimising the direct impacts of GMOs. For transgenic crop plants, the measures suggested (UNEP 1995) are: (i) elimination or disarmament of the biological vectors once transgenics are produced, to arrest the potential for further vector-mediated gene transfer; (ii) deletion of the pathogenic sequences contained in some vectors such as *Agrobacterium tumefaciens* Ti plasmid to preclude the possibility of their survival and transmission; and (iii) control of dispersal of the transgenic crop beyond the test site through reproductive isolation. For microbial GMOs, the measures suggested for minimising their direct impacts differ because the possibility of horizontal gene transfer cannot always be excluded. The measures suggested relate mainly to minimising dispersal of the transgene. The Cartagena Protocol on Biosafety signed in Montreal in January 2000 is the first global regulatory proposal to focus exclusively on Biotechnology and was endorsed by 135 of the 168 signatories of the CBD. The statement of paramount importance agreed upon was: 'Modern Biotechnology has great potential for human well-being if developed and used with adequate safety measures for the environment and human health'.

Another direct impact of biotechnology could be episodic genetic erosion, which would threaten the genetic diversity on which this technology depends. For instance, micropropagation and the consequent production of identical clones discourage perpetuation of genetic diversity through evolutionary adaptations.

GMOs can cause direct impacts on target organisms and indirect impacts on non-target organisms. A clear case of this duality of impact was recently reported by Watkinson *et al.* (2000). GM sugar-beet plants were raised in such a way that the weed *Chenopodium album*, a menace in sugar-beet fields, would be eliminated. But this has indirectly drastically reduced the population of a particular species of frugivorous bird whose main food is the seed of this weed. Pesticidal proteins produced by GMOs may have effects indirectly through bioaccumulations; when predators consume prey items that contain pesticidal proteins they ingest the latter (Wolfenberger and Phifer 2000).

Indirect Impacts

As discussed earlier, the direct impacts of biotechnology on biodiversity operate through biological processes (UNEP 1995). The indirect impacts, however, are predominantly socioeconomic ones, operated through human economic and social systems. Indirect impacts of biotechnology are immense and of very great relevance to people in developing countries who rely directly on biodiversity for their daily sustenance.

The impacts themselves are the results of human responses to the changes in relative cost and prices of biotechnologically derived items. This is best illustrated by an example. Suppose biotechnological methods lead to the identification of a plant material for an important pharmaceutical use. This use would raise the value of that material, resulting in increased collection pressures on that plant, which in turn would lead to overexploitation and species loss. Value increases of biotechnologically-derived items can increase the interest and conflict over ownership of genetic resources, the rates at which these resources should be utilised, IPRs, and formulation of laws relating to access to the resources. Value increases have also resulted in intense moral/ethical debates.

Ecoterrorism

While it is true that biotechnological exploitation of biodiversity has both positive and negative impacts, a new threat is fast emerging to counter this use of bioresources by biotechnologists. This threat is currently termed **Ecoterrorism.** A recent case of ecoterrorism should be mentioned here. Fearing that genetically modified trees might cause havoc on natural ecosystems, tree chopping fringe activists cut and damaged transgenic trees growing in an experimental plot of Oregon State University (Corvallis, Oregon) maintained by Dr. H.S. Strauss. These trees were being raised for traits ranging from fast growth to herbicide tolerance, which the experimenter felt could solve environmental problems such as pollution and loss of wild forests. Similar cases of 'vandalism' have also been reported on transgenic plants in the UK and Canada.

11

BIODIVERSITY PROSPECTING AND INDIGENOUS KNOWLEDGE SYSTEMS

Bioprospecting

The Europeans had learned and benefited greatly from observations of how plants were used by indigenous peoples of the Old World much before the voyages of Columbus. The ancient Greeks and Arabs were importing large quantities of spices such as cassia (*Cinnamomum cassia*), cinnamon (*Cinnamomum zeylanicum*), cardamom (*Eletteria cardamomum*) etc. from the Orient as early as the 5th century B.C. There existed a trans-Asian **silk route** during the 1st century A.D. that further facilitated trade in several items between the Old World and Europe. Today just the volatile oils of taxa mentioned above are worth more than US $7 million per annum to the countries that produce them. Besides these bioresources, indigenous knowledge of plants and microbes has hardly been tapped. Unfortunately hundreds and hundreds of knowledgeable people have either died without divulging their knowledge on traditional food and medicinal plants or refuse to part with it for reasons best known to them. This has resulted in the absence/scarcity of records or documentation of indigenous knowledge in many countries of the world. Unravelling such knowledge merits top priority. Hence it is not surprising that research today is mainly aimed at (i) identifying novel plants and plant products that have untapped but substantial economic potential, (ii) popularising and modifying traditional techniques so as to conserve vulnerable taxa and habitats, and (iii) conserving traditional germplasms of useful species and their wild relatives for further improvement programmes. These three aspects are covered under **Bioprospecting** or **Biodiversity prospecting**.

Bioprospecting is also referred to as Gene Prospecting. Bioprospecting, in more common usage, is the exploration of biodiversity for every valuable genetic and/or biochemical resource that finds use in pharmaceutical, biotechnological and agricultural industries either through bioprocesses unique to them or through novel end or by-products (Eisner 1992; Reid, Laird *et al.* 1993; Sittenberg and Gámez 1993). Increased interest in bioprospecting is attributable to: (i) a decline in innovativeness in the chemical and pharmaceutical industries; (ii) the rise of biotechnology as a dominant economic sector, as mentioned in the previous chapter; (iii) concern over biodiversity loss; (iv) the invigorated effort by developing countries to search for new economic activities; and above all (v) advances in the techniques for bioprospecting. As indicated in the last chapter, recent biotechnological advances have strengthened the hands of scientists in analysing organisms at the molecular-genetic level and finding ways of producing commercially viable products and processes at enhanced quantities/qualities using genetically modified organisms. Expectations from these new biotechnologies have further stimulated biodiversity prospecting which, according to some, will in turn stimulate profit-motivated steps towards active conservation of bioresources (Joyce 1994). However, one should not disregard the fact that in the absence of strict enforcement of 'legislation' and lack of suitable equitable benefit-sharing mechanisms, this 'gene rush' will not only deteriorate our ecosystems (including their species), but also provide less, if any, benefit sharing to indigenous people—the real owners of such traditional biodiversity knowledge. One of the striking causalities of bioprospecting is *Meytenus buchananni*, the important source of the anticancer compound, Meytansine; its entire population was completely lost from Kenya in a single expedition by the National Cancer Institute, USA. Already several pharmaceutical multinationals are screening the plants of tropical forests of several developing countries, which are rich in biodiversity, such as Costa Rica, Brazil, India, Colombia, Chile, Micronesia etc. Bioprospecting is also increasingly practised in forest habitats of temperate countries and 'hypothermal vents deep under the sea' (UNEP 1995).

It is very difficult to correctly estimate the commercial value of bioprospecting in pharmacy, industry, agriculture, and forestry. Tables 6.1 to 6.3 (Chapter 6) provide some data on the approximate value of some traditional natural products exploited in very recent years. Companies that produce seeds and agrochemicals substantially benefit from the free flow of germplasms from traditional societies and the market value of seed germplasms using land races is estimated at $ 50 million per year in the US alone (RAFI 1994). For many more bioresources obtained from indigenous societies, the value has not been estimated to date but is definitely substantial (see Chapter 6). It should, however, be mentioned that to date only a small proportion of biodiversity has been exploited; nevertheless there are sufficient indications that the economic potential of bioresources yet to be exploited is enormous. In view of the aforesaid, bioprospecting is rapidly growing into a new industry on its own merit.

Essentially three methods have been used until now in the choice and collection of plants for novel uses/compounds. The first is the **random method**, which involves the random collection of plants found in a given area of study for analysis of their economic potential. The second is the **phylogenetic method**, which involves the collection and analysis of all members of those plant families in which some taxa are already known to be good sources of useful products (e.g. Solanaceae). The third method, much superior to the first two, is the **ethno-directed method**. In this method, attention is particularly focused on plants which, based on traditional knowledge, are known to be used by tribal people/indigenous community, but not yet popularised. In other words, it is ready-made knowledge that is sure to yield the desired results in addition to involving less research and development costs; it is also less time-consuming. About three-quarters of the biologically active plant-derived compounds presently in use globally have been discovered through follow-up research on folk and ethnomedicinal uses (see Cotton 1996).

Indigenous Knowledge Systems

It is difficult to define 'indigenous people/ community'. The dictionary definition of 'indigenous' is **residence in any area for a very long time**. The widely acknowledged legal definition of 'indigenous people' is **politically weak people/community belonging to a culturally distinct ethnic entity and having an identity different from the mainstream-national society and deriving subsistence from local resources**. The World Bank additionally characterises them as a **social group vulnerable**

to domination. International labour organisations as well as the CBD expect such people to self-identify themselves as indigenous, which entitles them to their lawful rights. According to the International Working Group on Indigenous Affairs based at Copenhagen, there are approximately 300-500 million indigenous people the world over. Nearly half live in India and China. The knowledge these people have accumulated over several hundreds of years is termed traditional or indigenous knowledge.

It should be clear from the foregoing that traditional indigenous knowledge concerning biodiversity is of paramount importance. Today, traditional societies throughout the world possess a wealth of knowledge accumulated through trial and error for several hundreds of years during their prolonged and close interaction with Nature. Within a very short time after its inception, the science of biodiversity has become a very important tool in highlighting the significance of **Indigenous (= Traditional) Knowledge Systems (IKS** or **TKS)**. IKS is also known by many other terms such as **ITK (Indigenous Technical Knowledge), IAK (Indigenous Agricultural Knowledge), RPK (Rural Peoples' Knowledge), TBK (Traditional Botanical Knowledge), TEK (Traditional Ecological Knowledge)** etc. (for more details, see Cotton 1996).

The IKS, in its most modern concept, includes all the information, knowledge, wisdom, practices, beliefs and philosophies of traditional societies built and accumulated by members of these societies through several generations in close contact with Nature. The following are the important categories of IKS (D.D. Posey 1997): (i) sacred property (images, sounds, knowledge, material, culture) or anything considered sacred; (ii) knowledge of present, past and future potential uses of biodiversity elements as well as of soils and minerals; (iii) knowledge of preparations and processes as well as of storage/ preservation of useful taxa; (iv) knowledge of formulations involving more than one ingredient in food/beverage/medicine; (v) knowledge of individual species, such as planting, caring, cultivation and harvesting of each; (vi) knowledge of habitats and ecosystems; and (vii) classificatory systems of knowledge, such as traditional taxonomies. Three broad categories of TEK were identified by Inglis (1994): (i) knowledge about the specific components of environment such as plants, animals, soil etc.; (ii) development, evolution and use of appropriate technologies for farming, forestry etc.; and (iii) understanding the intimate relationship of traditional societies with the environment as a whole.

Indigenous knowledge is often possessed within a tribal/traditional society by all its members or may be held only by specialists, elders or specific gender groups (Warren *et al.* 1995). Within any traditional/tribal society, there are a number of social factors such as age, class and social status that can affect how knowledge is distributed or held among specific individuals. Usually the specialists/elders possess the right to hold such knowledge by virtue of their superiority over other common members of the society. The other most common social factor is gender. In most societies only men have been privy to such knowledge, in certain others only women and in yet others certain categories of knowledge were held by men and certain others by women. For example, among the Aguarura of Peru, details pertaining to manioc (manihot) cultivars were definitively the prerogative of only women in the past and still are even today (Boster 1985).

The importance of indigenous knowledge, especially with reference to bioprospecting, has been grossly underestimated. Even today more than 300-500 million people across the world are dependent on traditional modes of biodiversity utilisation. Advances in scientific documentation and understanding of traditional knowledge have resulted not only in recognising wild resources for human use and areas rich in them, but also in bringing into focus the sacred bondage that existed between traditional societies and nature (especially biodiversity). Such knowledge is often difficult for western science to understand and appreciate. The sacred bondage between Nature and Society was recognised by UNESCO in its 1972 Convention on the Protection of the World Cultural and Natural Heritage, often called the World Heritage Conference, and a new category of World Heritage Sites called 'Cultural Landscapes' was established (D.D. Posey 1997). Such an effort becomes all the more significant since the world's richest biodiversity areas, in terms of hot spots and megadiversity centres, have been maintained and defended by several diverse indigenous societies in the tropics as well as in temperate regions against destruction for all these years (Posey and Dutfield 1997); likewise the several land races of agricultural crops.

Biopiracy

The recognition of 'Cultural Landscapes' has very important implications for ownership and hence for **Intellectual Property Rights** (IPRs). Although wild species are products of nature and human societies can lay no special claim to them, any knowledge about them (including values) should logically be covered by an IPR. Moreover, many species/landscapes have been moulded/modified by indigenous societies who, therefore, are entitled to a special claim on them. This 'enlarges the implications of IPR for communities beyond the recognised categories of use, extraction, processing and preservation' (D.D. Posey 1997). Since ethno-directed bioprospecting will significantly reduce research and development costs by using traditional knowledge extracted from IKS, loss of bits of such knowledge will lead to the traditional societies losing control over information that has been theirs many many years. The existing mechanisms are also highly inadequate for effectively protecting the genuine rights of traditional societies (D.A. Posey 1996; Posey and Dutfield 1996,1997). It is very sad to note that only a miniscule proportion of profits (often less than 0.001%) earned from bioproducts has been returned to the indigenous peoples from whom much of the knowledge was extracted (D.A. Posey 1990). In several instances the indigenous society has received no return whatsoever. Only in recent years have a number of contracts have been set up between industrial organisations, academic researchers and traditional communities in biodiversity prospecting (Laird 1993). One instance may be mentioned, namely the agreement reached between INBio, Costa Rica's National Biodiversity Institute, and the US-based Merck & Co. Ltd. This agreement provides US $1.135 million to INBio from Merck for screening of the biodiversity elements of Costa Rica for prospective bioactive chemicals of economic potential, in addition to royalties. The royalties will be used by INBio to further its inventory and research and to support a fund for the management of Costa Rica's National Parks (Reid *et al.* 1993).

Biopiracy or 'gene robbing' may be defined as any effort by biodiversity prospectors to 'steal' or 'rob' traditional societies of their knowledge about biodiversity and use that knowledge to earn money without reimbursement to the owner-society. Until recently the biotechnologically rich developed countries and multinational companies have proprietorially exploited the rich bioresources of developing countries through the powerful tool of gene technology. These prospectors assumed ecosystems and their components to be 'wild' and 'part of the common heritage of humankind' and, therefore, 'leads' provided by indigenous knowledge systems need not be covered under patent rules; based on such leads, the prospectors make 'protectable products or processes'. Such piracies have been brought to public attention in recent years through various organisations such as RAFI (Rural Advancement Foundation International) (RAFI 1994). Cases of biopiracy are voluminous but only three classic examples are given below.

(i) The Urueu-Wau-Wau tribe of the Amazon region has long used an anticoagulant called 'tiki uba' and this fact was published in 1989 in an article written by McIntyre in a renowned magazine (McIntyre 1989). Based on this information, Merck Pharmaceuticals did further research and found that the extract of this plant was very helpful during heart surgery; the company subsequently patented its 'discovery' with no reference to the traditional knowledge provided by the Urueu-Wau-Wau tribe (Jacobs *et al.* 1990; D.A. Posey *et al.* 1995).

(ii) Indian people have for centuries used various parts of the neem plant for several purposes, including protection of crops against diverse pests. They found through experience that neem products are not only environment-friendly but also cheap. The active component of neem, azadiractin, was patented by two USA companies—W.R. Grace and Agrodyne Technologies—with no acknowledgement whatsoever that the traditional knowledge system of the indigenous people of India led to the isolation of azadiractin. However, on 30th September 1997, the European patent office delivered a favourable judgement against the patent given to W.R. Grace & Co for using neem-oil products as fungicides. This was due to the efforts of Indian NGOs who undertook the Neem Campaign.

(iii) The 'Endod' case is particularly interesting. Ethiopians have for countless years used the berries of endod, an indigenous African soapberry plant (*Phytolacca dodecandra*), as a detergent. It was later accidentally discovered that the mollusc carrying the pathogen causing the dreaded North African disease, **schistosomiasis,** was killed in streams where endod was used as a detergent. Ethiopian and Netherland scientists did further collaborative research, identified the elite E-44 variety of this plant taxon and obtained lemmatoxin, the active chemical involved in killing the mollusc that harbours the pathogen. This was patented by the Netherland and Ethiopian scientists who worked on the project with no benefit given to the local people who had been aware of this knowledge for a very long time (see Posey and Dutfield 1996, 1997).

IPRs and Ownership of Traditional Knowledge

Attention has already been drawn to the problem concerning protection of IKS by IPRs. Here the intricacies of the problem are elaborated and ways and means of solving them suggested. The three cases presented above and several others known of biopiracy raise the following two major issues (D.D. Posey 1997):

(i) Indigenous people and societies cannot obtain legal protection for their traditional knowledge and resources for various reasons (discussed below), while industries/academics somehow obtain legal rights for their application of traditional knowledge and resources, thus depriving the genuine beneficiaries of any benefit.

(ii) As already mentioned on page 186, the technologically poor but biodiversity-rich countries of the world lack the capacity/facilities and money to adequately exploit the commercial potential of their traditional knowledge or to defend such knowledge from being usurped by industries/academics of affluent countries.

As of today, IPRs are the main legal mechanisms available for protecting knowledge, discoveries, innovations, inventions and practices. Patents are the best known of these

IPRs. Patents, however, are of very limited value to most traditional people/societies as there are several difficulties in documenting such knowledge and in identifying the actual inventor or discoverer. In many instances, indigenous knowledge is considered in the 'public domain' and hence 'uniqueness' as required in Patent Rules also becomes problematic. In many traditional societies, even if technical requirements for patenting knowledge were satisfied, there are problems in meeting the costs of filing the Patent Papers, maintaining and monitoring their legal claims, and legal implementation and enforcement once protection is granted (D.D. Posey 1997)..

The other difficulties posed by IPRs to traditional societies are as follows (see D.D. Posey 1997): (i) IPRs benefit a society usually through the granting of rights to a small group/ individuals but not to the whole indigenous society. (ii) IPRs can protect only information that results from a specific historic act of 'discovery', whereas traditional knowledge is passed on through several generations; hence such knowledge falls in the 'public domain' and cannot be protected. (iii) While the basic motive of IPRs is to encourage commercialisation of knowledge and discovery, indigenous societies are not concerned with commercialisation; in fact, they prohibit it, thereby restricting the use of such knowledge to only within the society. (iv) IPRs invariably recognise only market value, while indigenous knowledge is often linked to spiritual, aesthetic and cultural values; at best, local economic value (within the society) is recognised. (v) IPRs are expensive, complex in nature, and time-consuming to obtain and defend.

The UN conference on Environment and Development and its Convention on Biological Diversity (CBD) have highlighted the importance of indigenous knowledge and the close dependence of traditional societies on bioresources. Article 8.j. of CBD addresses the different States of the world to 'respect, preserve and maintain knowledge, innovations and practices of indigenous and local communities' and 'encourage the equitable sharing of the benefits arising from the utilisation of such knowledge....' Article 18.4 of CBD, however, characterises such knowledge as 'traditional and indigenous technologies' and that those relevant to biodiversity should be transferred and used through the creation of a Clearing-House Mechanism (CHM) (Article 18.3). CHM is problematic in the sense that it expects the use of technologies to be regulated by an IPR. On the other hand, indigenous societies point out the problems of existing IPRs listed above in protecting IKS, and are adamant that 'additional, alternative or *sui generis* systems' should be created in place of IPRs to ensure their rights in profit-sharing. Unfortunately, the CBD provides no mechanism for protecting the rights of IKS (D.D. Posey 1997). Also, under Article 27.3b, Section 5, of the TRIPS agreement States are allowed to make exemptions from the patenting of plants, if an effective *sui generis* protection system for plant varieties also exists (Leskien and Filtner 1997). Given the foregoing, traditional societies demand implementation of one of the following three proposals: (i) modify the existing IPR to cover the previously unprotected IKS; (ii) formulate an entirely new IPR to protect IKS as well as the existing IPRs; or (iii) propose an alternative to IPR that will exclusively cover IKS. Most traditional societies favour the third proposal. A number of declarations to this effect have been made by indigenous and local communities. The most important are the Kari Oca Declaration, the Mataatua Declaration, the Santa Cruz Declaration, the Leticia Declaration and Plan and Action, the Treaty for Life Forms Patent Free Pacific, the Ukupseni Kuna Yala Declaration, the Jovel Declaration on Indigenous communities, the Chiapas Declaration etc.

Traditional Resource Rights

TRR is a process that respects the requirements of the indigenous people; in addition, it favours the development of appropriate *sui generis* systems for protecting IKS. TRR is one of the two alternative ways of implementing the idea of benefit sharing, that is, the internalisation of existing or expected benefits. TRR is considered more a process than a product. TRRs form around four processes (D.D. Posey 1997): (i) Identification and expression of **'bundles of rights'**. The 'bundles of rights' include human rights, rights to self-determination, land and territorial rights, right to development, environmental integrity rights, cultural heritage rights, IPRs, right to privacy, religious freedom etc. (ii) Fast evolving **'soft law'** primarily 'based on recognition of customary practice and legally non-binding agreements, declarations and covenants'. Traditional societies have their own system of justice with respect to sharing of knowledge and rights and responsibilities attached to possession of such knowledge. (iii) **'Harmonisation'** of all existing legally binding international agreements by obtaining signatures of all States; the States should sign such agreements only after ascertaining that the agreements are compatible with the rights of their traditional societies. (iv) **'Equitizing'** to ensure a very effective participation of indigenous and traditional societies in all phases of the implementation process.

TRR-guided efforts can open up a conducive atmosphere for new partnerships based on increased respect for IKS. This would then enable every State to implement its international obligations on trade and sustainable development. In the long run, such efforts would also open up IKS on biodiversity for more biotechnological exploitations.

Local Efforts to Date

Indigenous societies have recently become better -informed, organised and more articulate. Either they have started to assert their rights forcefully or have preferred to generate, control, distribute and use their knowledge through their own networks, databases and organisations. They have also made their voices heard by the developed countries, which either lack or are very poor in IKS. There is also a growing alliance between indigenous people and those who profess environmentalism and biosafety. Thus the theory and practice of local participation in rural/tribal development have become well established. Several conservationists and NGOs have recently begun to vigorously utilise this concept. These activities have now come to be known as **participatory approaches** to biodiversity management and utilisation (Cohen and Uphof 1977; Gadgil *et al.* 1993; Paul 1987; Salmen 1987). There are five major areas in which traditional societies and their people can participate: Information-gathering, consultation, decision-making, initiating actions and evaluation. A few instances of local efforts towards participatory management are provided below.

Territorial Demarcation of Traditional Societies

The most important requirement for traditional societies is legal title to their lands and territories where they practise their own land-use patterns, watershed management, cultivation, forest management etc. Sacred groves, for example, are areas of cultural and botanical significance and so should be covered by legal acts. Many cultural landscapes are often very difficult to detect and so must be properly documented and mapped. Such an act would indicate to outsiders that these are not 'wild' but 'private' and protected.

Self-demarcation of traditional territories has already been followed by several indigenous societies across the world, for example the Ye'kuana in Venezuela.

Community Forest Management

In many parts of South and South-east Asia, communities ruled and regulated forests and their resources in the historic past, thus ensuring equitable distribution of forest lands and products for agriculture, hunting and gathering of timber, fuel and minor forest produce.

However, in the last one to three hundred years the governance of forests has fallen into the hands of the government in many countries. In the last two decades or so, empowerment of communities as protectors of forests has again come into operation, especially in India, in order to facilitate the sustainable maintenance of their productivity. The concept of Joint Forest Management has gained ground in India, Thailand and a few other oriental countries, with local communities becoming partners in protecting forest resources and in sharing the benefits. This was initially tried in West Bengal and encouraged by the success of this system there, other States in India felt encouraged to follow suit. There are now nearly 10,000 forest protection committees managed by participating communities and approximately two million hectares of forest are managed under Joint Forest Management. A recent study in Thailand has similarly revealed nearly 12,000 community forest management initiatives.

The Asian experience on forest management has highlighted the fact that elaborate policies and heavily funded projects are no longer needed to protect our forests.

Indigenous People and Protected Areas

People living in or near protected areas were earlier denied derivation of benefits from them soon after the areas were declared protected. However, it was soon felt that such people should be provided adequate means to continue to derive benefits from protected areas without affecting/degrading them. One such measure was to create buffer and core zones in the protected areas and allow the buffer zones to be used for sustainable exploitation of bioresources, leaving the core zone highly protected (Wells *et al.* 1992). In many cases economic incentives are often additionally provided to the local community to exclusively exploit the buffer zones (for more details, see Chapter 8)

Community Biodiversity Registers

This is a 'bottoms-up system' of recording data/information on wild and domesticated biodiversity (Gadgil *et al.* 1995). These have been developed in India in an effort to secure community control over TEK. People of traditional societies are encouraged to document all the known plant and animal species, with all available details on their uses. CBRs will contain four separate sections (Gadgil *et al.* 1995): (i) **Background information**: This will cover two modules. Module 1 will delineate the total land and water areas, settlements and human communities. Module 2 will cover the local ecological history, which will include accounts of major changes that affect(ed) the landscape and waterscape elements, people, cultivated crops, livestock etc., the driving factors behind these major changes, and the establishment of 'Historical Benchmarks' (such as famine, year of completion of a dam etc.). As part of this effort, sufficient awareness must be created among local people regarding CBR and its purpose, along with preparation of landscape-waterscape maps and toposheets. (ii) **Practical ecological knowledge**: This should be collected through adequate surveys and documented. All knowledge relating to biodiversity status and utilisation should be collected and documented. (iii) **Claims**: This will enlist and record all the claims of local people regarding properties used and processes relating to bioresources irrespective of whether such uses and properties have been proved or not. (iv) **Scientific knowledge**: All other scientific information provided by local people regarding biodiversity elements should be documented. Ideally, there should be one CBR per village/society settlement in the preparation of which one or more educational institutions of that area may be involved. All members outside that particular community are then refused normal access to the details contained in the register; conditions are then set to allow others access to such data. Community registers can be advantageously used as documentary evidence of indigenous knowledge if there are legal disputes involving biopiracy.

Databases and Networks on IKS

Some progressive-thinking traditional societies have already established databases on their IKS. They essentially control these databases for access and use of knowledge. The most notable traditional society in this regard is the Canadian Inuit of Nunavik and the Dene (Simon and Brooke 1996).

Traditional societies are increasingly using electronic networks to exchange knowledge and information. A few such efforts are described here. The **Indigenous People's Biodiversity Network** (IPBN) is a global network of organisations of indigenous people working in the area of biodiversity conservation, especially in protecting IKS relating to bioresources. IPBN has already formulated suitable policies, laws and programmes in the above area of conservation; especially significant is its development of *sui generis* systems for protecting IKS on genetic resources. Working Groups have been established in the Americas, Asia and Africa, which have agreed to exchange ideas, share knowledge, and raise issues relating to protection of their IKS.

The other organisation worthy of mention is the Society for Research and Initiatives for Sustainable Technologies and Institutions (SRISTI), initiated by Prof. Anil Gupta of the Indian Institute of Management. SRISTI has already established communications with about 300 villages of India. SRISTI's main objectives are capacity-building at grassroots level in biodiversity conservation, protecting IPRs, adding more information to their IKS through further efforts and experiments, development of entrepreneurial enterprises among traditional people using IKS and enrichment of their cultural and institutional base. SRISTI also offers technical, developmental and legal/counselling help not only to the innovative efforts of traditional people, but also to their IKS relating to genetic resources. Village people are also trained in sustainable developmental activities. SRISTI has already established a computer database on IKS.

Community Controlled Research

CCR (Community Controlled Research) fulfils the objectives and methodologies formulated by indigenous peoples themselves. For instance, the Kuna tribe of Panama and the Inuit tribe of Canada have not only established CCRs,but have decided to allow only such research in their territories. According to D.D. Posey (1997), the Proyecto de Estudio para el Manejo de Areas Silvestres de Kuna Yala (AEK) of Panama has prepared an Information Manual for researchers of their area. External collaborative research is allowed, and even encouraged, provided such research is 'designed to provide the Kuna with information useful to them and under their control'. Similarly, an article 'Negotiating Research Relationship in the North' has been prepared which delineates the ethical principles of Inuit communities to be followed by researchers of that community. For more information on research principles for CRR suggested by this community, see D.D. Posey (1997).

Centre for Farmers' Rights

The term 'Farmers Right' was developed in the 1983 forum of the International Commission on Plant Genetic Resources of FAO chaired by Professor M.S. Swaminathan to denote 'the rights arising from the past, present and future contributions of farmers in conserving, improving and making available plant genetic resources', in particular those in the centres of origin/diversity (Swaminathan 1997). Subsequently a dialogue was organised in 1990 at Madras (now called Chennai) in the

M.S.Swaminathan Research Foundation (MSSRF) in collaboration with the Keystone Centre of US, wherein it was agreed to create a fund for Farmers' Rights. A draft legislation was also framed 'for converting the **know-how** relating to Farmers' Rights into **do-how'**. Now there is absolutely no difference of opinion about the need to recognise and reward the contributions of traditional farmers and indigenous tribals to the conservation as well as improvement of germplasms of crop plants. The concept of Farmers' Rights aims at benefit-sharing through a process of **compensation** for the use of traditional knowledge systems, in contrast to the system of **internalisation of benefits**, already referred to on page 189. The charter of Farmers' Rights includes the following rights (Shiva and Ramprasad 1993): (i) right to land; (ii) right to conserve, reproduce, and modify seed and plant material; (iii) right to feed, to ensure food security and to save the country; (iv) right to just agriculture prices and public support for sustainable agriculture; (v) right to information; (vi) right to participatory research; (vii) right to natural resources; and (viii) right to safety and health.

The Plant Variety Protection Acts introduced in advanced countries generally conform to the provisions stipulated by the UPOV Act of 1991 version (see Chapter 9). Because of this Act, it has become necessary to classify the Farmers' Rights into two distinct categories (Swaminathan 1997): (i) Farmer cultivators: Farmers cultivating new varieties of crops by buying seeds should have unrestricted rights to keep seeds for raising crops in their own fields for successive generations. They should also be allowed to enter into a very limited sale/ exchange of seeds with their immediate neighbours. (ii) Farmer conservers: Those indigenous people who have preserved land races and varieties raised by them through their own selection belong to this category. It is for this category of Farmers that protection should be given under 'Farmers' Rights' There was a strong recommendation, and rightly so, in the UPOV that a balance should be struck between 'homogeneity and heterogeneity in the genetic make-up of new cultivars' (Swaminathan 1997). As a sequel to this strong proposition, a Resource Centre for Farmers' Rights was organised in MSSRF Chennai, India. The government of India has planned the establishment of a National Community Gene Fund for rewarding farmers' efforts in maintaining agrobiodiversity. This constitutes the first legal recognition and reward of Farmers' Rights, based on remuneration rights.

This Resource Centre is involved in the following four activities: (i) Farmers' Rights Information Service (FRIS). FRIS is a collection of several component databases such as intellectual property rights database (with four modules, respectively Tribal contributions, Ethnobotanical features, Sacred groves, and rare Angiosperms) and multimedia database on the ecological farmers of India. (ii) Community Gene Bank: this is meant for storing seeds of land races, traditional cultivars and folk varieties. (iii) Community Genetic Resources Herbarium: this Herbarium is meant to serve as a reference centre for the identification of rare and threatened flowering plant species, economically useful plants and traditional cultivars. (iv) Lastly, creation of an Agrobiodiversity Conservation Corps to train young tribal and rural women and men in *in-situ* and *ex-situ* conservation techniques and seed technology.

Role of Women

It was pointed out earlier in this chapter that IKSs in several indigenous societies were held by women only or shared by both men and women. Women have traditionally played a key role in seed and plant selection and preservation, especially in PGRs. Women in some Indian and other Old World traditional societies, are responsible for the selection of viable and healthy seeds. Statistics show that women contribute to more than half the food grown every year in the developing countries (Damodaran 1997). It is also true that women and children are the first and worst victims of biodiversity degradation. If participation of the entire traditional society is needed in biodiversity development and conservation, 50% of the population should be women empowered for that purpose. It has been repeatedly demonstrated in the past that women's participation has been successful in solving pressing and acute environmental problems,e.g.: (i) the role of tribal women in the Chiku dam issue in the Philippines; (ii) the Chipko movement in the Himalayas; (iii) the now-raging Narmada Valley agitation spearheaded by Medha Patkar; (iv) the struggle of women for recognition as farmers in Burkino Faso, Kenya, Nigeria and Zambia. Efforts should be made immediately to see that women establish themselves as managers of natural resources through education, awareness and action and society should ensure every assistance possible, including financial, to women in all these endeavours.

Problems and Prospects of Participatory Management of Biodiversity

Reconciling economic growth with biodiversity conservation has become a major and acute problem in sustainable development in remote indigenous societies and traditional farming communities. While on the one hand these societies and the countries housing them lack adequate funds to venture into conservation actions, there is the problem of not only protecting their IKS from bioprospecting piracies, but also rectifying the lack of equitable sharing of benefits accruing from such knowledge-sharing. As a result of these concerns, an increasing number of efforts are underway by the respective governments, NGOs, international organisations and the indigenous communities themselves to link biodiversity conservation, protecting their IKS and equitable sharing of benefits through internalisation of existing or expected benefits. A number of projects have been launched towards this goal of participatory management of biodiversity (see UNEP 1995).

One of the major problems in such participatory management efforts is the poor response of the local people/tribal societies themselves, either because of the short duration of the projects executed thus far, or because of the failure of the projects to elicit the required participation of locals. More efforts should be focused on educating the traditional people about the likely advantages of participatory approaches in sharing benefits equitably and at the same time in a sustainable manner. Secondly, such participatory programmes have not reached very many indigenous societies, especially in the developing countries, in spite of efforts by such organisations as GEF, which had already committed as of 1997 a sum of $300 million to more than 50 developing countries. Thirdly, at least some of the NGOs who are entrusted with this job of familiarising participatory approaches to rural communities by national and international funding organisations have not satisfactorily carried out the missions entrusted to them or have mismanaged them.

It is true, for the reasons mentioned above, that the participatory approach to biodiversity management, conservation and sustainable use has yet to clearly demonstrate its effectiveness. Participation by indigenous sources can facilitate a more co-operative and harmonious relationship between protected areas and other useful ecosystems and local people who depend on them, thereby 'making law enforcement more humane and acceptable' (UNEP 1995).

REFERENCES

Adams, L.W. 1994. **Urban Wildlife Habitats**. Univ. Minnesota Press, Minneapolis, MN.

Adams, R.P. 1993. The conservation and utilisation of genes from endangered and extinct plants: DNA bank-net. In: Gustafson, J.P. Appels, R. and Raven, P. (Eds.). **Proceedings Twentieth Stadler Symposium: Gene Conservation and Exploitation**, Plenum Press, New York, NY, pp. 35-52.

Adams, R.P., Miller, J.S., Goldberg, E.M. and Adams, J.E. 1994. **Conservation of plant genes II**. Monographs in systematic botany from the Missouri Botanical Garden, Vol. 48., Missouri Botanical Gardens, MO.

Akeroyd, J and Synge, H. 1992. Higher Plant Diversity. In: Groombridge, B. (Ed.). **Global Biodiversity-Status of the Earth's Living Resources**. Chapman & Hall, London, pp. 64-87.

Alcorn, J.B. 1984. **Huastec Mayan Ethnobotany**. Univ. Texas Press, Austin, TX.

Aldred, J. 1994. Existence value, welfare and altruism. Environ. Values **2**: 381-402.

ALICE 1990. **ALICE: A Bio-diversity Database System**. ALICE software partnership.

Allan, D.J. and Flecker, A.S. 1993. Biodiversity Conservation in running waters: identifying the major factors that threaten destruction of riverine species and ecosystems. Bioscience **43**: 32-43.

Allen, J.M. (Ed.) 1963. **The Nature of Biological Diversity**. McGraw Hill Inc. New York, NY.

Altaba, C.R. 1996. Counting species names. Nature **380**: 488-489.

Altieri, M.A. and Anderson, M.K. 1992. Peasant Farming Systems, Agricultural Modernization, and the Conservation of Crop Genetic Resources in Latin America. In: Fiedler, P.L. and Jain, S. K. (Eds.). **Conservation Biology**. Chapman & Hall, New York, NY, , pp. 49-64.

Alvarez, L.W., Alvarez, W., Asaw, F. and Michel, H.V. 1980. Extraterrestrial cause for the Cretaceous-Tertiary extinction. Science **108** : 1095-1108.

Amaral, W. 2001. Characterization, evaluation and conservation of forest genetic resources: The potential and limitation of new biotechnology tools. In: Uma Shaanker, R., Ganeshaiah, K.N. and Bawa, K.S. (Eds.). **Forest Genetic Resources: Status, Threats and Conservation Strategies**. Oxford & IBH Publishing Co. Pvt. Ltd., New Delhi, pp. 115-125.

Angermeier, P.L. 1994. Does biodiversity include artificial biodiversity? Conser. Biol. **8**: 600-602.

Appasamy, P. 1993. Role of non-timber forest products in a subsistence economy: the case of a joint forestry project in India. Econ. Bot. **47**: (cited from UNEP 1995).

Apps, M.J., Kurz, W.A., Luxmore, R.J. Nilsson, L.O., Sedjo, R.A., Schmidt, R., Simpson, L.G. and Vinson, T.S. 1993. Boreal forests and tundra. Water, Air and Soil Pollution **70**: 39-53.

Archibold, O.W. 1989. Seed banks and vegetation processes in coniferous forests. In: Leck, M.A., Thomas, V.T. and Simpson, R.L. (Eds.). **Ecology of Soil Seed Banks**. Academic Press, London, (Cited from Given 1996).

Aronson, R.B.1990. Onshore - offshore patterns of human fishing activity. Palaios **5** : 88-93.

Arrhenius, O. 1921. Species and area. J. Ecol. **9**: 95-99.

Arrow, K.J. and Fisher, A.C. 1974. Environmental preservation, uncertainty, and irreversibility. Quart. J. Econ. **88**: 312-319.

Arroyo, M.T.K., Squeo, F., Armesto, J. and Villargran, C. 1988. Effects of aridity on plant diversity in the northern Chile Andes. Ann. Missouri. Bot. Gard. **75**: 55-78.

Arumuganathan and Earl 1991. Plant Mol. Biol. Rep. **9**: 208 - 218. (cited from Westhoff *et al.* 1998).

Aselmann, I., and Crutzen, P.J. 1989. Global distribution of natural fresh water wetlands and rice fields, their net primary productivity, seasonality and possible methane emissions. Jour. Atmos. Chem. **8**: 307-358.

Ashton, P.S. 1992. Species Richness in Plant Communities. In: Fiedler, P.L. and Jain, S.K. (Eds.). **Conservation Biology.** Chapman & Hall, New York, NY, pp. 3-22.

Atlay, G.L., Ketner, P. and Duvigneaud, P. 1979. Terrestrial primary production . In: Bolin, B. (Ed.). **The Global Carbon Cycle.** John Wiley and sons, Chichester, UK, pp. 129-182.

Avise, J.C. 1994. **Molecular Markers, Natural History and Evolution**. Chapman & Hall, London.

Ayensu, E. and 24 others. 1999. International ecosystem assessment. Science **286**: 685-686.

Aylward, B.A. 1993. The economic value of pharmaceutical prospecting and its role in biodiversity conservation. LEEC discussion paper. Inst. Environ. Develop., London, pp. 93-105.

Aylward, B.A. and Barbier, E.B. 1992. Valuing environmental functions in developing countries. Biodiv. Conser. **1**: 34-50.

Bachmann, K. 1994. Tansley Review No. 63: Molecular markers in plant ecology. New Phytol. **126**: 403-418.

Bailey, R.G. 1989a. **Ecoregions of the Continents**. U.S. Dept. Agri. Forest Service, Washington DC.

Bailey, R.G. 1989b. Explanatory supplement to ecoregions map of the continents. Environ. Conser. **16**: 307-309.

Balick, M. and Mendelsohn, R.O. 1992. Assessing the economic value of traditionsl medicines from tropical forests. Conser. Biol. **6**: (cited from UNEP 1995)

Balick, M.J. 1990. Ethnobotany and the identification of therapeutic agents from the rainforest. CIBA Foundation Symposia **154**: 22-39.

Balmford, A., Mace, G.M. and Ginsberg, J.R. 1998. The challenges to conservation in a changing world: putting process on the map. In: Mace, G.M., Balmford, A. and Ginsberg, J.R. (Eds.). **Conservation in a Changing World.** Cambridge Univ. Press, Cambridge, pp. 1-28.

Barbier, E.B., Adams, W.M. and Kimmage, K. 1991. Economic valuation of wetland benefits: Hadejia-jam'are floodplain, Nigeria. LEEC discussion paper. Inst. Environ. Develop. London, pp. 91-102.

Barbier, E.B., Burgess, J.C. and Folke, C. 1994. **Paradise Lost? The Ecological Economics of Biodiversity**. Earthscan, London.

Barbier, E.B., Burgess, J.C., Swanson, T.M. and Pearce, D.W. 1990. **Elephants, Economics, and Ivory.** Earthscan, London.

Barthlott, W., Lauer, W., and Placki, A. 1996. Global distribution of species diversity in vascular plants: *Towards a world map of phytodiversity.* Erdkunde **50**: 317 - 327.

Becker, J. and Heun, M. 1995. Barley microsatellite allele variation and mapping. Plant Mol. Biol. **27**: 838-845.

Bellamy, D. 1979. The role of the media in conservation. In: Synge, H and Townsend, H. (Eds.). **Survival or Extinction**. Royal Botanic Gardens, Kew, England, pp. 165-170.

Bennet, M.D. and Smith, J.B..1991. Nuclear DNA amounts in angiosperms. Phil. Trans. Royal Soc. (London) Ser. B. **334** : 309-345.

Bennett, E. (Ed.) 1968. **Record of the FAO/IBP Technical Conference on the Exploration, Utilisation and Conservation of Plant Genetic Resources, 1967.** FAO, Rome.

BGCS 1989. **The Botanic Gardens Conservation Strategy.** WWF, IUCN, Botanic Gardens Conservation Secretariat, Kew, England.

Bhat, C.P. 1987. The *Chipko Andolan*. In: Agarwal, A., D'Monte, D., and Samanth, U. (Eds.). **The Fight for Survival—People's Action for Environment**. Centre for Science and Environment, New Delhi, pp. 43-56.

Bisby, F.A. 1994. Global master species databases and biodiversity. Biol. Inter. **29**: 33-40.

Bisby, F.A. 2000. The quiet revolution: Biodiversity Information and Internet. Science **289**: 2309-2312.

Bisby, F.A., Russell, G.F. and Pankhurst, R.J. (Eds.). 1993. **Designs for a Global Plant Species Information System.** Oxford Univ. Press, Oxford.

Boster, J.S. 1985. Requiem for omniscient informant: there's life in the old girl yet. In: Dougherty, J. (Ed.). **Directions in Cognitive Anthropology.** Univ. Illinois press, Champaign, IL, pp. 177-198.

Bowes, M.D. and J.V. Krutilla 1989. **Multiple-use Management : The Economics of Public Forestlands.** Resources for the Future, Washington DC.

Bowles, M.L. and Whelan, C.J. (Eds.). 1996. **Restoration of Endangered species.** Cambridge Univ. Press, Cambridge.

Boyle, T.J.B. and Lenne, J.M. 1997. Defining and Meeting Needs for Information: Agriculture and Forestry Perspective. In: Hawksworth, D.L., Kirk, P.M. and Dextre Clarke, S. (Eds.). **Biodiversity Information: Needs and Options.** CAB International, Wallingford, UK, pp. 31-53.

Breiman, A., Bogher, M., Sternberg, H. and Graur, D. 1991. Variability and uniformity of mitochondrial DNA in populations of putative diploid ancestors of common wheat. Theoret. Appl. Gen. **52:** 145- 157.

Briffa, K.R., Bartholin, T.S., Ecksetein, D., Jones, P.D., Karlen, W. and Schweingruber, F.H. 1990. A 1,400-year tree ring record of summer temperatures in Fennoscandia. Nature **346:** 434-439.

Brink, K.H. 1993. The coastal ocean process (CoOP) effort. Oceanus **36:** 47-49.

Briscoe, J., Furtado de Castro, P., Griffin, C., North, J. and Olsen, O. 1990. Toward equitable and sustainable rural water supplies: a contingent valuation study in Brazil. World Bank Econ. Rev. **4:** 115-134.

Bromley, D.W. 1986. The common property challenge. In: **Proceedings, Conference on Common Property Resource Management,** National Academy of Sciences, Washington DC, pp. 1-5.

Brown, A.H.D. 1989. Core collections: a practical approach to genetic resources management. Genome **31:** 818-824.

Brown, A.H.D. and Schoen, D.J. 1992. Plant Population genetic structure and biological conservation. In: Sandlund, O.T., Hindar, K. and Brown, A.H.D. (Eds.). **Conservation of Biodiversity for Sustainable Development.** Scandinavian Univ. Press, Oslo, pp. 88-104.

Brown, G.M. 1990. Valuation of genetic resources. In: Orians, G.H., Brown, G.M., Kunin W.E. and Swierbinski, J.E. (Eds.). **Preservation and Valuation of Biological Resources.** Univ. Washington Press, Seattle, WA, pp. 203-228.

Brown, G.M. and W. Henry 1993. The viewing value of elephants. In: Barbier, E.B. (Ed.). **Economics and Ecology. New Frontiers and Sustainable Development.** Chapman & Hall, London, pp. 146-155.

Brown, J.S. 1996. Restoration ecology: living with the prime directive. In: Bowles, M.L. and Whelan, C.J. (Eds.). **Restoration of Endangered Species.** Cambridge Univ. Press, Cambridge, pp. 355-380.

Brown, K. and G.G. Brown, 1992. Habitat alteration and species loss in Brazilian forests. In: Whitmore, T.C. and Sayer, J.A. (Eds.). **Tropical Deforestation and Species Extinction.** Chapman & Hall, London, pp. 119-142.

Browne, J. 1983. **The Secular Ark: Studies in the History of Biogeography.** Yale Univ. Press, New Haven, CT.

Brummitt, R.K. 1992. **Vascular Plant Families and Genera.** Royal Botanic Gardens, Kew, England.

Bruner, A.G., Gullison, R.E., Rice, R.E. and da Fonseca, G.A.B. 2001. Effectiveness of parks in protecting tropical biodiversity. Science **291:** 125-128.

Burley, J.R., Scott, P.R., and Speedy, A.W. 1997. Biodiversity: The role of Information Technology in distributing Information. In: Hawksworth, D.L., Kirk, P.M. and Dextre Clarke, S. (Eds.). **Biodiversity Information: Needs and Options.** CAB International, Wallingford, UK, pp. 157-171.

Burnett, J. 1993. IOPI: Genesis of GPSIS? In: Bisby, F.A., Russell, G.F. and Pankhurst, R.J. (Eds.). **Designs for a Global Plant Species Information System.** Oxford Univ. Press, Oxford, pp. 334-342.

Busby, J.R. 1997. Management of Information to support conservation decision making. In: Hawksworth, D.L., Kirk, P.M. and Dextre Clarke, S. (Eds.). **Biodiversity Information: Needs and Options.** CAB International, Wallingford, UK, pp. 105-114.

Cabrera, A. 1970. **Flora de la Provincia de Buenos Aires: Gramineas**. Instituto Nacional de Tecnologia Agropecuara, Buenos Aires.

Cairns, J. Jr. 1986. Restoration, reclamation and regeneration of degraded or destroyed ecosystems. In. Soulé, M.J. (Ed.). **Conservation Biology**. Sinauer Associates, Sunderland, pp. 465-484.

Callicott, J.B. 1986. On the intrinsic value of nonhuman species. In: Norton, B.G. (Ed.). **The Preservation of Species: The Value of Biological Diversity.** Princeton Univ. Press, Princeton, NJ, pp. 138-172.

Canhos, D.A.L., Canhos, V.P. and Kirsop, B.E. (Eds.). 1994. **Linking Mechanisms for Biodiversity Information**. (UNEP-sponsored workshop)-Tropical Foundation, Champinas, Brazil.

Canhos, V.P., Manifio, G.P. and Canhos D.A.L. 1997. Networks for distributing Information. In: Hawksworth, D.L., Kirk, P.M. and Dextre Clarke, S. (Eds.). **Biodiversity Information: Needs and Options**. CAB International, Wallingford, UK, pp. 147-156.

Chakraverty, R.K. and Mukhopadhyaya, D.P. 1990. **A Directory of Botanic Gardens and Parks in India.** Botanical Survey of India, Kolkata, India.

Chalmers, K.J., Sprent, J.L., Simons, A.J., Waugh, R. and Powell, W. 1992. Patterns of genetic diversity in a tropical tree legume (*Gliricidia*) revealed by RAPD markers. Heredity **69**: 465-472.

Chang, T.T. 1995. Rice. In: Smartt, J. and Simmonds, N.W. (Eds.). **Evolution of Crop Plants.** Longmans, London, pp. 147-155.

Charlesworth, D. and Charlesworth, B. 1987. Inbreeding depression and its evolutionary consequences. Ann. Rev. Ecol. Syst. **18**: 237-268.

Child, B. 1990. Assessment of wild life utilisation as a land-use option in the semi-arid rangeland of Southern Africa. In: Kiss, A. (Ed.). **Living with Wildlife: Wildlife Resource Management with Local Participation in Africa**. Tech. Paper no.10. World Bank, Washington DC.

Chilean Forest Service 1989. **Red List of Chilean Terrestrial Flora**. Chilean Forest Service (CONAF), Santiago.

Chin, H.F. and Pritchard, H.W. 1988. **Recalcitrant seeds—a Status Report, Including Bibliography, 1979-1987**. IBPGR, Rome.

Chopra, K. 1993. The value of non-timber forest products: an estimate from India. Econ. Bot. **47**: 251-257.

Clegg, M.T. and Durbin, M.L. 1990. Molecular approaches to the study of plant biosystematics. Austral. Syst. Bot. **3** : 1-8.

Club of Earth, 1990. **Loss of Biodiversity Threatens Human Future**. Dept. Biol. Sci., Stanford Univ, Stanford, CA.

Cohen, J.M. and Uphof, N.T. 1997. Rural development and participation: concepts and measures for project design, inplementation and evaluation. **Rural Development Committee Monograph 2**. Centre for International Studies, Cornell Univ., Ithaca, NY.

Cole, J.R., Cascarelli, A.L., Mohn, W.W. and Tiedje, J.M. 1994. Isolation and characterization of a novel bacterium growing via reductive dehalogenation of 2-chlorophenol. Appl. Environ. Microbiol. **60**: 3536-3542.

Connor, E.P. and McCoy, E.D. 1979. The statistics and biology of the species - area relationship. Am. Nat. **113**: 791-833.

Cotton, C.M. 1996. **Ethnobotany, Principles and Applications**. John Wiley and Sons, Chichester, UK.

Crawley, M.J. and Harral, J.E. 2001. Scale dependence in plant biodiversity. Science **291**: 864-868.

Cromarty, A.S., Ellis, R.H. and Roberts, E.H. 1982. **The Design of Seed Storage Facilities for Genetic Conservation.** IBPGR, Rome.

Crozier, R.H. 1992. Genetic diversity and the agony of choice. Biol. Conser. **61**: 11-15.

D'Arrigo, R., Jocoby, G.C. and Fung, I.Y. 1987. Boreal forests and atmosphere-biosphere exchange of carbon dioxide. Nature **329**: 321-323.

Dahl, T.E. 1990. **Wetland Losses in the United States 1780s to 1980s.** US Department of the Interior, Fish and Wildlife Service, Washington DC.

Dale. P.J. 1994. The impact of hybrids between genetically modified crop plants and their related species: general considerations. Mol. Ecol. **3**: 31-36.

Damodaran, V.K. 1997. Biodiversity Conservation, Environment Regulations, Impact of Technologies and the Need for Empowering Women in Developing Countries. In:

Pushpangadan, P., Ravi, K. and Santhosh, V. (Eds.). **Conservation and Economic Evaluation of Biodiversity,** vol 2. Oxford & IBH Publishing Co. Pvt. Ltd. New Delhi, pp. 491-500.

Darwin, C. 1859. **On the Origin of Species by means of Natural Selection**. John Murry, London.

Davis, S.D., Heywood, V.H. and Hamilton, A.C. (Eds.). 1994-1995. **Centres of Plant Diversity. A Guide and Strategy for their Conservation. Vol.1. Europe, Africa, South West Asia and the Middle East, vol. 2. Asia, Australia and the Pacific**. WWF and IUCN, IUCN Publications Unit, Cambridge, UK.

de Klemm, C. 1990. **Wild Plant Conservation and the Law.** IUCN, Gland.

del Tredici, P., Ling, H. and Yang, G. 1992. The *Ginkgos* of Tian Mu Shan. Conserv. Biol. **6**: 202-209.

Denisiuk, Z. 1990. Lowland grasslands in Poland—their natural resources, managment and protection. In: **Lowland Grassland of Eastern Europe**. IUCN, Gland, Switzerland.

Dennis, J.G. and Ruggiero, M. A. 1996. Biodiversity Inventory: Building an Inventory at scales from local to global. In: Szaro, R.C. and Johnston, D.W. (Eds.). **Biodiversity in Managed Landscapes.** Oxford Univ. Press, Oxford, pp. 149-156.

de Rosnay, J. 1996. Biodiversity in the twenty-first century. In: di Castri, F. and Younès,T. (Eds.). **Biodiversity, Science and Development. Towards a New Partnership**. CAB International & IUBS, Wallingford, UK, pp. 596-598.

Diamond, J. 1986. The design of a nature reserve system for Indonesian New Guinea. In: Soulé, M.J., (Ed.). **Conservation Biology**. Sinauer Associates, Sunderland, pp.485-503.

Diamond, J.M. 1972. Biogeographic kinetics: estimation of regulation times for avifauna of southwest Pacific islands. Proc. Wash. Acad. Sci., USA. **69**: 3199-3203.

di Castri, F. 1995. Una silla de cuatro patas. In: **Las Ultimas Noticias**. Santiago, Chile, 29th April 1995 (cited from di Castri and Younès 1996).

di Castri, F. and Younès, T. 1996. Introduction: Biodiversity, the Emergence of a New Scientific field—Its Perspectives and Constraints. In: di Castri, F. and Younès, T. (Eds). **Biodiversity Science and Development. Towards a New Partnership**. CAB International & IUBS, Wallingford, UK, pp. 1-11.

di Castri, F., Verntus, J.R. and Younès, T. (Eds.). 1992. **Inventorying and Monitoring Biodiversity: A Proposal for an International Network.** Biol. Internat., Special issue No. 27.

Dickie, J.B., Linington, S. and Williams, J.T. (Eds.). 1984. **Seed Management Techniques for Gene Banks**. IBPGR, Rome.

Dinerstein, E. and Wikramanayake, E.D. 1993. Beyond hotspots: How to prioritize investments to conserve biodiversity in the Indo-Pacific region. Conser. Biol. 7: 53-65.

Dinerstein, E. Wikramanayake, E.D. and Forney, M. 1995. Conserving the reservoirs and remnants of tropical moist forest in the Indo-Pacific. In: Primack R. and Lovejoy T.E. (Eds.). **Ecology, Conservation and Management of Southeast Asian Rainforest**. Yale Univ. Press, New Haven, CT,pp. 140-175.

Dinerstein, E., Olson, D.M. , Graham, D.J., Webster, A.L. Bookbinder, M.A., and Ledec, G. 1995b. **A Conservation Assessment of the Terrestrial Ecoregions of Latin America and Caribbean.** The World Bank, Washington DC.

Dixon, D.A. and Sherman, P.B. 1990. **Economics of Protected Areas: A New Look at Benefits and Costs**. Island Press, Washington DC.

Dixon, J.A., Carpenter, R.A., Fallon, L.A., Sherman, P.B. and Manipomoke, S. 1988. **Economic Analysis of the Environmental Impact of Development Projects**. Earthscan, London.

Dixon, P.S. 1982. Rhodophycota. In: Parker, S.P. (Ed.) **Classification of Living Organisms**. McGraw-Hill, New York, NY, pp. 62-79.

Doebley, J., Stec, A. and Gustus, C. 1995. *Teosinte branched 1* and the origin of maize: evidence for epistasis and the evolution of dominance. Genetics **141**: 333-346.

Dong, J. and Wagner, D.B. 1993. Taxonomic and population differentiation of mitochondrial diversity in *Pinus banksiana* and *Pinus contorta*. Theoret. Appl. Gen. **86**: 573-578.

Drury, W.H. 1974. Rare species. Biol. Conser. **6**: 162-169.

Dudash, M.R. 1990. Relative fitness of selfed and out-crossed progeny in a self-compatible, protandrous species, *Sabatia angularis* L. (Gentianaceae): A comparison in three environments. Evolution **44**: 1129-1139.

Duff, F. 1997. Overview of the UNEP\GEF Biodiversity Data Management Project (BDM). In: Hawksworth, D.L., Kirk, P.M. and Dextre Clarke, S. (Eds.). **Biodiversity Information: Needs and Options**. CAB International, Wallingford, UK, pp. 115-123.

Dugan, P.J. 1990. **Wetland Conservation: a Review of Current Issues and Required Action**. IUCN, Gland, Switzerland.

Duque, E., Haidour, A., Godoy, F. and Ramos, J.L. 1993. Construction of *Pseudomonas* hybrid strain that mineralizes 2,4,6-trinitrotoluene. J.Bacteriol. **175**: 2278-2283.

Duthie, D. 1997. The Science of Conservation Biology and the extinction crisis. In: Pushpangandan, P., Ravi, K. and Santhosh, V. (Eds.). **Conservation and Economic Evaluation of Biodiversity,** vol.1. Oxford & IBH Publishing Co., Pvt. Ltd., New Delhi, pp. 31-46.

Duvick, D.N. 1984. Genetic contributions to yield gains of US hybrid maize, 1930 to 1980. In: Fehr, W.R. (Ed.). **Genetic Contributions to Yield Gains of Five Major Crop Plants.** Crop Science Society of America, Special Publication 7, Madison, WI, pp. 15-47.

Edge, T. 1994. Genetically modified organisms. In: **Biodiversity in Canada: A Science Assessment**.156. Environment Canada, Canadian Wildlife Service, Ottawa.

Edwards, C. (Ed.). 1990. **Microbiology of Extreme Environments**. Open Univ. Press, Milton Keynes, UK.

Edwards, J.L., Lane, M.A. and Nielsen, E.S. 2000. Interoperability of biodiversity data bases: Biodiversity Information on every desktop. Science **289**: 2314-2318.

Ehrlich, P.R. 1986. Extinction: what is happening now and what needs to be done. In: Elliott, D.K. (Ed.). **Dynamics of Extinction**. Wiley-Interscience, New York, pp.157-164.

Ehrlich, P.R. 1988. The loss of diversity. In: Wilson E.O. and Peters, F.M. (Eds.). **Biodiversity**. National Academy Press, Washington DC, pp. 21-27.

Ehrlich, P.R. and Ehrlich, A.H. 1981. **Extinction: The Causes and Consequences of the Disappearance of Species.** Random House, New York.

Ehrlich, P.R. and Wilson, E.O. 1991. Biodiversity studies: Science and Policy. Science **253**: 758-762.

Eisner, T. 1992. Chemical prospecting: a proposal for action. In: Bormann, F.H. and Kellert, S.R. (Eds.). **Ecology, Economics and Ethics: The Broken Circle**. Yale Univ. Press, New Haven, CT.

Eldridge, K., Davidson, J., Harwood, C. and Van Ivyk, G. 1993. **Eucalypt Domestication and Breeding.** Clarendon Press, Oxford.

Elliot, R. 1992. Intrinsic value, environmental obligation and naturalness. Monist **75**: 138-160.

Ellis, R.H., Hong, T.D. and Roberts, E.H. 1985. **Handbook of Seed Technology for Gene Banks** (2 vols.). IBPGR, Rome.

Engel, J.R. and J.G. Engel (Eds.) 1990. **Ethics of Environment and Development**. Belhaven, London.

English Nature, 1993. **RECORDER Specification.** English Nature, Peterborough, UK. .

Erickson, B.D. and Mondello, F.J. 1993. Enhanced biodegradation of polychorinated biphenyls after site-directed mutagenesis of a biphenyl dioxygenase gene. Appl. Environ. Microbiol. **59**: 3858-3962.

Erwin, D.H. 1996. The geological history of diversity. In: Szaro, R.C. and Johnston, D.W. (Eds.). **Biodiversity in Managed Landscapes.** Oxford Univ. Press, Oxford, pp. 3-16.

Erwin, T. 1982. Tropical forests: their richness in Coleoptera and other arthropod species. Coleopterists Bull. **36**: 74-75.

Esquirel, M. and Hammer, K. 1992. The Cuban homegarden 'Conuco': A perspective environment for evolution and *in situ* conservation of plant genetic resources. Gen. Resour. Crop Evol. **39** : 9-22.

Falk, D.A. 1987. Endangered species in botanic gardens. In. Elias, T.S. (Ed.). **Conservation and Management of Rare and Endangered Plants.** California Native Plant Society, Sacramento, CA, pp. 553-562.

FAO (Food and Agricultural Organization), 1975. **Reports of the Sixth Session of the FAO Panel of Experts on Plant Exploration and Introduction.** FAO, Rome.

FAO, 1988. **An Interim Report on the State of Forest Resources in Developing Countries**. FAO, Rome.

FAO, 1990. **Outline for a Draft on an International Convention on Conservation and Sustainable**

Use of Biological Diversity. Draft 6, August 1990. FAO, Rome.

FAO, 1991a. **Interim Report on Forest Resources Assessment 1990 Project.** Committee on Forestry, Tenth Session. FAO, Rome.

FAO, 1991b. **Second Interim Report on the State of Tropical Forests by Forest Resources Assessement 1990 Project.** Tenth World Forestry Congress, Sep 1991, Paris. FAO, Rome.

FAO, 1991c. Strategies for the establishment of a network of *in situ* conservation areas. Forest Genetic Resources Information **19**: 3-8. FAO, Rome.

FAO, 1993. **Tropical Forest Resources Assessment.** FAO, Rome.

FAO, 1995. **Report of the Sixth Session of the Commission on Plant Genetic Resources.** 19 - 30 June 1995, Document CPGR-6/95 REP. FAO, Rome.

FAO, 1996a. **The State of the World's Plant Genetic Resources for Food and Agriculture. Background Documentation Prepared for the International Technical Conference on Plant Genetic Resources.** Leipzig, Germay, 17-23 June 1996. FAO, Rome.

FAO, 1996b. **Report of the International Technical Conference on Plant Genetic Resources.** Leipzig, Germany, 17-23 June 1996. Document : ITCPGR/96/REP.. FAO, Rome.

FAO, 1997. **Background Documentation Provided by the International Union for the Protection of New Varieties of Plants** (UPOV). Commission on Genetic Resources for Food and Agriculture, CGRFA-7/97/Inf.5, Rome.

FAO/UNEP, 1981. **Tropical Forest Resources Assessment Project (in the framework of GEMS).** FAO, Rome.

Farnsworth, N.R. 1988. Screening plants for new medicines. In: Wilson, E.O. and Peters, F.M. (Eds.). **Biodiversity.** National Academy of Sciences, Washington DC, pp. 212-216.

Favre, D.S. 1989. **International Trade in Endangered Species: A Guide to CITES.** Martinus Nijhoff, Dordrecht.

Fiedler, P.L. 1986. Concepts of rarity in vascular plant species, with special reference to the genus *Calochortus* Pursh (Liliaceae). Taxon **35** : 502 - 518.

Fiedler, P.L., and Ahouse, J.J. 1992. Hierarchies of cause: Toward an understanding of rarity in vascular plant species. In: Fiedler, P.L. and Jain, S.K. (Eds.). **Conservation Biology.** Chapman & Hall, New York, NY, pp. 23-47.

Finlayson, M. and Moses, M. (Eds.). 1991. **Wetlands.** Facts on File Limited, Oxford.

Finnerty, W.R. 1994. Biosurfactants in environmental biotechnology. Curr. Opinions Biotech. **5**: 291-295.

Fitter, R. 1986. **Wildlife for Man: How and Why We Should Conserve Our Species.** Collins, London.

Flannery, K.V. 1969. Origins and ecological effects of early domestication in Iran and the Near East. In: Ucko, P.J. and Dimbleby, G.W. (Eds.). **The Domestication and Exploitation of Plants.** Duckworth, London, pp. 73-100.

Fliermans, C.B. and Balkwill, D.L. 1989. Microbial life in deep terestrial surface. BioScience **39** : 370-377.

Florention, J.M., Maurrasse, R., and Sen, G. 1991. Impocts, tsunamis, and the Haitian Cretaceous-Tertiary boundary layer. Science **252** : 1690-1693.

Folch, R. 1996. Biodiversity in urban and peri - urban zones. In: di Castri, F. and Younès, T. (Eds.). **Biodiversity, Science and Development. Towards a New Partnership.** CAB International & IUBS, Wallingford, pp. 539-542.

Folke, C., Hammer,M. and Jansson, A.M. 1991. Life-support value of ecosytems: a case study of the Baltic region. Ecol. Econ. **3**: 123-137.

Ford-Lloyd, B. and Jackson, M. 1986. **Plant Genetic Resources: An Introduction to their Conservation and Use.** Edward Arnold, London.

Forman, R.T.T. and Godron, M. 1986. **Landscape Ecology.** John Wiley and Sons, New York, NY.

Forrero, E. 1988. Botanical exploration and phytogeography of Colombia: Past, present and future. Taxon **37** : 561-566.

Foster, K.R., Vecchia, P. and Repachali, M.H. 2000. Science and the precautionary principle. Science **288**: 979-981.

Fowler, C. 1994. **Unnatural Selection: Technology, Politics and Plant Evolution.** Gordon and Breach Science Publishers, Yverdon, Switzerland.

Fowler, C. and Mooney, P. 1990. **Shattering: Food, Politics, and the Loss of Genetic Diversity.** Univ. Arizona Press, Tucson, AZ.

Frankel, O.H. 1970. Variation - the essence of life. Proc. Linn. Soc., New South Wales, 95: 158-169.

Frankel, O.H. and Bennett, E. 1970. **Genetic Resources in Plants—Their Exploration and Conservation**. IBP Handbook No.11. Blackwell Scientific Publications, Oxford.

Frankel, O.H. and Brown, A.H.D. 1984. Current plant genetic resources - a critical appraisal. **Proc. XV Intl. Congr. Genetics**. Oxford & IBH Publishing Co., New Delhi, pp. 3-11.

Frankel, O.H., Brown, A.H.D. and Burdon, J.J. 1995. **The Conservation of Plant Biodiversity**. Cambridge Univ. Press, Cambridge.

Frankel, O.H. and Soulé, M.E. 1981. **Conservation and Evolution.** Cambridge Univ. Press, Cambridge, UK.

Franklin, I.R. 1980. Evolutionary change in small populations. In: Soulé, M. (Ed.). **Conservation Biology: An Evolutionary Ecological Perspective.** Sinauer Associates, Sunderland, pp. 135-150.

Franklin, J.F. 1988. Structural and functional diversity in temperate forests. In: Wilson, E.O. and Peters, F.M. (Eds.). **Biodiversity**. National Academy Press, Washington DC, pp. 166-175.

Freedman, B. 1989. **Environmental Ecology : The Impacts of Pollution and Other Stresses in Ecosystem Structure and Function**. Academic Press Inc., San Diego, CA.

Friis-Hansen, E. 1994. Conceptualizing *in situ* conservation of land reces. In: Krattiger, A.P., McNeely, J., Lesser, W., Miller, K., St. Hill, Y. and Senanayake, R. (Eds.). **Widening Perspectives**. IUCN/IAE, Gland, Switzerland, Chapter 5-3.

Fuerst, J.A. and Hugenholtz, P. 2000. Micro organisms should be high on DNA preservation list. Science **290**: 1503.

Fuller, W.A. 1987. Synthesis and recommendation. In: Fitter, R. and Fitter, M. (Eds.). **The Road to Extinction.** IUCN/UNEP, Gland, Switzerland.

Furukawa, K., Hirose, J., Suyama, A., Zaiki, T. and Hayashida, S. 1993. Gene components responsible for discrete substrate specificity in the metabolism of biphenyl (bph operon) and tolune (tod operon). J. Bacteriol. **175**: 5224-5232.

Furtado, J.I. 1987. Some Problems. In: Fitter, R. and Fitter, M. (Eds.). **The Road to Extinction**. IUCN/UNEP, Gland, Switzerland, pp. 62-63.

Gadgil, M., Berkes, F. and Folke, C. 1993. Indigenous knowledge for biodiversity conservation. Ambio **22**: 151-156.

Gadgil, M., Devasia, P. amd Seshagiri Rao, P.R. 1995. A comprehensive framework for nurturing practical ecological knowledge. Centre for Ecological Science, Indian Institute of Science, Bangalore, Karnataka, pp. 1-74.

Ganeshaiah, K.N., Chandrashekara, K., and Kumar, A.R.V. 1997. Avalanche Index: A measure of biodiversity based on biological heterogeneity of the communities. Curr. Sci. **73**: 128-133.

Ganeshaiah, K.N., Uma Shaanker, R. and Bawa, K.S. 2001. Conservation of forest genetic resources of a region: combining species-centered and ecosystem-based approaches. In: Uma Shaanker, R., Ganeshiah, K.N. and Bawa, K.S. (Eds.). **Forest Genetic Resources: Status, Threats and Conservation Strategies**. Oxford & IBH Publishing Co. Pvt. Ltd., New Delhi, pp. 273-281.

Garrod, G. and Willis, K. 1991. Some empirical estimates of forest amenity value. Working paper 13, Countryside Change Centre, Univ. New Castle-upon-Tyne, England.

Gaston, K.J., Williams, P.H. Eggketib, P. and Humphries, C.J. 1995. Large scale patterns of biodiversity: Spatial variation in family richness. Proc. Roy.Soc. Lond. **260** : 149-154.

Gentry, A.H. 1988. Changes in plant community diversity and floristic composition on environmental and geographical gradients. Ann. Missouri Bot. Gard. **75**: 1-34.

Gentry, A.H. 1991. Biological extinction in Western Ecuador. Ann. Missouri Bot. Gard. **75** : 273-295.

Gentry, A.H. 1996. Species extirpations and current extinction rates: A review of the evidence. In: Szaro, R.C. and Johnston, D.W. 1996. (Eds.). **Biodiversity in Managed Landscapes** . Oxford Univ. Press, Oxford, pp. 17-26.

Gepts, P. and Clegg, M.T. 1989. Genetic diversity in pearl millet [*Pennisetum glaucum* (L.) R. Br.] at the DNA sequence level. J. Heredity. **80**: 203-208.

Gibbons, A. 1992. Conservation biology in the fast lane. Science **255**: 20-22.

Gilpin, M.E. and M.J. Soulé 1986. Minimum viable populations: Processes and species extinctions. In: Soulé M.J. and Wilcox, B.A. (Eds.). **Conservation Biology**. Sinauer, Associates, Sunderland, Mass., USA pp. 19-34.

Given, D.R. 1984. Monitoring and science— the next stage in threatened plant conservation in New Zealand. In: Given, D.R. (Ed.). **Conservation of Plant Species and Habitats.** Nature Conservation Council, Wellington, New Zealand, pp. 83-102.

Given, D.R. 1996. **Principles and Practice of Plant Conservation.** Timber Press Inc., Portland, OR (2nd ed.).

Glaszmann, J.C. 1987. Isozymes and classification of Asian rice varieties. Theoret. Appl. Gen. **74:** 21-30.

Glaszmann, J.C. 1988. Geographical pattern of variation among Asian rice native cultivars (*Oryza sativa*) based on fifteen isozyme loci. Genome **30** : 782-792.

Glowka, L., Burhenne-Guilmin, F. and Synge, H. 1994. **A Guide to the Convention on Biological Diversity.** IUCN, Gland, Switzerland.

Good, R.B. and Leigh, J.H. 1986. **Guidelines for the Formulation of Uniform Flora Legislation in all States.** Report to Australia CONCOM Ad Hoc Working Group on Endangered Flora.

Goode, D. 1993. Local authorities and Urban conservation. In: Goldsmith, F.B. and Warren, A. (Eds.). **Conservation in Progress.** John Wiley and Sons, Chichester, UK, pp. 335-345.

Gove, J.N., Patil, G.P., and Talle, C. 1996. Diversity measurement and comparison with examples. In: Szaro, R.C., and Johnston, D.W. (Eds.). **Biodiversity in Managed Landscapes.** Oxford Univ. Press, Oxford, pp. 157-175.

Grabherr, G., Gottfriend, M. and Pauli, H. 1994. Climatic effects on mountain plants. Nature **369:** 448.

Grassle, J.F. and Maciolek, N.J. 1992. Deep-sea species richness: regional and local diversity estimates from quantitative bottom samples. Amer. Natur. **193:** 313-341.

Grassle, J.F., Patil, G.P., Smith, W.K. and Talle, C. (Eds.) 1979. **Ecological Diversity in Theory and Practice. Statistical Ecology:** Vol 6. International Co-operative Publishing House, Burtonsville.

Gray, J. 1985. The microfossil record of early land plants: advances in understanding of early terrestrialization. 1970-1984. Phil. Trans. Roy. Soc. Lond. **309** : 167-195.

Green, D.G. and Croft, J.R. 1994. Proposal for implementing a biodiversity information network. In: **Linking Mechanisms for Biodiversity Information.** Proceedings Workshop for the Biodiversity Information Network, Base de Dados Tropical, Campinas, São Pauli, Brazil.

Gregg, W.P. Jr. 1988. On wilderness, national parks and biosphere reserves. In: **Fourth World Wilderness Congress on Worldwide Conservation.** Proceedings of the Symposium on Biosphere Reserves. U.S. Dept. of Interior, Washington DC.

Gren, I.M., Folke, C., Turner, R.K. and Bateman, I.J. 1994. Primary and secondary values of wetland ecosystems. Envir. Res. Econ. **4:** 55-74.

Groom, M.J. and N. Schumaker. 1993. Evaluating landscape change: Patterns of worldwide deforestation and local fragmentation. In: Kareiva, P.M., Kingsolver, J.G. and Huey, R.B. (Eds.). **Biotic Interaction and Global Change.** Sinauer Associates, Sunderland, Mass, USA, pp. 24-44.

Groombridge, B. (Ed.). 1992. **Global Biodiversity— Status of the Earth's Living Resources.** Chapman & Hall, London.

Groves, R.H. 1981. **Australian Vegetation.** Cambridge Univ. Press, Cambridge.

Hamann, O. 1991. The joint IUCN-WWF Plants Conservation Programme and its interest in medicinal plants. In: Akerele, O., Heywood, V. and Synge, H. (Eds.). **The Conservation of Medicinal Plants.** Cambridge Univ. Press, Cambridge, pp. 13-22.

Hammer, M., Jansson, A.M. and Jansson, B.O. 1993. Diversity, change and sustainability: implications for fisheries. Ambio **22:** 97-105.

Hammond, P.M. 1992. Species inventory. In: Groombridge, B. (Ed.). **Global Biodiversity- Status of the Earth's Living Resources.** Chapman & Hall, London, pp. 17-39.

Hammond, P.M. 1995. Described and estimated species number: an objective assessment of current knowledge. In: Allsopp, D., Colwell, R.R. and Hawksworth, D.L. (Eds.). **Microbial Diversity and Ecosystem Function.** CAB International, Wallingford, UK, pp. 29-71.

Hamrick, J.L. and Godt, M.J.W. 1989. Allozyme diversity in plant species. In: Brown, A.H.D., Clegg, M.T., Kahler, A.L. and Weir, B.S. (Eds.).

Plant Population Genetics, Breeding, and Genetic Resources. Sinauer Associates, Sunderland, Mass, USA, pp.43-63.

Hanks, J. (Ed.). 1984. **Traditional Life-styles, Conservation and Rural Development.** Commission on Ecology Paper No. 7. IUCN, Gland, Switzerland.

Hanley, N. 1989. Valuing rural recreation benefits: and empirical comparison of two approaches. J. Agric. Econ. **40**: 361-374.

Hanson, J. 1985. **Procedures for Handling Seeds in Gene Banks.** IBPGR, Rome.

Hardon, J.J., Vosman, B. and Van Hintum, Th.J.L. 1994. Identifying genetic resources and their origin: the capabilities and limitations of modern biochemical and legal systems. Commission on Plant Genetic Resources. Background Study Paper No.3. FAO, Rome.

Harlan, J.R. 1971. Agricultural origins: centres and non-centres. Science **174**: 468-474.

Harlan, J.R. 1975a. **Crops and Man.** American Society of Agronomy, Madison, WI.

Harlan, J.R. 1975b. Our vanishing genetic resources. Science 188: 618-621.

Harlan, J.R. 1976. Genetic resources in wild relatives of crops. Crop Sci. **16**: 329-333.

Harlan, J.R. and de Wet, J.M.J. 1971. Toward a rational classification of cultivated plants. Taxon **20**: 509-517.

Harper, J.L. and Hawksworth, D.L. 1994. Biodiversity: Measurement and estimation. Preface. Phil. Trans. Roy. Soc., Lon. **345**: 5-12.

Harris, D.R. 1969. Agricultural systems, ecosystems and the origin of agriculture. In: Ucko, P.J., and Dimbleby, G.W. (Eds.). **The Domestication and Exploitation of Plants and Animals.** Duckworth, London, pp. 3-15.

Harris, D.R. and Hillman, G.C. 1989. Introduction. In: Harris, D.R. and Hillman, G.C. (Eds.). **Foraging and Farming: the Evolution of Plant Exploitation.** Unwin Hyman, London, pp.1-8.

Harris, L.D. and G. Silva-Lopez 1992. Forest fragmentation and the conservation of biological diversity. In: Fiedler, P.L. and Jain, S.K. (Eds.). **Conservation Biology.** Chapman & Hall, New York, pp. 197-237.

Harwood, M. 1982. Myth of extinction. Audubon Mag. **84**: 18-21.

Hawkes, J.G. 1970. Potatoes. In: Frankel, O.H. and Bennett, E. (Eds.). **Genetic Resources in Plants—Their Exploration and Conservation.** IBP Handbook No.11. Blackwell Scientific Publications, Oxford, pp. 311-319.

Hawkes, J.G. 1983. **The Diversity of Crop Plants.** Harvard Univ. Press, Cambridge, MA.

Hawkes, J.G. 1987. **A Feasibility Study for the Preparation of a List of Economic Plants for Conservation and Development.** Report to IUCN/WWF.

Hawksworth, D.L. 1991. The fungal dimension of biodiversity: magnitude, significance, and conservation. Mycol. Res. **95**: 641-655.

Hawksworth, D.L. (Ed.). 1994. Biodiversity: measurement and estimation. Phil. Trans. Roy. Soc., Lon. **B. 345**: 1-136.

Hayden, B.P., Ray, G.C. and Dolan, R. 1984. Classification of coastal and marine environments. Environ. Conser. **11**: 199-207.

Heip, C. 1996. Biodiversity in marine sediments. In: di Castri, F. and Younès, T. (Eds.). **Biodiversity, Science and Development. Towards a New Partnership.** CAB International & IUBS, Wallingford, UK, pp. 139-148.

Hellawell, J.M. 1991. Development of a rationale for monitoring. In: Goldsmith, F.B. (Ed.). **Monitoring for Conservation and Ecology.** Chapman & Hall, London, pp. 1-14.

Heywood, V.H. , 1987. The changing role of botanic gardens. In: Bramwell, D., Hamann, O., Heywood, V.H. and Synge, H., (Eds.). **Botanic Gardens and the World Conservation Strategy.** Academic Press, London, pp. 3-18.

Heywood, V.H. 1990. Botanic gardens and the conservation of plant resources. Impact of Science on Society **158**: 121-132.

Heywood, V.H. 1991a. Botanic gardens and the conservation of medicinal plants. In: Akerele, O., Heywood, V. and Synge, H. (Eds.). **The Conservation of Medicinal Plants.** Cambridge Univ. Press, Cambridge, pp. 213-218.

Heywood, V.H. 1991b. The needs for stability of nomenclature in conservation. In: Hawksworth, D.L. (Ed.). **Improving the Stability of Names: Needs and Options** (Regnum Vegetabile No. 123). Koeltz Scientific Books, Konigstein, Germany, pp. 53-58.

Heywood, V.H. 1992. Conservation of germplasm of wild plant species. In: Sandlund, O.T., Hindar, K. and Brown, A.H.D. (Eds.). **Conservation of Biodiversity for Sustainable Development**. Scandinavian Univ. Press, Oslo, pp. 189-203.

Heywood, V.H. 1996. The importance of urban environments in maintaining biodiversity. In: di Castri, F. and Younès, T. (Eds.). **Biodiversity, Science and Development. Towards a New Partnership.** CAB International, Wallingford, UK, pp. 543-550.

Heywood, V.H. 1997a. Information needs in biodiversity assessment - From genes to ecosystems. In. Hawksworth, D.L., Kirk, P.M. and Dextre Clarke, S. (Eds.). **Biodiversity Information : Needs and Options**. CAB International, Wallingford, UK, pp. 5-20.

Heywood, V.H. 1997b. Biodiversity conservation: An eco-biogeographical approach. In: Pushpangadan, P., Ravi, K. and Santhosh, V. (Eds.). **Conservation and Economic Evaluation of Biodiversity**, vol. 1. Oxford & IBH Publ. Co. Pvt. Ltd., New Delhi, pp. 17-30.

Heywood, V.H. and Stuart, S.N. 1992. Species extinctions in tropical forests. In: Whitmore , T.C. and Sayer, J.A. (Eds). **Tropical Deforestation and Species Extinction**. Chapman & Hall, London, pp. 91-118.

Highet, K. 1992. The legal odyssey of the continental shelf: Is it a shelf? Is it a slope? Is it only a legal concept? Oceanus 35: 6-8.

Hobbs, R.J. 1992. The role of corridors in conservation: solution or bandwagon. Trends Ecol. Evol. 7: 389-392.

Hodgson, G. and G. Dixon. 1988. Measuring economic losses due to sediment pollution: logging versus tourism and fisheries. Trop. Coastal Area Management, pp. 5-8.

Holdridge, L. 1967. **Life Zone Ecology**. Tropical Science Centre, San Jose, Costa Rica.

Hollingsworth, M.L., Bailey, J.P. Hollingsworth, P.M., and Ferris, C. 1999. Chloroplast DNA variation and hybridization between invasive populations of Japanese knotweed and giant knotweed (*Fallopia*, Polygonaceae). Bot. J.Linn. Soc. 129: 139-154.

Hornby, R.J. 1992. Grasslands.. In: Groombridge, B. (Ed.). **Global Biodiversity, Status of the Earth's Living Resources**. Chapman & Hall, London, pp. 280- 292.

Houghton, R.A., Lefkowitz, D.S. and Skole, D.L. 1991. Changes in the landscape of Latin America between 1850 and 1985. I. Progressive loss of forests. For. Eco. Mgmt. 38: 143-172.

Huff, D.R., Peakall, R., and Smouse, P.E. 1993. RAPD variation within and among natural populations of outcrossing buffalo grass [*Buchloe dactyloides* (Nutt.) Englem.]. Theoret. Appl. Gen. 86: 927-934.

Hughes J.B., Daily G.C. and Ehrlich P.R. 1997. Population diversity: its extent and extinction. Science 278: 689-691.

Hughes, J.D. and Chandran, M.D.S. 1998. Sacred groves around the earth: an overview. In: Ramakrishnan, P.S. (Ed.). **Conserving the Sacred for Biodiversity Management**. Oxford & IBH Publishing Co. Pvt. Ltd., New Delhi, pp.69-85.

Hughes, J.D. and Swan, J. 1986. How much of the earth is sacred space? Environ. Rev. 10: 247-259.

Huntley, B.J. and B.H. Walker, 1982. Ecology of tropical savannas. In: Huntley, B.J. and Walker, B.H. (Eds.). **Ecological Studies** 42, Springer-Verlag, Heidelberg.

Hurka, H. 1994. Conservation genetics and the role of botanical gardens. In. Loeschcke, V., Tomiuk, J. and Jain, S.K. (Eds.). **Conservation Genetics.** Birkhausen Verlag, Basel, pp. 371-380.

Hurlbert, S.H. 1971. The non-concept of species diversity: A critique and alternative parameters. Ecology 52: 577 - 586.

Huttel, B. 1996. **Mikrosatelliten als Molekulare marker in der Kichererbs** (*Cicer arietinum* L.). Ph.D. thesis, University of Frankfurt, Germany (cited from Weising *et al.* 1998).

Hymowitz, T. 1972. The trans-domestication concept as applied to gaur. Econ. Bot. 26: 49-69.

IBP (International Biological Programme), 1966. **Plant Gene pools**. IBP News 5: 48-51.

IBPGR (International Bureau of Plant Genetic Resources), 1985. **Cost-effective Long-term Seed Stores**. IBPGR, Rome.

IBPGR, 1991. **Descriptors for Beta**. IBPGR/CGN, Rome.

Inglis, I.R. 1994. Nat. Res. 30: 2-4 (cited from Posey, D.D. 1997).

Inouye, D.W. and McGuire, A.D. 1991. Effect of snowpack on timing and abundance of flowering in *Delphinium nelsonii*: implication for climatic change. Am. J. Bot. **78**: 997-1001.

International Environment Education Program 1985 (cited from Given 1996)

ITF (International Transfer Format), 1987. **The International Transfer Format for Botanic Garden Plant Records. Plant Taxonomic Database Standards. No.1.** Hunt Institute for Botanical Documentation, Pittsburgh, PA.

ITTO (International Tropical Timber Organization), 1998a. **ITTO Criteria and Indicators for Sustainable Management of Natural Tropical Forests**. ITTO, Yokohama.

ITTO, 1998b. **Manual for the Application of Criteria and Indicators for Sustainable Management of Natural Tropical Forests (Parts A and B)**. ITTO, Yokonama.

ITTO 1990. **ITTO Guidelines for the Sustainable Management of Natural Tropical Forest**. ITTO Policy Development Series 1. ITTO, Yokohama. 18 pp.

ITTO, 1992. **Criteria for the Measurement of Sustainable Tropical Forest Management.** ITTO Policy Development Series 3. ITTO, Yokohama. 5 pp.

ITTO, 1993. **ITTO Guidelines on the Conservation of Biological Diversity in Tropical Production Products**. ITTO Policy Development Series 5. ITTO, Yokohama. 18 pp.

IUCN (International Union for the Conservation of Nature and Natural Resources), 1980. **World Conservation Strategy**. IUCN (in collaboration with WWF and UNEP), Gland, Switzerland.

IUCN, 1992. **Protected Areas of the World: A Review of National Systems**(4 vols.). WCMC, Cambridge and IUCN Commission on National Parks and Protected Areas. IUCN, Gland, Switzerland.

IUCN, 1993. **Draft IUCN Red List Categories.** IUCN, Gland, Switzerland.

IUCN, 1994a. **United Nations List of National Parks and Protected Areas**. WCMC, Cambridge and IUCN Commission on National Parks and Protected Areas. Gland, Switzerland.

IUCN, 1994b. **Guidelines for Protected Area Management Categories**. WCMC, Cambridge and IUCN Commission on National Parks and Protected Areas. Gland, Switzerland.

IUCN, 1995. **IUCN Red List Categories**. IUCN, Gland, Switzerland.

IUCN/UNEP/WWF. 1990. **Caring for the World. A Strategy for Sustainability** (second draft). IUCN/UNEP/WWF, Gland.

Jablonski, D. 1986. Background and mass extinctions: The alteration of macroevolutionary regimes. Science **213** : 129-133.

Jackson, L.L. 1992. The role of ecological restoration in conservation biology. In: Fiedler, P.L.and Jain, S. K. (Eds.). **Conseravtion Biology**. Chapman & Hall, New York, NY, pp. 433 - 451.

Jackson, L.L., McAuliffe, J.R. and Roundy, B.A. 1991. Desert restoration. Restoration and Management Notes **9** : 71-79.

Jackson, P.S.W. (Ed.) 1989. **Resolutions of the Second International Botanic Gardens Conservation Congress.** IUCN Botanic Gardens Congress, Kew Green, Richmond, UK.

Jackson, P.S.W. 1991. Developing a world network of botanic gardens. In: **Proceedings of the Conference—Protective Custody?—*Ex situ* Plant Conservation in Australia**. Australian National Botanic Gardens, Canberra.

Jacobs, J.W., Petroski, C., Friedman, P.A. and Simpson, E. 1990. Characterization of the anticoagulant activities from a Brazilian arrow poison. Thrombosis and Haemostasis **63**: 31-35.

Jacobs, M. 1988. **The Tropical Rain Forest: A First Encounter**. Springer-Verlag, Berlin.

Jana, S. 1993. Utilisation of biodiversity from *in situ* reserves, with special reference to wild wheat and barley. In: Damania, A.B. (Ed.). **Biodiversity and Wheat Improvement**. John Wiley and sons, Chichester. (cited from Maxted *et al.* 1997).

Janzen, D.H. 1986. The eternal threat. In: Soulé M.J. (Ed.). **Conservation Biology**. Sinauer Associates, Sunderland, Mass, USA, pp. 286-303.

Janzen, D.H. 1988. Tropical dry forests - the most endangered major tropical ecosystem. In: Wilson, E.O. and Peters, F.M. (Eds.). **Biodiversity.** National Academy Press, Washington DC, pp. 130-137,

Janzen, D.H. 1993a. Taxonomy: Universal and essential infrastructure for development and management of tropical wildland biodiversity.

In: Sandlund, O.T. and Schei, P.J. (Eds.). **Proc. Norway/UNEP Expert Conference on Biodiversity**, Trondheim, Norway, pp. 100-112.

Janzen, D.H. 1993b. Taxonomy: Universal and essential infrastructure for development and management of tropical wildland biodiversity. In: Sandlund, O.T., Hindar, K. and Brown, A.H.D. (Eds.). **Conservation of Biodiversity for Sustainable Development**. Scandinavian Univ. Press, Oslo, pp.100-113.

Janzen, D.H. and Gámez, R. 1997. Assessing information needs for sustainable use and conservation of biodiversity. In: Hawksworth, D.L., Kirk, P.M. and Dextre Clarke, S. (Eds.). **Biodiversity Information: Needs and Options**. CAB International, Wallingford, UK, pp. 21-29.

Jenkins, M. 1992. Species Diversity: An Introduction. In: Groombridge, B. (Ed.). **Global Biodiversity, Status of the Earth's Living Resources**. Chapman & Hall, London, pp. 40-46.

Jimenez, J.A. Hughes, K.A., Alaks, G., Graham, L. and Lacy, R.C. 1994. An experimental study of inbreeding depression in a natural habitat. Science **226**: 271-273.

John, R.W. and Tothill, J.C., 1985. Ecology and managment of world savannahs. Definition and geographic outline of savannah lands. In: Tothill, J.C. and Mott, J.C. (Eds.). **Ecology and Management of World Savannahs**. Australian Academy of Science, Canberra, pp. 1-13.

Johns, R.J. and Bellamy, A. 1979. **The Ferns and Fern Allies of Papua New Guinea**. Papua New Guinea Forestry College, Papua New Guinea.

Jones, T. and Cook, M.A. 1993. Proceedings First BioNET International Consultation. CAB International, London.

Joyce, C. 1994. **Earthly Goods: Medicine-Hunting in the Rainforest**. Little Brown & Co., Boston, MA.

Jutro, P.R. 1993. Human influence on ecosystems: dealing with biodiversity. In: McDonnel, M.J. and Pickett, S.T.A. (Eds.). **Humans as Components of Ecosystems**. Springer-Verlag, New York, pp. 246-256.

Kapos, V. 1989. Effects of isolation on the water status of forest patches in the Brazilian Amazon. J. Trop. Ecol. **5**: 173-185.

Keller, B.D. and Jackson, J.B.C. (Eds.). 1993a. **Long-term Assessment of the Oil-spill at Bahia Las Minas Panama. Synthesis Report**, vol. I.. **Executive Summary**. U.S. Department of the Interior, Minerals Management Service, New Orleans, LA.

Keller, B.D. and Jackson, J.B.C. (Eds.). 1993b. **Long-term Assessment of the Oil-spill at Bahia Las Minas Panama. Synthesis Report**, vol. II. **Technical Report**. U.S. Department of the Interior, Minerals Managements service, New Orleans, LA.

Keystone Center, 1991. **Final Consensus Report of the Keystone Policy Dialogue on Biological Diversity on Federal Lands**. The Keystone Center, Keystone, Brandon, Canada .

Khoshoo T.N. 1991. **Indian Biosphere and Geosphere**. Har-Anand Publications, New Delhi.

Khoshoo, T.N. 1995. Census of India's biodiversity: Tasks ahead. Curr. Sci. **69**: 14-17.

Khoshoo, T.N. 1996. Biodiversity in the Developing Countries. In: di Castri, F. and Younés, T. (Eds.). **Biodiversity, Science and Development. Towards a New Partnership**. CAB International, Wallingford, UK, pp. 304-311.

Khoshoo T.N. 1998. Sustaining development in developing countries. Curr. Sci. **75**: 652-660.

Kirsop, B.E.,and A. Doyle, (Eds.).1991. **Maintenance of Microorganisms**. Academic Press, London (2nd ed.).

Knoll, A.H. 1986. Patterns of change in plant communities through geological time. In: Diamond, J. and Case, T.J. (Eds.). **Community Ecology**. Harper and Row, New York, pp. 126-141.

Komen, J. 1991. Screening plants for new drugs. Biotech. Devt. Monitor **9**: 4-6.

Koopowitz, H. and Kaye, H. 1990. **Plant Extinction. A Global Crisis**. Christopher Helm, London (2nd ed.).

Krebs, C.J. 1972. **Ecology**. Harper and Row Publishers Inc., New York, NY.

Krishnamurthy, K.V. 1999. The *Nandavanas* of medieval India. Lecture, C.P.R. Foundation, Chennai, Tamil Nadu, India (mimeographed).

Krishnamurthy, K. V. and Upreti, D. K. 2001. Reproductive Biology of Lichens. In: Johri, B.M. and Srivastava, P.S. (Eds.) **Reproductive Biology of Plants**. Narosa Publishing House, New Delhi. pp. 127-147.

Krutilla, J.V. 1967. Conservation reconsidered. Amer. Econ. Rev. **57**: 778-786.

Kushalappa, C.G. and Bhagwat, S.A. 2001. Sacred groves: Biodiversity, threats and conservation. In: Uma Shaanker, R., Ganeshaiah, K.N. and Bawa, K.S. (Eds.). **Forest Genetic Resources: Status, Threats and Conservation Strategies**. Oxford & IBH Publishing Co. Pvt. Ltd., New Delhi, pp. 21-29.

Laird, S.A. 1993. Contracts for biodiversity prospecting. In: Reid W.V., Laird, S.A., Meyer, C.A., Gámez, R., Sittenfeld, A., Janzen, D.H., Gollin, M.A. and Juma, C. (Eds.) **Biodiversity Prospecting: Using Genetic Resources for Sustainable Development**. World Resources Institute Publications, Washington DC, pp. 99-130.

Lakshmikumaran, M., Srivastava, P.S. and Singh, A. 2001. Application of molecular marker technologies for genome analysis and assessment of genetic diversity in forest tree species. In: Uma Shaanker, R., Ganeshaiah, K.N. and Bawa, K.S. (Eds.). **Forest Genetic Resources: Status, Threats and Conservation Strategies**. Oxford & IBH Publishing Co. Pvt. Ltd., New Delhi, pp. 153-181

Lande, R. 1988. Genetics and demography in biological conservation. Science **241**: 1455-1460.

Lande, R. 1996. The meaning of quantitative genetic variation in evolution and conservation. In : Szaro, R.C. and Johnston, D.W. (Eds.). **Biodiversity in Managed Landscapes.** Oxford Univ. Press, Oxford, pp. 27-40.

Lande, R. and Barrowclough, G.F. 1987. Effective population size, genetic variation, and their use in population management. In: Soulé, M.J. (Ed.). **Viable Populations for Conservation.** Cambridge Univ. Press, Cambridge, pp.87-124.

Lanly, J.P., Singh, K.D., and Janz, K. 1991. FAO's 1990 reassessment of tropical forest cover. Nature and Resources **27**: 21-26.

la Riviere, J.W.M. 1989. Threats to the world's water. Sci. Amer. Sep. 1989: 80-94.

Larsen, K., Morley, B. and Ern, H. 1987. The role of the International Association of Botanic Gardens (IABG) in conservation world-wide. In: Bramwell, D., Hamann, O., Heywood, V.H. and Synge, H., (Eds.). **Botanic Gardens and the World Conservation Strategy.** Academic Press, London, pp. 277-284.

Lechowicz, M.J. 1987. Resistance of the Carbou lichen *Cladonia stellans* (Opiz.) Brodo to growth reduction by simulated acidic rain. Water, Air and Soil Pollution **34**: 71-77.

Ledig, F.T. 1992. Human impacts on genetic diversity in forest ecosystems. Oikos **63**: 87-108.

Leskien, D. and Flitner, M. 1997. **Intellectual Property Rights and Plant Genetic Resources: Options for a *Sui Generic* System**. Issues in Genetic Resources No.6. IPGRI, Rome.

Li, W. and Graur, D. 1991. **Fundamentals of Molecular Evolution.** Sinauer Associates, Sunderland, Mass, USA..

Liu, Z. and Furnier, G.R. 1993. Comparison of allozyme, RFLP and RAPD markers for revealing genetic variation within and between trembling aspen and bigtooth aspen. Theoret. Appl. Gen. **87**: 97-105.

Lopez-Moreno, I. (Ed.). 1993. **Ecologia Urbana Aplicada a la citudad de Xalapa**. Instituto de Ecologia, Xalapa, Mexico.

Loreau, M., Barbault, R., Rawanabe, H., Higashi, M., Alvarez-Buylla, E. and Renaud, F. 1995. Dynamics of biodiversity at the community and ecosystem level. In: UNEP. **Global Biodiversity Assessment**. Cambridge Univ. Press, Cambridge, pp. 245-274.

Lovejoy, T.E. 1980. A projection of species extinctions. In: **The Global 2000 Report to the President. Entering the Twenty-first Century**, vol. 2. Council on Environmental Quality, Washington DC, pp. 328-331.

Lovejoy, T.E. 1986. Species leave the arc one by one. In: Norton, B.G. (Ed.). **The Preservation of Species.** Princeton Univ. Press, Princeton, NJ, pp. 13-17.

Lovejoy, T.E., Bierregaard, R.O., Rylands, A.B., Malcolm, A.B., Quintela, C.E., Harper, L.H., Brown, K.S., Jr., Powell, A.H., Powell, G.V.N., Shubart, H.O.R. and Hays, M.B. 1986. Edge and other effects of isolation in Amazon forest fragments. In: Soulé, M. (Ed.). **Conservation Biology: The Science of Scarcity and Diversity**. Sinauer Associates, Sunderland,Mass, USA, pp. 257-285.

Lovelock, J.E. 1988a. The Earth as a living system. In: Wilson, E.O. and Peters, F.M. (Eds.). **Biodiversity.** National Academy Press, Washington DC, pp. 486-489.

Lovelock, J.E. 1988b. **The Ages of Gaia : A Biography of Our Living Earth.** W.W. Norton, New York.

Lovely, D.R., Woodward, J.C. and Chapelle, F.H. 1994. Stimulated anoxic biodegradation of aromatic hydrocarbons using Fe (III) ligands. Nature **370**: 128-131.

Lucas, G. and Synge, H. 1978. **The IUCN Plant Red Data Book.** IUCN, Gland, Switzerland.

Lucas, G.V. 1976. Conservation: recent developments in international co-operation and legislation. In: Simmons, J.B., Beyer, R.I., Brandham, P.E., Lucas, G.V. and Parry, T.H. (Eds.). **Conservation of Threatened Plants**. Plenum Press, New York, NY, pp. 271-277.

Lugo, A.E. 1988a. Diversity of tropical species: questions that elude answers. Biology International. Special Issue 19, p. 37.

Lugo, A.E. 1988b. Estimating reductions in the diversity of tropical forest species. In: Wilson, E.O. and Peters, F. M. (Eds.). **Biodiversity.** National Academy Press, Washington DC, pp. 58-70.

Lugo, A.E. 1996. Monitoring Biodiversity at Global scales. In: di Casttri, F and Younés, T. (Eds.). **Biodiversity, Science and Development. Towards a New Partnership**. CAB International & IUBS, Wallingford, UK, pp. 189-196.

Lugo, A.E. and Brown, S. 1996. Management of land and species richness in the tropics. In: Szaro, R.C. and Johnston, D.W. (Eds.). **Biodiversity in Managed Landscapes.** Oxford Univ. Press, Oxford, pp. 280-295.

Lugo, A.E. Brown, S. and Brinson, M.M. 1990. Concepts in wetland ecology. In: Lugo, A.E. Brinson, M. and Brown, S. (Eds.). **Ecosytems of the world 15.** Elsevier, Amsterdam, pp. 53-85.

Lugo, A.E., Parratta, J. and Brown, S. 1993. Reduction of species due to tropical deforestatiion and their recovery through management. Ambio **22**: 106-109.

Lyster, S. 1985. **International Wildlife Law**. Grotius, Cambridge, UK.

Mabey, R. 1973. **The Unofficial Countryside**. Collins, London.

MacArthur, J. 1997. The economic valuation of biodiversity, its implications and importance in bioresource planning, and initiations for its regular use in planning conservation projects in India. In: Pushpangandan, P., Ravi, K. and Santhosh, V. (Eds.). **Conservation and Economic Evaluation of Biodiversity.** Vol.2. Oxford & IBH Publ. Co.Pvt. Ltd., New Delhi, pp. 335-354.

MacArthur, R.H. 1985. Patterns of species diversity. Biol. Rev. **40**: 510-533.

MacArthur, R.H. and Wilson, E.O. 1963. An equilibrium theory of insular zoogeography. Evolution **17**: 373-387.

MacArthur, R.H. and Wilson, E.O. 1967. **The Theory of Island Biogeography**. Princeton Univ. Press, Princeton, NJ.

MacFie, C. 1987. In: Longman Paul (Ed.). **Touch: Nature Awareness Activities for Teachers, Leaders and Parents**. Auckland (cited from Given 1996).

Mackinnon, J. 1994. **A Method for Evaluating and Classifying Habitat Importance for Biodiversity Conservation**. WCMC/WCI Meeting on Identification of Habitat Criteria, 11-12 October 1994, Cambridge (cited from UNEP 1995).

Mackinnon, J. and Mackinnon, K. 1986. **Review of the Protected Areas System in the Afrotropical Realm.** IUCN, Gland, Switzerland.

Mackinnon, J., Mackinnon, K., Child, G. and Thorsell, J. 1986. **Managing Protected Areas in the Tropics**. IUCN/UNEP, Gland, Switzerland.

Magurran, A.E. 1988. **Ecological Diversity and its Measurement.** Princeton Univ. Press, Princeton, NJ.

Malhotra, K.C. 1998. Anthropological dimensions of sacred groves in India: An overview. In: Ramakrishnan, P.S. (Ed.). **Conserving the Sacred for Biodiversity Management**. Oxford & IBH Publishing Co. Pvt. Ltd., New Delhi, pp. 423-438.

Mangelsdorf, P.C. 1966. Genetic potentials for increasing yields of food crops and animals. Proc. Nat. Acad. Sci. USA **56**: 370-375.

Manilal, K.S. 1997. Taxonomy teaching in Universities and Colleges in India. Rheedea **7**: 51-55.

Mares, M.A. 1992. Neotropical mammals and the myth of Amazonian biodiversity. Science **255** : 976-979.

Margules, C.R., and Nicholls, A.O. 1988. Selecting network of reserves to maximise biological diversity. Biol. Conserv. **43**: 63-76.

Margules, C.R., Higgs, A.J. and Rafe, R.W. 1982. Modern biogeography theory: are there any lessons for nature reserve design? Biol. Conserv. **24**: 115-128.

Margulis, L. and Schwartz, K.V. 1988. **Five Kingdoms: An Illustrated Guide to the Phyla of Life on Earth**. W.H. Freeman and Co., San Francisco, CA.

Mathews, E. 1983. Global vegetation and land use: new high resolution data bases for climate studies. J. Climate Appl. Meterol. **23**: 474-487.

Matthews, E. and Fung, I. 1987. Methane emission from natural wetlands: global distribution, area and environmental characteristics of sources. Global Biogeochem. Cycles **1**: 61-86.

Maunder, M. 1992. Plant reintroduction : an overview. Biodiver. Conser. **1**: 51-61.

May, R.M. 1975. Island biogeography and the design of wildlife reserves. Nature **254**: 177-178.

May, R.M. 1992a. Bottoms up for the oceans. Nature, London. **357** : 278-279.

May, R.M. 1992b. How many species inhabit the Earth? Sci. Amer. pp. 18-24.

May, R.M. 1992c. Past efforts and future prospects towards understanding how many species there are. In: Solbrig, O.T., van Emden, H.M. and Van Oordt, P. G. W. J. (Eds.). **Biodiversity and Global Change.** IUBS Press, Paris, pp. 71-81.

May, R.M. 1994. Biological diversity : Differences between land and sea. Phil. Trans. Roy. Soc. Lond. Ser. B. **343**: 105-111.

May, R.M. 1995. Conceptual aspects of the quantification of the extent of biological diversity. In: Hawksworth, D.L. (Ed.). **Biodiversity Measurement and Estimation.** Chapman & Hall in Association with the Royal Society, London, pp. 13-20.

May, R.M. 2002. The future of biological diversity in a crowded world. Curr. Sci. **82**: 1325-1331.

May, R.M., Lawton, J.H. and Stork, N.E. 1995. Assessing extinction rates. In: Lawton, J.H. and May, R.M. (Eds.). **Extinction Rates.** Oxford Univ. Press, Oxford, pp. 1-24.

May, R.M. and Nee, S. 1995. The species alias problem. Nature **378**: 447-448.

May, R.M. and Tregonning, K. 1998. Global Conservation and UK government policy. In: Mace, G.M., Balmford, A. and Ginsberg, J.R. (Eds.). **Conservation in a Changing World.** Cambridge Univ. Press, Cambridge, pp. 287-301.

May, R.M. and Stumpf, M.P.H. 2000. Species-area relations in tropical forests. Science **290:** 2084-2086.

McCarty, P.L. 1993. *In situ* bioremediation of chlorinated solvents. Curr. Opinions Biotech. **4**: 323-330

McCouch, S. 1998. Toward a plant genomics initiative: thoughts on the value of cross-species and cross-genera comparison in the grasses. Proc. Nat. Acad. Sci. USA **95**: 1983-1985.

McIntyre, L. 1989. Last days of Eden. National Geographic **174**: 800-817.

McNeely, J.A. 1988. **Economics and Biological Diversity: Developing and Using Economic Incentives to Conserve Biological Resources.** IUCN, Gland, Switzerland.

McNeely, J.A. 1994. Protected areas for the 21st century: Working to provide benefits to society. Biodiver. Conser. **3** : 390-405.

McNeely, J.A. 1996. Conserving Biodiversity: The key political, economic and social measures. In. di Castri, F. and Younès, T. (Eds.). **Biodiversity, Science and Development. Towards a New Partnership**. CAB International & IUBS, Wallingford, UK, pp. 264-281.

McNeely, J.A., Miller, K.R. and Thirsell, J.W. 1987. Objectives, selection and management of protected areas in tropical forest habitats. In: Marsh, C. and Mittermeier, R.A. (Eds.). **Primary Conservation in the Tropical Rain Forest**. Alan R. Liss, Inc., New York, NY, pp. 181-204.

McNeely, J.A., Miller, K.R., Reid, W.V., Mittermeier, R.A. and Werner, T.B. 1990. **Conserving the World's Biological Diversity**, IUCN, Gland, Switzerland.

Meadows, D.H. 1990. Biodiversity: The key to saving life on earth. Land Stewardship Letter (summer) 4-5 (cited from Meffe and Carroll 1994).

Meffe, G.K. and Carroll, C.R. 1994. **Principles of Conservation Biology**. Sinauer Associates, Sunderland, Mass, USA .

Menaut, J. –C. 1983. The vegetation of African savannas. In: Bouliere, F. (Ed.). **Ecology and**

Conservation of Grassland Birds. Technical Publication No 7., ICBP, Cambridge, UK.

Menges, E.S. 1991. The application of minimum viable population theory to plants. In: Falk, D.A. and Holsinger, K.E. (Eds.). **Genetics and Conservation of Rare Plants**. Oxford Univ. Press, Oxford, pp. 45-61.

Merriam, G. 1991. Corridors and Connectivity: annual populations in heterogeneous environments. In: Saunders, D. and Hobbs, R. (Eds.). **Nature Conversation: The Role of Corridors**. Surrey Bealty & Sons, Chipping Norton, New South Wales, pp. 133-142.

Milchunas, D.G. and Lauenroth, W.K. 1993. Quantitative effects of grazing on vegetation and soil over a global range of enviroments. Ecol. Monographs **63**: 327-366.

Miller, C.I. and Tanksley, S.D. 1990. RFLP analysis of phylogenetic relationships and genetic variation in the genus *Lycopersicon*. Theoret. Appl. Gen. 80: 437-448.

Mittermeier, R.A. 1988. Primate diversity and the tropical forest: Case studies from Brazil and Madagascar and the importance of the megadiversity countries. In: Wilson, E.O. and Peters, F.M. (Eds.). **Biodiversity.** National Academic Press, Washington DC, pp. 145-154.

Mittermeier, R.A. and Werner, T.B. 1990. Wealth of plants and animals unites "Megadiversity Countries". Tropics **4**: 4 -5.

Mohn, W.W. and Tiedje, J.M. 1992. Microbial reductive dehalogenation. Microbiol. Rev. **56**: 482-507.

Mondor, C. and Kun, S. 1982. The long struggle to protect Canada's vanishing Prairie. Ambio **2**: 286-291.

Mooney, H.A. 1988. Lessons from Mediterranean-climate regions. In: Wilson, E.O. and Peters, F.M. (Eds.). **Biodiversity.** National Academy Press, Washington DC, pp. 157-165.

Mooney, H.A. 1996. Biotic Interactions and the Ecosystem Function of Biodiversity. In: di Castri, F. and Younès, T. (Eds.). **Biodiversity Science and Development. Towards a New Partnership.** CAB International & IUBS, Wallingford, UK, pp.153-165.

Moore, R.M. 1970. **Australian Grasslands.** Alexander Bros., Melbourne.

Mori, S.A. 1992. The Brazil nut industry: past, present and future. In: Plotkin, M.J. and Famolare, L.M. (Eds.). **Sustainable Harvest and Marketing of Rain Forest Products**. Island Press, Washington DC (cited from UNEP 1995).

Mosseler, A., Egger, K.N. and Hughes, G.A. 1992. Lower levels of genetic diversity in red pine confirmed by random amplified polymorphic DNA markers. Can. J. Forest Res. **22**: 1332-1337.

Moyle, P.B. and Leidy, P.B. 1992. Loss of biodiversity in aquatic ecosytems: Evidence from fish faunas. In: Fiedler, P.L. and Jain, S. K. (Eds.). **Conseravtion Biology**. Chapman & Hall, New York, NY, pp. 127 - 169.

Murphy, R.W. Sites, J.W. Jr., Buth, D.G. and Haufler, C.H. 1990. Proteins. I. Isozyme electrophoresis. In: Hillis, D.M. and Moritz, C. (Eds.). **Molecular Systematics.** Sinauer Associates, Sunderland, Mass., USA, pp. 45-126.

Myers, N. 1979. **The Sinking Arks: A New Look at the Problem of Disappearing Species**. Pergamon Press, New York, NY.

Myers, N. 1980a. The problem of disappearing species: What can be done? Ambio **9**: 229-235.

Myers, N. 1980b. **Conservation of Tropical Moist Forests.** National Academy of Sciences, Washington DC.

Myers, N. 1985a. A look at the present extinction spasm . In: R.J. Hoage, (Ed.). **Animal extinctions: What Everyone Should Know**. Smithsonian Institution Press, Washington DC, pp. 47-57.

Myers, N. (Ed.). 1985b. **The Gaia Atlas of Planet Management.** Pan, London.

Myers, N. 1988a. Threatened biotas : "Hotspots" in tropical forest. The Environmentalist **8(3)** : 187-208.

Myers, N. 1988b. Tropical forests and their species—going , going ...? In: Wilson. E.O. and Peters, T. (Eds.). **Biodiversity.** National Academy Press, Washington DC, pp. 218-235,

Myers, N. 1989. **Deforestation Rates in Tropical Countries and Their Climatic Implications.** Friends of the Earth, Washington DC.

Myers, N. 1990. The biodiversity challenge: Expanded hotspots analysis. The Environmentalist **10** : 243-256.

Myers, N. 1992. **Future Operational Monitoring of Tropical Forests: An Alert Strategy**. Joint

Research Centre, Commission of the European Community, Ispra, Italy.

Myers, N. 1994. Global Biodiversity II. Losses. In: Meffe, G.K. and Carroll, C.R. (Eds.). **Principles of Conservation Biology**. Sinauer Associates, Sunderland, Mass., USA, pp. 110-140.

Myers, N. 1998. Global biodiversity priorities and expanded conservation policies In: Mace, G.M., Balmford A. and Ginsberg B.R. (Eds.). **Conservation in a Changing World.** Cambridge Univ. Press, Cambridge, pp. 273-285.

Myers, N. 2000. The new millennium: An ecology and an economy of hope. Curr. Sci. 78: 686-693.

NASC (National Academy of Sciences), 1975. **Underexploited Tropical Plants With Promising Economic Value.** NASC, Washington DC.

Naresh Kumar 1998. Shifting paradigms of Industrial Microbiology. In: Sahai, S. (Ed.) **Microorganisms and Intellectual Property Rights.** Gene Campaign, New Delhi.

Navid, D. 1994. The legal development of the convention on wetlands: getting it right; or the importance of proper legal drafting 1-4, Ramsar Newsletter Special Issue, April. 1994.

Nayar, M.P. 1997. Conservation of rare and endangered species of Indian Flora: Strategies for botanic gardens. In: Pushpangadan, P., Ravi, K. and Santhosh, V. (Eds.). **Conservation and Economic Evaluation of Biodiversity.** Oxford & IBH Publishing Co. Pvt. Ltd., New Delhi, pp. 47-57.

Nayar, M.P. and Sastry, A.R.K. (Eds.). 1987, 1988, 1989. **Red Data Book of Indian plants** (vols. 1-3). Botanical Survey of India, Kolkata.

Nei, M. 1973. Analysis of gene diversity in subdivided populations. Proc. Nat. Acad. Sci., U.S.A. 70: 3321-3323.

Newbury, H.J. and Ford-Lloyd, B.V. 1997. Estimation of genetic diversity. In: Maxted, N., Ford-Lloyd, B.V. and Hawkes, J. G. (Eds.). **Plant Genetic Conservation—the *in situ* Approach.** Chapman & Hall, London, pp. 192-206.

Nicholson-Lord, D. 1987. **The Greening of the Cities.** Routledge and Kegan Paul, London.

Niklas, K.J., Tiffney, A.H. and Knoll, A.H. 1985. Patterns in vascular land plant diversification: analysis at the species level. In: Valentine, J.W. (Ed.). **Phanerozoic Diversity Patterns: Profiles in Macroevolution.** Princeton Univ. Press, Princeton, NJ, pp. 97-128.

Nisbet, L.J. and Fox, F.M. 1991. The importance of microbial biodiversity to biotechnology. In: Hawksworth, D.L. (Ed.). **The Biodiversity of Microorganisms and Invertebrates: Its Role in Sustainable Agriculture.** CAB International, Wallingford, UK, pp. 229-244.

Norse, E.A. (Ed.) 1993. **Global Marine Biodiversity. A Strategy for Building Conservation into Decision Making.** Island Press, Washington DC.

Norse, E.A. and McManus, R.E. 1980. Ecology and living resources biological diversity. In: **Environmental quality 1980.** Eleventh annual report of the council on environmental quality, Washington DC, pp. 31-80.

Norse, E.A., Rosenbaum, K.L., Wilcove, D.S., Wilcox, B.A., Romme, W.H., Johnston, D.W. and Stout, M.L. 1986. **Conserving Biological Diversity in our National Forests.** The Wilderness Society, Washington DC.

Norton, B.G. 1986. On the inherent danger of undervaluing species. In: Norton, B.G. (Ed.). **The Preservation of Species.** Princeton Univ. Press, Princeton, NJ, pp. 110-137.

Norton, B.G. 1987. **Why Preserve Natural Variety?** Princeton Univ. Press. Princeton.

Norton, B.G. 1991. **Towards Unity among Environmentalists.** Oxford Univ. Press, New York, NY.

Norton, B.G. 1994. Thoreau and Leopold on scientific values. In: Kim, C.K. and Weaver, R.D. (Eds.). **Biodiversity and Landscapes.** Cambridge Univ. Press, Cambridge, pp. 31-46.

Noss, R.F. 1990. Indicators for monitoring biodiversity: A hierarchical approach. Conser. Biol. 4: 355-364.

Noss, R.F. 1992. Issues of scale in conservation biology In: Fiedler, P.L. and Jain, S.K. (Eds.). **Conservation Biology.** Chapman & Hall, New York, NY, pp. 239-250.

Noss, R.F. 1994. Hierachial indicators for monitoring changes in biodiversity. Essay 4A. In: Meffe, G.K. and Carroll, C. R. (Eds.). **Principle of Conservation Biology.** Sinauer Associates, Sunderland, Mass., USA, , pp. 79- 80.

Noss, R.F. 1996. Conservation of Biodiversity at the landscape scale. In: Szaro, R.C., and Johnston,

D.W. (Eds.). **Biodiversity in Managed Landscapes.** Oxford Univ. Press, Oxford, pp. 574 - 589.

NRC (National Research Council), 1980. **Research Priorities in Tropical Biology.** National Academy of Sciences, Washington DC.

NRC, 1992. **Restoration of Aquatic Ecosystems.** National Academy Press, Washington DC.

NRC, 1993. **A Biological Survey for the Nation.** National Academy Press, Washington DC.

NRC, 1995. **Understanding Marine Biodiversity: A Research Agenda for the Nation.** National Academy Press, Washington DC.

O'Neal, M. and Walter, K.S. 1989. The BG-BASE users Manual: designing a computer-software application to meet the plant-record needs of the Arnold Arboretum. Arnoldia 49: 42-53.

Oakeson, R.J. 1986. A model for the analysis of common property probelms: In: **Proceedings Conference on Common Property Resource Management,** Washington DC. National Academy of Sciences, New York, NY, pp. 13-30.

Odum, E.P. 1992. Great ideas in ecology for the 1990s. BioScience 42: 542-545.

Odum, H.T. 1983. **Systems Ecology: An Introduction.** John Wiley and Sons, New York, NY.

OECD (Organization for Economic Cooperation and Development). 1996, **Intellectual Property, Technology Transfer and Genetic Resources. An OECD Survey of Current Practices and Policies.** OECD, Paris.

Oldfield, S. 1988. **Buffer Zone Management in Tropical Moist Forests : Case Studies and Guidelines.** IUCN, Gland, Switzerland.

Oldfield, S. 1992. Plant use. In: Groombridge, B. (Ed.). **Global Biodiversity-Status of the Earth's Living Resources.** Chapman & Hall, London, pp. 331-358.

Olembo, R.J. 1996. Bridge-building for Biodiversity. In: di Castri, F and Younès, T. (Eds.). **Biodiversity, Science and Development. Towards a New Partnership.** CAB International, Wallingford, UK, pp. 30-34.

Olivieri, S., Harrison, J. and Busby, J.R. 1995. Data and Information management and communication. In: UNEP, **Global Biodiversity Assessment.** Cambridge Univ. Press, Cambridge, pp. 607-670.

Olsen, J.S., Watts, J.A. and Allison, L.J. 1983. **Carbon in Live Vegetation of Major World Ecosystems.** Oak Ridge National Laboratory Technical Report ORNL - 5862. Oak Ridge, TN, USA .

OTA, 1987. Office of Technology Assessment. **Technologies to Maintain Biological Diversity.** OTA-F-330, Washington DC.

Palmer, J.D. 1991. Plastid chromosomes: Structure and evolution. **In:** Bogorad, L.and Vasil, I. K. (Eds.). **Cell Culture and Somatic Cell Genetics of Plants,** vol. VIIIA. Academic Press, San Diego, CA, pp. 5-53.

Parilla, G., Lavin, A., Bryden,H., Garcia, M. and Millard, R. 1994. Rising temperatures in the subtropical North Atlantic Ocean over the past 35 years. Nature **369:** 48-51.

Paroda, R.S., Kapoor, P., Arora, R.K., Mal, B. (Eds.) 1988. **Life Support Plant Species. Diversity and Conservation.** NBPGR, New Delhi.

Pastor, J. and Johnston, C.A. 1992. Using simulation models and geographic information systems to integrate ecosystem and landscape ecology. In: Naiman, R.J. (Ed.). **Watershed Management: Balancing Sustainability with Environmental Change.** Springer-Verlag, New York, pp. 324-246.

Pastor, J. and Post, W.M. 1988. Response of northern forests to CO_2 induced climate change. Nature **334:** 55-58.

Patterson, B, D. 1996. The 'Species alias' problem. Nature **380:** 589.

Paul, S. 1987. **Community Participation in Development Projects: the World Bank Experience. World Bank Discussion Paper 6.** World Bank, Washington DC.

Peacock, W.J. 1989. Molecular biology and genetic resources In: Brown, A.H.D., Frankel, O.H., Marshall, D.R. and Williams, J.T. (Eds.). **The Use of Plant Genetic Resources.** Cambridge Univ. Press, Cambridge, pp. 363-376.

Pearce, D.W. 1990. An economic approach to saving the tropical forest. In: **LEEC Discussion paper, London Environmental Economics Centre,** London, pp. 90-95.

Pearce, D.W. and D. Moran 1994. **The Economic Value of Biological Diversity.** Earthscan, London.

Perlin, J. 1989. **A Forest Journal: The Role of Wood in the Development of Civilization.** W.W. Norton and Co., New York, NY.

Perrings, C. 1997. The value of biodiversity for sustainable economic development. In: Pushpangadan, P., Ravi, K. and Santhosh, V. (Eds.). **Conservation and Economic Evaluation of Biodiversity**, vol.2. Oxford & IBH Publishing Co. Pvt. Ltd., New Delhi, pp. 317-334.

Perrings, C., Müler, K.-G., Folke, C., Holling, C.S. and Jansson, B.-O. (Eds.). 1994. **Biodiversity Loss: Ecological and Economic Issues.** Cambridge Univ. Press, Cambridge.

Peters, C., A. Gentry and R. Mendelsohn, 1989. Valuation of an Amazonian rainforest. Nature **339**: 655-656.

Pickersgill, B. 1989. Cytological and genetic evidence on the domestication and diffusion of crops within the Americas. In: Harris, D.R. and Hillman, G.C. (Eds.). **Foraging and Farming: the Evolution of Plant Exploitation.** Unwin Hyman, London, pp. 426-439.

Pielou, E.C. 1974. **Population and Community Ecology.** Gardon and Breach, New York, NY.

Pielou, E.C. 1975. **Ecological Diversity**. John Wiley and Sons, New York, NY.

Pielou, E.C. 1977. **Mathematical Ecology**. John Wiley and Sons, New York, NY.

Pimentel, D., Harvey, C., Resosudarmo, P., Sinclair, K., Kurz, D., McNair, M., Crist, S., Shpritz, L., Fitton, L., Saffouri, R., and Blair, R. 1994. **Environmental and Economic Costs of Soil Erosion and Conservation Benefits**. College of Agriculture and Life Sciences, Cornell Univ., Ithaca, New York (draft).

Pimm, S.L., Russell, G.J., Gittleman, J.L. and Brooks, T.M. 1995. The future of biodiversity. Science **269**: 347-350.

Pitt, D. 1987. Outline for a book on education and botanic gardens. Unpublished report to IUCN, Gland (cited from Given 1966).

Platt, R.H., Rowntree, R.A. and Muick, P.C. (Eds.). 1994. **The Ecological City. Preserving and Restoring Urban Biodiversity**. Univ. Massachusetts Press, Amherst, MA.

Posey, D.A. 1990. Intellectual Property Rights and just compensation for indigenous knowledge. Arthropology Today **6**: 13-16.

Posey, D.A. 1996. **Traditional Resources Rights: International Instruments for Protection and Compensation for Indigenous Peoples and Local Communities.** IUCN and International Books, Gland, Switzerland.

Posey, D.A. and Dutfield, G. 1996. **Beyond Intellectual Property: Towards Traditional Resource Rights for Indigenous Peoples and Local Communities.** IDRC, Ottawa, Canada.

Posey, D.A. and Dutfield, G. 1997. **Indigenous Peoples and Sustainability. Cases and Actions.** International Books, Utrecht.

Posey, D.A., Dutfield, G. and Plenderleith, K. 1995. Collaborative research and intellectual property rights. Biodiver. Conser. **4**: 892-902.

Posey, D.D. 1997. Wider use and applications of indigenous knowledge, innovations and practices. Information systems and ethical concerns. In: Hawksworth, D.L., Kirk, P.M. and Dextre Clarke, S. (Eds.). **Biodiversity Information—Needs and Options.** CAB International, Wallingford, UK, pp. 69-97.

Postgate, J. 1994. **The Outer Reaches of Life.** Cambridge Univ. Press, Cambridge.

Prance, G.T. 1977. Floristic inventory of the tropics: Where do we stand? Ann. Missouri Bot. Gard. **64**: 659-684.

Prance, G.T. 1997. The role of botanic gardens in the conservation of biodiversity. In: Pushpangadan, P., Ravi, K. and Santhosh, V. (Eds.). **Conservation and Economic Evaluation of Biodiversity,** vol.1. Oxford & IBH Publishing Co. Pvt. Ltd., New Delhi, pp. 3-16.

Prescott-Allen, R. and Prescott-Allen, C. 1982. **What's Wildlife Worth** ? Earthscan, London.

Prescott-Allen, R. and Prescott-Allen, C. 1988. **Genes from the Wild: Using Wild Genetic Resources for Food and Raw Materials.** Earthscan, London (2nd ed.).

Preston, F.W. 1962. The canonical distribution of commonness and rarity: Part 1. Ecology **43**: 185-215.

Primack, R.B. 1992. Tropical community dynamics and conservation biology. Bioscience **42**: 818-621.

Principle, P.P. 1991. Valuing the biodiversity of medicinal plants. In: Akerele, O., Heywood, V. and Synge, H. (Eds.). **The Conservation of Medicinal Plants**. Cambridge Univ. Press, Cambridge, pp. 79-124.

Qualset, C.O., Damania, A.B., Zanatta, A.C.A. and Brush, S.B. 1997. Locally based crop plant conservation. In: Maxted, N., Ford-Lloyd, B.V. and Hawkes, J.G., (Eds.). **Plant Genetic Conservation.** Chapman & Hall, London, pp. 160 - 175.

Rabinowitz, D. 1981. Seven forms of rarity. In: Synge, H. (Ed.). **The Biological Aspects of Rare Plant Conservation.** John Wiley and Sons, New York, NY, pp. 205-217.

Rabinowitz, D., Cairns, S. and Dillon, T. 1986. Seven forms of rarity and their frequency in the flora of the British Isles. In: M.E. Soulé, (Ed.). **Conservation Biology: The Science of Scarcity and Diversity.** Sinauer Associates, Sunderland, Mass., USA, pp. 182-204,

RAFI (Rural Advancement Foundation International), 1994. **Conserving Indigenous Knowledge: Integrating Two Systems of Innovation.** An independent study by RAFI commissioned by the United Nations Development Program, New York, NY.

Rahul and Jacob Nellithanam. 1998. Return of the Native Seeds. The Ecologist **28**: 29-33.

Ramakrishnan, P.S. 1992. International sustainable biosphere initiative: a participatory research agenda for India. Curr. Sci. **63**: 127-131.

Randall, A. 1991. Total and non-use values. In: Braden, J.B. and Kolstad, C.D. (Eds.). **Measuring the Demand for Environmental Quality.** Elsevier Science Publishers, Amsterdam, pp. 303-321.

Ratnam, W. 2001. Use of molecular markers to quantify genetic diversity parameters of forest trees. In: Uma Shaanker, R., Ganeshaiah, K.N., and Bawa, K.S. (Eds.). **Forest Genetic Resources: Status, Threats and Conservation Strategies.** Oxford & IBH Publishing Co. Pvt. Ltd., New Delhi, pp. 127-139.

Raup, D.M. 1978. Cohort analysis of genetic survivorship. Palaeobiol. **4**: 1-15.

Raup, D.M. 1979a. Biases in the fossil record of species and genera. Bull. Carnegie Museum Nat. Hist. 13: 85-91.

Raup, D.M. 1979b. Size of the Permo-Triassic bottleneck and its evolutionary implications. Science **206**: 217-218.

Raup, D.M. 1991a. **Extinction. Bad Genes or Bad Luck?** W.W. Norton & Co., New York, NY.

Raup, D.M. 1991b. A kill curve for phanerozoic marine species. Paleobiol. **17**: 37-48.

Raup, D.M. and Sepkoski, J.J. Jr. 1982. Mass extinctions in the marine fossil record. Science **215**: 1501-1503.

Raven, P.H. 1976. Ethics and attitude. In: Simmons, J.B. *et al.* (Eds.). **Conservation of Threatened Plants.** Plenum Press, New York, pp. 155-179 (cited from Gentry 1996).

Raven, P.H. 1987. The scope of the plant conservation problem world-wide. In: Bramwell, D., Hamann, O., Heywood, V.H. and Synge, H. (Eds.). **Botanic Gardens and the World Conservation Strategy.** Academic Press, London, pp. 19-29.

Raven, P.H. 1988a. Our diminishing tropical forests. In: Wilson, E.O. and Peters, F.M. (Eds.). **Biodiversity.** National Academy Press, Washington DC, pp. 119-122.

Raven, P.H. 1988b. Tropical floristics tomorrow. Taxon **37**: 549-560.

Raven, P.H. 1990. The politics of preserving biodiversity. Bioscience **40**: 769-774.

Ravi, K. and Pushpangadan, P. 1997. Application of environmental valuation techniques for economic evaluation biodiversity: A critical investigation: In: Pushpagadan, P., Ravi, K. and Santhosh, V. (Eds.). **Conservation and Economic Evaluation of Biodiversity,** vol. 2. Oxford & IBH Publishing & Co. Pvt. Ltd., New Delhi, pp. 355-369.

Ray, G.C. 1988. Ecological diversity in coastal zone and oceans. In: Wilson, E.O. and Peters, F.M. (Eds.). **Biodiversity.** National Academy Press, Washington DC, pp. 36-50.

Ray, G.C. 1996. Conservation of coastal-marine biological diversity. In. di Castri, F. and Younès, T. (Eds.). **Biodiversity, Science and Development. Towards a New Partnership.** CAB International & IUBS, Wallingford, UK, pp.224 - 225.

Ray, G.C., Hayden, B.P., Bulger, Jr., A.J. and McCormick-Ray, M.G. 1992. Effects of global warming on the biodiversity of coastal-marine zones. In. Peters, R.L. and Lovejoy, T.E. (Eds.) **Global Warming and Biological Diversity.** Yale Univ. Press, New Haven, CT, pp. 91-104.

Ray, G.C. and McCormick-Ray, M.G. 1989. Coastal and marine biosphere reserves. In. Gregg, Jr.

W.P., Krugman, S.L. and Woods, Jr. J.D. (Eds.). **Proc. Symp. Biosphere Reserves, Fourth World Wilderness Conference**. US National Park Service and US MAB, Washington DC, pp. 68-78.

Redford, K.H., Taber, A. and Simonetti, J.A. 1990. There is more to biodiversity than the tropical rain forests. Conserv. Biol. **4**: 328-330.

Reid, W.V. 1992. How many species will there be? In: Whitmore, T. and Sayer. J. (Eds.). **Tropical Deforestation and Species Extinction.** Chapman & Hall, London, pp. 55-74.

Reid, W.V., Laird, S.A., Meyer, C.A., Gámez, R., Sittenfeld, A., Janzen, D.H., Gollin, M.A., and Juma, C. 1993. **Biodiversity Prospecting: Using Genetic Resources for Sustainable Development**. WRI, Washington DC.

Reid, W.V., McNeely, J.A., Tunstall, D.B., Bryant, D.A. and Winograd, M. 1993. **Biodiversity Indicators for Policymakers.** WRI, Washington DC.

Reid, W.V. and Miller, K.R. 1989. **Keeping Options Alive: The Scientific Basis for Conserving Biodiversity.** World Resources Institute, Washington DC.

Reynolds, J. 1993. **National Biodiversity Data Bank: Software Guide and User Manual**. Makerere Univ. Institute of Environment and National Resources, Uganda.

Rick, C.M. 1982. The potential of exotic germplasm for tomato improvement. In: Vasil, I.K., Scowcroft, W.R. and Frey, K.J. (Eds.). **Plant Improvement and Somatic Cell Genetics**. Academic Press, New York, NY, pp. 1-28.

Risser, P.G. 1988. Diversity in and among grasslands. In: Wilson, E.O. and Peters, F.M. (Eds.). **Biodiversity.** National Academy Press, Washington DC, pp. 176-180

Ritland, K.R. 1990. Inference about inbreeding depression based on changes of the inbreeding coefficient. Evolution **44**: 1230-1241.

Roberts, E.H. 1975. Problems of long-term storage of seed and pollen for genetic resources conservation. In: Frankel, D.H. and Hawkes, J.G. (Eds.). **Crop Genetic Resources for Today and Tomorrow.** International Biological Programme 2. Cambridge Univ. Press, Cambridge, pp. 269-295.

Roberts, E.H. 1989. Seed storage for genetic conservation. Plants Today **2**: 12-17.

Roche, L. and Dourjeanni, M.J. 1984. **A Guide to In Situ Conservation of Genetic Resources of Tropical Woody Species**. Forest Resources Division Report No. FORGEN/Misc84/2. FAO, Rome. .

Ruffie, J. 1982. **The Population Alternative: A New Look at Competition and the Species**. Pantheon Books, New York, NY.

Ruitenbeek, H.J. 1989a. **Economic Analysis of Issues and Projects Relating to the Establishment of the Proposed Cross River National Park (Oban division) and Support Zone**. WWF, London.

Ruitenbeek, H.J. 1989b. **Social Cost-benefit Analysis of the Korup Project, Cameroon**. Prepared for the World Wildlife Fund for Nature and the Republic of Cameroon, London.

Ruitenbeek, H.J. 1992. **Mangrove Management. An Economic Analysis of Management Options with a Focus on Bintuni Bay, Irian Jaya. Environmental Management Development in Indonesian Projects**. Environmental Reports, no.8, Djakarta.

Runge, C.F. 1986. Common property and collective action in economic development. In: **Proceedings Conference on Common Property Resource Management.** National Academy of Sciences, Washington DC, pp. 31-60.

SA, 1994a. **Systematics Agenda 2000**. Charting the Biosphere. The Natural History Museum, London SW7 5BD, UK

SA, 1994b. **Systematics Agenda 2000**. Charting the Biosphere, Technical Report. The Natural History Museum, London SW7 5BD, UK

Saenger, P., Hegerl, E.J. and Davie, J.D.S. 1983. **Global Status of Mangrove Ecosystems**. Commission on Ecology Papers, No. 3. IUCN, Gland, Switzerland.

Saenger, S.C. 1986. Traditional uses of South American mangrove resources and the socio-economic effect of ecosystem changes. In: Kunstadler, P., Bird, E.C.F. and Sabhasri, S. (Eds.). **Proceedings Workshop on Man in the Mangroves**. United Nations Univ. Tokyo, pp. 104-112.

Sala, O.E. and 18 others. 2000. Global biodiversity scenarios for the year 2100. Science **287**: 1770-1774.

Salmen, L.F. 1987. **Listen to the People: Participant-Observer Evaluation of Development Projects**. Oxford Univ. Press, New York,NY.

Saunders, D.A., Hobbs, R.J. and Margules, C.R. 1991. Biological consequences of ecosystem fragmentation: a review. Conser. Biol. **5**: 18-32.

Schalk, P.H. 1992. Computer-aided Taxonomy. Binary **4**: 124-126

Scheiner, S.M. 1992. Measuring pattern diversity. Ecology **73**: 1860-1867.

Schneckenburger, S. 1991. **Neukaledonien Pflanzenwelt einer Pazifikinsel. Palmengarten**. Sonderheft16. Palmengarten, Frankfurt.

Schwanitz, F. 1966. **The Origin of Cultivated Plants**. Harvard Univ. Press, Cambridge, MA.

Scott, D.A. and Carbonell, M. 1986. **A Directory of Neotropical Wetlands**. IUCN, Cambridge and IWRB, Slimbridge, Gloucestershire, UK.

Scott, D.A. and Poole, C.M. 1989. **A Status Overview of Asian Wetlands**, no. 53.. AWB, Kuala Lumpur, Malaysia.

Scott, J.M., Csuti, B. Jacobi, J.D. and Estes, J.E. 1987. Species richness: A geographic approach to protecting future biological diversity. BioScience **37**: 782-788.

Scott, M.P. and Williams, S.M. 1994. Measuring reproductive success in insects. In: Scherwater, B., Streit, B., Wagner, G.P. and DeSalle, R. (Eds.). **Molecular Ecology and Evolution: Approaches and Applications.** Birkhauser Verlag, Basel.

Sekar, S. and Kandavel, D. 2002. Patenting Microorganisms: Towards creating a policy framework. J. Intellectual Property Rights **7**: 211-221.

Sepkoski, J.J., Jr. 1988. Alpha, beta or gamma: where does all the diversity go? Paleobiol. **14**: 221-234.

Sepkoski, J.J., Jr. 1992. Phylogenetic and ecological patterns in the phanerozoic history of marine biodiversity. In: Eldridge, N. (Ed.). **Systematics, Ecology, and the Biodiversity Crisis**. Columbia Univ. Press, New York, NY, pp. 77-100.

Sepkoski, J.J. Jr. and Raup, DM. 1986. Periodicity in marine extinction events. In: Elliott, D.K. (Ed.). **Dynamics of Extinction.** John Wiley and Sons, New York, NY, pp. 3-36.

SER (Society for Ecological Restoration), 1991. Program and Abstracts, 3rd Annual Conference 1991. Orlando, FL, pp. 18-23.

Shaffer, M.L. 1981. Minimum population sizes for species conservation. Bioscience **31**: 131-134.

Shaffer, M.L. 1987. Minimum viable populations: Coping with uncertainty. In: Soulé, M.E. (Ed.). **Viable Populations for Conservation**. Cambridge Univ. Press, Cambridge, pp. 59-68.

Shaffer, M.L. 1990. Population viability analysis. Bioscience **31**: 131-134.

Shaffer, M.L. 1995. Population viability analysis—determining Nature's share. Essay 7B, In: Meffe, G.K. and Carroll, C.R. (Eds.). **Principles of Conservation Biology.** Sinauer Associates, Sunderland, Mass., USA, pp. 195-196.

Shantz, H.L. 1954. The place of grasslands in the earth's cover of vegetation. Ecology **35**: 142-145.

Shiva, V., Anderson, P., Schiicking, H., Gray, A., Lohman, L. and Cooper, D. 1991. **Biodiversity: Social and Ecological Perspectives**. Zed Books, Atlantic Highlands, NJ.

Shiva, V., and Ramprasad, V. 1993. **Cultivating Diversity**. Research Foundation for Science, Technology and Natural Resources Policy. Dehra Dun, UA, India.

Shugart, H.H., Leemans, R. and Bonan, G.B. (Eds.). 1992. **A Systems Analysis of the Global Boreal Forest**. Cambridge Univ. Press, Cambridge.

Signor, P.W. 1990. The geological history. Ann. Rev. Ecol. Syst. **21**: 509-539.

Silva, P.C. 1982. Chlorophycota, In: Parker, S.P. (Ed.). **Classification of Living Organisms**. McGraw-Hill, New York NY, pp. 133-161.

Simberloff, D. 1986. Are we on the verge of a mass extinction in tropical rain forests? In: D.K. Elliot, (Ed.). **Dynamics of Extinctions**. John Wiley and Sons, New York, NY, pp. 165-180.

Simberloff, D., Farr, J.A., Cox, J. and Mehlman, D.W. 1992. Movement corridors: Conservation bargains or poor investments. Conser. Biol. **6**: 493-504.

Simmonds, N.W. 1979. **Principles of Crop Improvement**. Longmans, London.

Simmons, J.B., Beyer, R.I., Brandham, P.E., Lucas, G.L. and Parry, V.T.H. (Eds.). 1976. **Conservation of Threatened Plants**. Plenum Press, New York, NY.

Simon, M.M. and Brooke, L. 1996. Inuit Science: Nunavik's experience in Canada. In: **Indigenous Conservation in the Modern World: Case**

Studies in Resource Exploitations. Traditional Practice and Sustainable Development. IUCN, Gland, Switzerland.

Singh, S.P. 2002. Balancing the approaches of environmental conservation by considering ecosystem services as well as biodiversity. Curr. Sci. **82**: 1331-1335.

Sittenberg, A. and Gámez, R. 1993. Biodiversity prospecting by INBio. In: Reid, W., Laird, S.A., Meyer, C.A., Gámez, R., Sittenfeld, A., Janzen, D.H., Gollin, M.A. and Juma, C. (Eds.). **Biodiversity Prospecting: Using Genetic Resources for Sustainable Development.** World Resources Institute, Washington DC.

Skole, D. and Tucker, C. 1993. Tropical deforestation and habitat fragmentation in the Amazon: Satellite data from 1978-1988. Science. **260**: 1905-1910.

Smartt, J. 1990. **Grain Legumes—Evolution and Genetic Resources.** Cambridge Univ. Press, Cambridge.

Smith, Adam. 1776. **Wealth of Nations—An Inquiry into the Nature and Causes of Wealth of Nations.** Strahan and Cadell, UK.

Smith, D.S. and Hellmund, P.C. (Eds.). 1993. **Ecology of Greenways. Design and Function of Linear Conservation Areas.** Univ. Minnesota Press, Minneapolis, MN.

Solbrig, O.T. (Ed.). 1991. **From Genes to Ecosystems: A Research Agenda for Biodiversity**, IUBS, Paris.

Solow, 1974. The economics of resources or the resources of economics. Amer. Econ. Rev. **64** (cited from Groombridge 1992).

Solow, A.R., Mound, L.A. and Gaston, K.J. 1995. Syst. Biol. **44**: 93-96 (cited from May 2002).

Soltis, D.E., Soltis, P.S. and Milligan, B.G. 1992. Intraspecific chloroplast DNA variation: Systematic and phylogenetic implications. In: Soltis, P.S., Soltis, D.E. and Doyle, J.J. (Eds.). **Molecular Systematics of Plants.** Sinauer Associates, Sunderland, Mass., USA, pp.117-150.

Sommer, A. 1976. Attempt at an assessment of the World's tropical moist forests. Unsylva **28 (112-113)** :5-24.

Soulé, M.E. 1980. Thresholds for survival: maintaining fitness and evolutionary potential. In: Soulé, M.E. and Wilcox, B.A. (Eds.). **Conservation Biology: An Evolutionary-Ecological Perspective.** Sinauer Associates, Sunderland, Mass., USA, pp. 119-133.

Soulé, M.E. 1985. What is conservation biology? BioScience **35**: 727-734.

Soulé, M.E. 1986. **Conservation Biology: The Science of Scarcity and Diversity.** Sinauer Associates, Sunderland, Mass., USA, .

Soulé, M.E. 1987a. **Viable Populations for Conservation.** Cambridge Univ. Press, Cambridge.

Soulé, M.E. 1987b. Introduction. In: Soulé, M.E. (Ed.). **Viable Populations for Conservation.** Cambridge Univ. Press, Cambridge. pp. 1-10.

Soulé, M.E. 1987c. Where do we go from here? In: Soulé, M.J. (Ed.). **Viable Populations for Conservation.** Cambridge Univ. Press, Cambridge, pp. 175-184.

Soulé, M.E. and Wilcox, B.A. (Eds.). 1980. **Conservation biology: An Evolutionary-Ecological Perspective.** Sinauer Associates, Sunderland, Mass., USA, .

Soulé, M.E., Wilcox, B.A. and Holtby, C. 1979. Benign neglect: a model of faunal collapse in the game reserves of East Africa. Biol. Conser. **15**: 259-272.

Spellerberg, I.F. 1991. Biogeographical basis for Conservation. In: Spellerberg, I.F., Goldsmith, F.B. and Morris, M.G. (Eds.). **The Scientific Management of Temperate Communities for Conservation.** Blackwell Scientific Publications, Oxford, pp. 293-322.

Stalker, H.T. 1980. Utilisation of wild species for crop improvement. Adv. Agron. **33**: 111-147.

Stanley, S. 1988. Paleozoic mass extinctions: shared patterns suggest global cooling as a common cause. Amer. J. Sci. **288**: 334-352.

Stevens, G.C. 1989. The latitudinal gradient in geographical range: How so many species coexist in the tropics. Am. Nat. **133**: 240-256.

Stevens, G.C. 1992. The elevational gradient in altitudinal range: an extension of Rapoport's latitudinal rule to altitude. Am. Nat. **140**: 893-911.

Stewart-Tull, D.E.S. and Sussman, M. (Eds.). 1992. **The Release of Genetically Modified Microorganisms-REGEM2 (Second International Conference).** Plenum Press, New York, NY.

Stohlgren, T.J. and Quinn, J.F. 1991. **Status of National Resources Data Bases in National Parks: Western Region**. National Park Service, Co-operative Park Studies Unit, Technical Report 44. Univ. California, Davis, CA.

Strahm, W. 1989. **Plant Red Data Book for Rodrigues**. Scientific Books, Koengstein, Germany.

Strauss, H.S. 1991. Lessons from chemical risk assessment. In: Levin, M. and Strauss, H. (Eds.). **Risk Assessment in Genetic Engineering: Environmental Release of Organisms.** McGraw-Hill, New York, NY, pp. 297-318.

Strauss, S.H., Hong, Y.-P. and Hopkins, V.D. 1993. High levels of population differentiation for mitochondrial DNA haplotypes in *Pinus radiata, muricata* and *attenuata*. Theoret. Appl. Gen. **86**: 605-611.

Stuart, S.N. and R.J. Adams, 1990. Biodiversity in Sub-Saharan Africa and Its Islands. Occasional Papers of the IUCN Species Survival Commission No.6. IUCN, Gland, Switzerland.

Swaminathan, M.S. 1997. Implementing the global biodiversity convention: IPR for public good. In: Pushpangadan, P., Ravi, K. and Santhosh, V.(Eds.). **Conservation and Economic Evaluation of Biodiversity**,vol. 2. Oxford & IBH Publishing Co. Pvt. Ltd., New Delhi, pp. 399-412.

Swaminathan, M.S. 2000. Government-Industry—Civil society Partnerships in integrated gene management. *Curr. Sci.* 78: 55-562.

Swamy, B.G.L. 1973. Sources for a history of plant sciences in India. I. Epigraphy. Indian Jour. Hist. Sci. **8**: 61-98.

Swofford, D.L. and Olsen, G.J. 1990. Phylogeny reconstruction. In: Hillis, D.M. and Moritz, C. (Eds.). **Molecular Systematics**. Sinauer Associates, Sunderland, Mass., USA, pp. 411-501.

Sykora, K., de Nijs, L. and Pelsma, T. 1994. Plant communities in road verges and their importance for the conservation of plant communities. Abstract TU/2A, **Symposium on Community Ecology and Conservation Biology**. Bern, Switzerland, 14-18 Aug 1994.

Synge, H. and Heywood, V.H. 1987. **IUCN Plant Information Programme**. IUCN Threatened Plants Unit, Kew, England.

Szaro, R. and Shapiro, B. 1990. **Conserving our Heritage: America's Biodiversity**. The Nature Conservancy, Arlington, VA.

Szaro, R.C. and Salwasser, H. 1991. The management context for conserving biological diversity. 10th World Forestry Congress (Paris, France, Sep. 1991). Revue Forestière Française Actes **2**: 530-535.

Tangley, L. 1988. Research priorities for conservation. BioScience **38** : 444-448.

Tanksley, S.D. and McGough, S.R. 1997. Seed banks and molecular maps: Unlocking genetic potential from the wild. Science **277**: 1063-1066.

Temple, S.A. 1995. Role of the University in conservation biology. Essay 1A. In: Meffe, G.K and Carroll, C.R. (Eds.) **Principles of Conservation Biology**. Sinauer Associates, Sunderland, Mass., USA, , pp. 5-6.

Temple, S.A., Bolen, E.G., Soulé, M.E., Brussard, P.F., Salwasser, H. and Teer, J.G. 1988. What's so new about conservation biology? In: McCabe, R.E. (Ed.). **Transactions of the 53rd North American Wildlife and Natural Resources Conference.** Wildlife Management Institute, Washington DC, pp. 609-612,

Templeton, A.R. 1996. Translocation in conservation. In: Szaro, R.C. and Johnston, D.W. **Biodiversity in Managed Landscapes**. Oxford Univ. Press, Oxford, pp. 314-325.

Teng, S.-C. 1927. The early history of forestry in China. Jour. Forestry **25**: 564-570.

Tenhunen, J.D., Lange, O.L., Hahn, S., Siegwolf, R. and Oberbauer, S.F. 1992. The ecosystem role of poikilohydric tundra plants. In: Chapin, F.S., II, Jefferies, R.L., Reynolds, J.F., Shaver, G.R. and Svoboda, J. (Eds.). **Arctic Ecosystems in a Changing Climate**. Academic Press, San Diego, CA, pp. 213-237.

Thom, B.G. 1982. Mangrove ecology : a geomorphological perspective. In: Clough, B.F. (Ed.). **Mangrove Ecosystem in Australia.** Australian National Univ. Press, Canberra, pp. 3-17.

Thomas, R. 1992. Genetic Diversity. In: Goombridge, B. (Ed.). **Global Biodiversity. Status of the Earth's Living Sources**. Chapman & Hall, London, pp. 1-6.

Tinker, P.B. 1996. Inventorying and monitoring biodiversity. In. di Castri, F. and Younès, T. (Eds.). **Biodiversity Science and Development. Towards a New Partnership**. CAB International & IUBS, Wallingford, UK, pp. 166-170.

TNC (The Nature Conservancy), 1975. **The Preservation of Natural Diversity: A Survey and Recommendations**. TNC, Washington DC.

TNC, 1992. **An Overview of the Biological and Conservation Data (BCD) System.** The Nature Conservancy, Washington DC.

Tobias, D. and R. Mendelsohn 1991. Valuing ecotourism in a tropical rain-forest reserve. Ambio **20** : 91-93.

Tomiuk, J. and Loescheke, V. 1993. Conditions for the establishment and persistence of populations of transgenic organisms. In: Woehrmann, K. and Tomiuk, J. (Eds.). **Transgenic Organisms: Risk Assessment of Deliberate Release.** Birkhäuser-Verlag, Basel. pp 117-134.

Tomlinson, P.B. 1986. **The Botany of Mangroves.** Cambridge Univ. Press, New York, NY.

Towill, L.E. 1985. Low temperature and freeze-/ vacuum-drying preservation of pollen. In: Kartha, K.K. (Ed.). **Cryopreservation of Plant Cells and Organs**. CRC Press, Boca Raton, FL, , pp. 171-197.

Trexler, M.C. and Kosloff, L.H. 1991. International Implementation: The Longest arm of the law? In: K.A. Kohm (Ed.). **Balancing on the Brink of Extinction : The Endangered Species Act and Lessons for the Future.** Island Press, Washington DC, pp. 114-133.

Turner, I.M., Tan, H.T.W., Wee, Y.C., Ali Bin Ibrahim, Chew, P.T. and Corlett, R.T. 1994. A study of plant species extinction: Lessons for the conservation of biodiversity. Conser. Biol. **8**: 705-712.

Turner, R.K. and Pearce, D.E. 1993. Sustainable economic development: Economic and ethical principles. In: Barbier, E.B. (Ed.). **Economics and Ecology**. Chapman & Hall, London, pp. 177-194.

Twilley, R.R., Chen, R.H. and Hargis, T. 1992. Carbon sinks in mangroves and their implications to carbon budget of tropical coastal ecosystems. Water, Air and Soil Pollution **64**: 265-288.

Udvardy, M.D.F. 1975. **A Classification of the Biogeographical Provinces of the World** (IUCN Occasional paper No. 18). IUCN, Morges, Switzerland.

UN (United Nations), 1989. **World Population Prospects**. United Nations, New York.

UNEP (United Nations Environment Programme), 1992. **Convention on Biological Diversity**, June 1992. UNEP, Nairobi.

UNEP, 1995. **Global Biodiversity Assessment.** Cambridge Univ. Press, Cambridge.

Urban, D.L., ONeill, R.V. and Shugart, H.H. 1987. Landscape Ecology. BioScience **37**: 119-127.

Valencia, R., Balslev, H. and Paz y Mino, G. 1994. High tree apha-diversity in Amazonian Ecuador. Biodiver. Conser. **3**: 21-28.

Vane-Wright, R.I. 1992. Systematics and diversity. In: Groombridge, B. (Ed.). **Global Biodiversity-Status. The Earth's Living Resources**. Chapman & Hall, London, pp. 7-12.

Vane-Wright, R.T, Humphries, C.J. and Williams, P.H. 1991. What to protect? Systematics and the agony of choice. Biol. Conser. **55**: 235-253.

Vavilov, N.I. 1926. Studies on the origin of cultivated plants. Bull. Appl. Bot. Gen. Plant Breed. **16**: 1-248.

Vavilov, N.I. 1949-50. The origin, variation, immunity and breeding of cultivated plants. Chronica Botanica **13**: 1-366.

Virchow, D. 1998. **Conservation of Genetic Resources**. Springer-Verlag, Berlin.

Voelker, T.A., Worrell, A.C., Anderson, L., Bleibaum, J., Fan, C., Hawkins, D.J., Radke, S.E. and Davies, H.M. 1992. Fatty acid biosynthesis redirected to medium chains in transgenic oil seed plants. Science **257**: 72-74.

Walker, B.H. 1992. Biodiversity and ecological redundancy. Conser. Biol. **6**: 18 - 22.

Warren, D.M., Slikkerveer, L.J. and Brokensha, D. (Eds.). 1995. **The Cultural Dimension of Development. Indigenous Knowledge Systems**. Intermediate Technology Publications, London.

Watkinson, A.R., Freckleton, R.P., Robinson, and Sutherland, W.J. 2000. Predictions of biodiversity response to genetically modified herbicide-tolerant crops. Science **289**: 1554-1557.

WCED (World Commission on Environment and Development), 1987. **Our Common Future.** Oxford Univ. Press, Oxford.

WCMC (World Conservation Monitoring Centre), 1992. **Global Biodiversity: Status of the Earth's Living Resources.** Chapman & Hall, London.

WCMC, 1994. **Biodiversity Data Source Book.** WCMC, Cambridge, UK.

Weil, O.S. 1986. Beyond the WCS: integrating development and environment in Latin America and the Caribbean. IUCN Bulletin **17**: 64.

Weising, K., Winter, P., Hüttel, B. and Kahl, G. 1998. Microsatellite marker for molecular breeding. J.Crop Production **1**: 113-143.

Weller, S.G. 1996. The relationship of rarity to plant reproductive biology. In: Bowles, M.L. and Whelan, C.J., (Eds.). **Restoration of Endangered Species.** Cambridge Univ. Press, Cambridge, pp.

Wells, M., Brandon, K. and Hannah, L. 1992. **People and Parks: Linking Protected Area Management with Local Communities.** World Bank, US Agency for International Development and WWF, Washington DC.

West, N.E. 1993. Biodiversity on rangelands. J. Range Manage. **46**: 2-13.

West, N.E. 1996. Strategies for maintenance and repair of biotic community diversity on rangelands. In: Szaro, R.C. and Johnston, D.W. (Eds.). **Biodiversity in Managed Landscapes—Theory and Practice.** Oxford Univ. Press, Oxford, pp. 326-346.

Western, D., Pearl, M.C., Pimm, S.L., Walker, B., Atkinson, I. and Woodruff, D.S. 1989. An agenda for conservation action. In: Western, D. and Pearl, M.C. (Eds.). **Conservation for the Twenty-First Century.** Oxford Univ. Press, New York, pp. 304-323.

Westhoff, P., Jeske, H., Jurgens, G., Kloppstech, K. and Link, G. 1998. **Molecular Plant Development.** Oxford Univ. Press, Oxford.

Westman, W. 1990. Managing for biodiversity. BioScience **40** : 26-33.

Westneat, D.F. and Webster, M.S. 1994. Molecular analysis of kinship in birds: interesting questions and useful techniques. In: Scherwater, B., Streit, B., Wagner, G.P. and DeSalle, R. (Eds.). **Molecular Ecology and Evolution : Approaches and Applications.** Birkhauser Verlag, Basel.

Westoby, M., Walker, B. and Noy-Meir, I. 1989. Opportunistic management for rangelands not at equilibrium. J. Range Manage. **42**: 266-274.

Wetzel, 1983. **Limnology.** Saunders, Philadelphia, PA.

Whitmore, T.C. 1980. The conservation of tropical rain forest. In: Soulé, M.J. and Synge, H. (Eds.). **Conservation Biology.** John Wiley and Sons, Chichester, UK, pp. 281-296.

Whitmore, T.C. and Sayer, J.A. (Eds.). 1992. **Tropical Deforestation and Species Extinction.** Chapman & Hall, London.

Whitson, P.D. and Massey, J.R. 1981. Information system for use in studying the population status of threatened and endangered plants. In: Morse, L.E. and Henifin, M.S. (Eds.). **Rare Plant Conservation: Geographical Data Organization.** New York Botanical Garden, New York, NY, pp. 217-236.

Whittaker, R.H. 1972. Evolution and measurement of species diversity. Taxon **21**:213-251.

Whittaker, R.H. and G.E.Likens, 1975. The biosphere and man. In: Leith, H. and Whittaker, R.H. (Eds.). **Primary Productivity of the Biosphere. Ecological Studies,** vol. 14. Springer-Verlag, New York, pp. 305-328.

Whyte, R.O. and Julen, G. 1963. **Proceedings of a Technical Meeting on Project Exploration and Introduction.** Genetica Agraria. **17** : 1- 573.

WIEWS (World Information and Early Warning System), 1996. Data Information from WIEWS. FAO, Rome.

Wilcox, B.A. and D.D. Murphy, 1985. Conservation strategy: The effects of fragmentation on extinction. Amer. Natur. **125**: 879-887.

Wilkes, G. 1989. Maize: domestication, racial evolution and spread. In: Harris, D.R. and Hillman, G.C. (Eds.). **Foraging and Farming: The Evolution of Plant Exploitation.** Unwin Hyman, London, pp. 440-455.

Williams, P.H. and Humphries, C.J. 1994. Biodiversity, taxonomic relatedness, and endemism in conservation. In: Forey, P.L., Humphries, C.J. and Vane-Wright, R.I. (Eds.). **Systematics and Conservation Evaluation.** Clarendon Press, Oxford, pp. 269-287.

Williams, P.H., Humphries, C.J. and Vane-Wright, R.I. 1991. Measuring biodiversity: taxonomic relatedness for conservation priorities. Austral. Syst. Bot. **4**: 665-679.

Williams, R.E., Alltred, B.W., Denio, R.M., and Paulsen, H.A., Jr. 1968. Conservation, development and use of the world's rangeland. J. Range Manage. **21**: 355-360.

Wilson, E.O. 1984. **Biophilia**. Harvard Univ. Press, Cambridge, MA.

Wilson, E.O. 1988a. The diversity of life. In: De Blij, H. J. (Ed.). **Earth's 88: Changing Geographic Perspective**. National Geographic Society, Washington DC, pp. 68-78.

Wilson, E.O. 1988b. The current state of biological diversity. In: Wilson, E.O. and Peters, F.M. (Eds.). **Biodiversity.** National Academy Press, Washington DC, pp. 3-18.

Wilson, E.O. 1992. **The Diversity of Life.** Belknap Press, Harvard Univ., Cambridge, MA.

Wilson, E.O. and Peters, F.M. (Eds.) 1988. **Biodiversity**. National Academy Press, Washington DC.

Winpenny, J.T. 1991. **Values for the Environment: A Guide to Economic Appraisal**. HMSO, London.

Withers, L.A. 1989. In vitro conservation and germplasm utilisation. In: Brown, A.H.D., Frankel, O.H., Marshall, D.R. and Williams, J.T. (Eds.). **The Use of Genetic Resources**. Cambridge Univ. Press, Cambridge, pp. 309-334.

Woese, C.R., Kandler,O. and Wheelis, M.L. 1990. Towards a natural system of organisms: Proposal for the domains Archaea, Bacteria, and Eucarya. Proc. Nat. Acad. Sci., USA **87**: 4576-4579.

Wolfenberger, L.L. and Phifer, P.R. 2000. The ecological risks and benefits of genetically engineered plants. Science **290**: 2088-2093.

Wolff, K., Rogstad, S.H. and Schaal, B.A. 1994. Population and species variation of minisatellite DNA in *Plantago*. Theoret. Appl. Genet. **87**: 733-740.

WFCC (World Federation for Culture Collections), 1990. **Guidelines for the Establishment and Operation of Collections of Cultures of Microorganisms**. WFCC, Champinas, Brazil, 16 pp.

WRI (World Resources Institute), 1991. **World Resources Report 1991-1992. A Guide to the Global Environment**. Oxford Univ. Press, New York.

WRI, 1992. **World Resources, 1992-1993: Toward Sustainable Development**. Oxford Univ. Press, New York.

WRI/IUCN/UNEP. 1992. **Global Biodiversity Strategy: Guidelines for Action to Save, Study, and Use Earth's Biotic Wealth Sustainably and Equitably**. WRI Publications, Baltimore, MD.

Wright, S. 1938. Size of population and breeding structure in relation to evolution. Science **87**: 430-431.

Wright, S.J. and Hubbel, S.P. 1983. Stochastic extinction and reserve size: a focal species approach. Oikos 41: 466-476.

Wu, K.R., Jones, R., Danneberger, L. and Scolnik, P.A. 1994. Detection of microsatellite polymorphisms without cloning. Nucleic Acids Res. **22**: 3257-3258.

WWF/IUCN/BGCS. 1989. **The Botanic Gardens Conservation Strategy**. WWF and IUCN, Gland, Switzerland.

Wynne, M.J. 1982. Phaeophyceae. In: Parker, S.P. (Ed.) **Classification of Living Organisms**. McGraw-Hill, New York, NY, pp. 115-125.

Yen, D.E. 1985. Wild plants and domestication in pacific islands. In: Misra, V.N. and Bellwood, P. (Eds.). **Recent Advances in Indo-Pacific Prehistory**. Oxford & IBH Publishing Co. Pvt. Ltd., New Delhi, pp. 315-316.

Yoon, C.K. 1993. Counting creatures great and small. Science **260**: 620-622.

Zarucchi, J.L., Winfield, P.J., Polhill, R.M., Hollis, S., Bisby, F.A. and Allkin, R. 1993. The ILDIS project on the world's legume species diversity. In. Bisby, F.A., Russell, G.F. and Pankhurst, R.J. (Eds.). **Designs for a Global Plant Species Information System**. Oxford Univ. Press, Oxford, pp. 131-144.

Zohary, D. 1989. Domestication of the Southwest Asian Neolitic Crop assemblage of cereals, pulses and flax: the evidence from the living plants. In: Harris, D.R. and Hillman, G.C. (Eds.). **Foraging and Farming: The Evolution of Plant Exploitation**, Unwin Hyman, London, pp. 358-372.

GLOSSARY

Abundance: *see* Species abundance.

Accession: A sample of a domesticated crop variety/cultivar or its wild relative collected at a particular place and time and stored/ maintained in a conservation facility; it can be a single specimen or more than one. It can be a whole plant or a part of it, such as the seed or pollen. In other words, it represents the smallest storable unit of a plant.

Active collections: Those seeds that are stored in a seed bank, and from which subsamples are periodically taken for experimentation, exchange, evaluation and display.

Adaptation: The activity of an organism to cope with any change in any environmental parameter through phenotypic or genetic responses; such changes enhance the fitness of the organism.

Advanced cultivar: Cultivars of recently evolved plants that invariably have homogeneous populations derived from careful individual selection and progeny testing. Also called **'Breeders' line** and **plus plant**.

Aesthetic value: Value of a component of biodiversity that satisfies the aesthetics of a person and appeals to him/her visually or otherwise.

Aggregated species: These species are generally characterised by very wide distribution, often with a wide variety of habitats, and with a large degree of morphological variation, which can be correlated with geographic and /or habitat variation.

Agrobiodiversity: 'That part of biodiversity which nurtures people and which is nurtured by people' (FAO definition). Functionally it is the diversity of existing domesticated plants.

Agroecosystem: An ecosystem comprised of agricultural activities (often indigenous) and the biological, genetic and cultural processes that support it.

Agroforestry: A system where woody perennials are deliberately grown in an area already subjected to cultivation of agricultural crops, either in some form of spatial adjustment or in temporal sequence.

Alien species: A species deliberately or inadvertently introduced into an area that does not historically belong to its range. Also known as **Introduced species** or **Exotic species**.

Alien invasive species: Alien species that not only becomes firmly established in an area, but also spreads very rapidly, causing problems for native taxa.

Allele: One or several alternatives of a gene at a particular genetic locus.

Allozyme: A subset of isozymes which are variants of polypeptides representing different allelic alternatives of the same gene locus, i.e., one of several possible forms of an enzyme.

Alpha richness: Refers to the number of species found in a small, homogeneous area. Also known as **Point richness**.

Amplification: Production of large amounts of a small segment of DNA molecule.

Aquaculture: Breeding and rearing of aquatic organisms which are of value to human beings.

Arrhenius equation or relationship: The species-area relationship derived by the equation log S = e + z log A, where S = No. of species; A= Area of study and e, z are constants.

Avalanche index: A measure of species diversity. Here, in addition to species number and frequencies, the biological and ecological differences among species comprising a community are also used.

Background extinction: Historical rates of extinction due to environmental cause alone; average background extinction rate of a species is estimated to be 4 million years.

Base collection: Seeds stored under optimum conditions in a seed bank and not interfered with until reduced viability is noticed through germination tests.

Baseline data: Fundamental data in a basic biodiversity inventory that are crucial for conservation planning and management.

Benthos: Organisms living at the bottom of a large aquatic body such as lakes or seas.

Bequest value: The value of knowing that others may benefit from the existence of a resource in future; it is a value defined by willingness to pay.

Beta richness: Rate of change in species composition across habitats in a landscape.

Biodiversity: Variously defined. The variety in living organisms considered at different bio-organisational levels—genes, species, ecosystems and landscapes. (Abbreviated form of **Biological Diversity**).

Biodiversity loss: Loss of biodiversity elements at its different scales—gene, species and ecosystem—due to stochastic or deterministic processes.

Biodiversity prospecting: *see* Bioprospecting.

Biogeography: A branch of science dealing with geographic distribution of organisms.

Biological control: Control of harmful pests/ microbes by using another living organism, which predates or parasitises on them. Also abbreviated as **biocontrol**.

Biological diversity: *see* Biodiversity.

Biological resource (also called **Bioresource**): A component of biodiversity of direct, indirect or potential value to human beings.

Biome: Any part of the living environment of a specific region characterised by a particular and dominant vegetation type; the biome is always under the constant influence of the local climatic conditions. Examples: grassland, coniferous forest.

Biopesticide: A pesticide obtained from a biological source and readily biodegradable.

Biophilia: A term first used by E.O.Wilson to describe the innate positive attitudes of human beings expressed as love for living organisms/nature.

Biopiracy (also called 'gene robbing'): Any effort by biodiversity prospectors to 'steal' or 'rob' indigenous knowledge about biodiversity from traditional societies and to use that knowledge to earn money with no payment whatsoever to the owner-society.

Bioprospecting: Abbreviated form of Biodiversity prospecting. Exploration of biodiversity for very valuable genetic and/or biochemical resources that find present use or may find future use in pharmaceutical, agricultural, food or other industries.

Biosphere reserve (also called **Bioreserve**): A protected area established under UNESCO's Man and Biosphere (MAB) Programme. The area may be a terrestrial, aquatic or coastal/ marine ecosystem meant not only to conserve the ecosystem and its biodiversity, but also to provide sustainable use of natural resources for the benefit of local communities. It is also meant for research, education and training activities (*see* also **Buffer Zone, Core Zone**).

Biotechnology: Variously defined. 'Any technological application that uses biological systems, living organisms or derivatives thereof, to make or modify products or processes for specific use' (CBD).

BIOTROP: An approach towards the establishment of research stations in areas believed to be important hot spots.

Bog: Peat-producing wetlands of moist climatic regions with sphagnum moss as the most dominant taxon.

Boreal forests: Circumpolar forests of peatland.

Botanic garden: Broadly defined as a place of collection of living plants grown for education, recreation, economic, medicinal, conservation or scientific purposes.

Breeder's line: seeAdvanced cultivar

Buffer zone: The peripheral region of a protected area covering the core region; the buffer zone is meant to sustain the requirements of local people who depend on the bioresources of the protected area, thus protecting the core region from exploitation. This area is also meant for non-destructive human activities such as ecotourism.

Bundle of rights: The set of rights that the indigenous people should be empowered with to protect their IKS as well as their society *per se*. These include human rights, rights to development, environmental integrity rights, cultural heritage rights, IPRs, right to privacy, religious freedom etc.

C value: The DNA quantity per genome, i.e., per chromosome set. The DNA content of a diploid nucleus is usually referred to as the 2C value.

Cambrian: The earliest period in the Palaeozoic era.

Candidate (species): Species whose status is being assessed and which are suspected but not yet definitely known to belong to any threatened category.

cDNA: Single-stranded complementary DNA, i.e., a DNA molecule synthesised from an RNA template as an exact template by reverse transcriptase.

Centre of diversity: An area with great species richness, recognised at the local, regional or global level.

Centres of genetic diversity: *see* Centres of origin

Centres of origin: Areas of ancient agricultural civilisation where the modern domesticated plant species had their origin. Also called **Vavilov centres** or **Centres of genetic diversity**.

Cerrado: *see* Ilano.

Character: Any trait of an organism that can be seen, felt, measured or otherwise assessed.

Characterisation (of Germplasms): Recording of heritable and easily identifiable characters, prevailingly of mature plants, that are expressed in all environments.

Charismatic species: Species that are significant from social, cultural or anthropomorphic standpoints, and /or usually attractive.

Chemotype: A taxon identified in terms of chemical/physiological characteristics but not necessarily accompanied by morphological changes.

Chipko Andolan: *see* Chipko movement.

Chipko movement: A social conservation movement initiated by the people, especially women, of the Himalayan region to protect trees from being felled for trade. The women hugged the trees when the trees were about to be cut. *Chipko* means to hug.

Clade: A set of taxa derived from a common ancestral taxon.

Cladistic: Pertaining to branching patterns. *See* also **clade, cladistics, cladogram.**

Cladistic prioritisation: A method to assess, and to some extent quantify, the distinctiveness between taxa considered for conservation; it is based on the phylogenetic relationship between species arrived at through cladistics.

Cladistics: A system of classification often based on phylogenetic divergence from a common ancestor.

Cladogram: A diagram representing cladistic relationships. It is supposed to be a rough estimate of a true genecological relationship among taxa.

Cline: This is a 'regional variation in one or more characters which vary unidirectionally over a significant geographical distance'.

Clones: Genetically identical organisms produced from one common organism.

Collapse: Reduction in population size of a species, leading to its extinction.

Community: All organisms belonging to a particular habitat and influencing each other by forming part of the food chain.

Community registers: Registers that contain all information about the biodiversity knowledge that an indigenous community has accumulated over the long years of its association with nature.

Competition: All activities of two or more organisms directed towards the same resource for utilisation.

Conservation: Variously defined; protection of biodiversity for sustainable utilisation.

Conservation biology: A scientific discipline that describes/promotes an integrated approach and management strategies to biodiversity conservation based on scientific principles of other disciplines as well as on experience.

Conservation, *ex situ*: The maintenance and conservation of a biodiversity element, say a species, away from its original habitat. Also called **off-site conservation.**

Conservation, *in situ:* Preservation of a species as a component of its original functioning ecosystem so that it continues to be subjected to selection pressures and adaptive evolution. Also called **on-site conservation.**

Conservation, *in vitro*: Conservation of plant germplasms through micropropagation or through storage of synseeds, embryos, buds or meristems.

Conservation potential: An approach to identifying protected areas for conservation action developed by Dinerstein and Wikramanayake (1993).

Consumptive use value: A type of direct use value that represents the value placed on a biodiversity component consumed directly without passing through a market.

Core collection. A minimal set of accessions that represents the genetic diversity present in the total collection of germplasms.

Core zone: The central region of a protected area where disruptive activities are prohibited.

Corridor: A connection between adjacent and similar habitats and protected areas.

Cost-benefit analysis: Human appraisal of any project, including all social and financial costs and benefits accruing to the project.

Cp DNA: The DNA present in the plastids, especially chloroplast. The Cp DNA is highly conserved.

Critically endangered (Species): A taxon facing an extremely high risk of extinction in the wild in the immediate future.

Cultivar: A cultivated variety of a domesticated crop plant. The individual precise race of a domesticated species. It should have DHS (or DUS). (i.e., Distinctiveness, Uniformity or Homogeneity and Stability).

Cultivated Species: *see* Domesticates.

Cultivation: *see* Domestication.

Cultural diversity: The diversity shown by different cultural elements and societies, placing different values, exerting different driving forces and influences and practising different measures for conserving and sustaining biodiversity.

Cultural landscape: A category of World Heritage Sites recognised by UNESCO, wherein a complex interrelationship between man and nature has existed from the historic past; rich in culture and biodiversity.

Demand value: Demand value occurs when a component of biodiversity provides satisfaction for some felt preferences;

commonly recognised in terms of Goods, Services and Information.

Demographic bottleneck: An episode of dramatic but usually temporary reduction in population size due to various reasons.

Demographic uncertainty: Random and chance populational events that influence survival in a small population. Example: extremely skewed sex ratio in dioecious plants.

Deterministic process: A cause-and-effect process responsible for loss of biodiversity. Example: Habitat fragmentation.

Direct use value: The value of biodiversity that is evident in actual use, especially in consumption.

Dispersal: Movement of an organism or plant away from its place of birth; in plants it takes place through various types of propagules/ diaspores.

Distribution: The spatial range of occupation of an organism.

DNA: Deoxyribonucleic acid, the macromolecule that controls all heritable characters. It is the blueprint that makes all living organisms.

DNA bank: A gene library in which samples of DNA are stored. It may comprise total genomes or individual cloned DNA fragments.

Domesticated species: *see* Domesticates.

Domesticates: Organisms that have undergone domestication.

Domestication: The process by which plants, animals and microbes are selected from the wild and adapted to habitats created by humans; also called **cultivation** in the case of plants and microbes.

Domestication syndrome: The co-ordinated morphological, structural, physiological, cytological, biochemical, reproductive and other changes undergone by a plant (or animal) during the process of domestication from the wild.

Earth summit: *see* Rio summit.

Ecological diversity: *see* Ecosystem diversity.

Ecosystem: Variously defined. A dynamic but complex system encompassing plants, animals and microbes interacting among themselves and with the abiotic environment; varies very widely in size as well as temporally and spatially.

Ecosystem diversity: Diversity at the habitat or ecological level.

Ecosystem restoration: *see* Restoration.

Ecotype: A morphologically distinctive form of a taxon restricted to an identifiably different habitat, often identified in terms of restriction to particular soils or climate regimes.

Edge effect: When a habitat is fragmented, the peripheral regions of fragmented bits become the edges; the edges do not possess the original features of the habitat. Edges will create a negative influence on the habitat's interior. This negative influence is called edge effect.

Effective size (of population): The size of an ideal population whose genome is affected by random drift to the same extent as is the genome of the real population under study.

Electrophoresis: A process by which gene products, especially proteins, of an organism are separated by applying an electrical field in a gel medium, and visualised, identified and classified using specific dyes.

Endangered species: A species in danger of extinction and whose survival is unlikely if the causal factors continue to operate.

Endemic: Any localised process or pattern but the term is usually applied to taxa restricted in distribution to a specified region.

Endemism: The phenomenon wherein a taxon is found only in a particular restricted area and nowhere else.

Environmental uncertainty: Any unpredictable environmental event that affects an organism's or habitat's survival.

Ethical value: The principle that forms the basis for private and social valuation of biological resources (UNEP 1995).

Ethnobotany: Studies, which concern the mutual relationships between plants and traditional people.

Ethno-directed method: One of the three-methods of bioprospecting wherein a bioprospector depends on the already available and tested traditional knowledge with respect to the usefulness of a biodiversity element or a bioprocess. This method is less time-consuming, cost-effective and sure to yield results.

Eukaryote: An organism whose DNA is enclosed in nuclear membranes.

Evaluation (of germplasm): A process by which germplasms of wild relatives and different populations of a cultivated taxon are assessed for their potential genetic diversity in terms of useful phenotypic or expressible traits.

Evenness (of genetic diversity): Refers to the frequency of the different genotypes present in the same population or a sample from it.

Evolutionarily Significant Unit (ESU): Also called Ecologically Significant Unit. A population that shows reproductive isolation from other conspecific populations and thereby represents an important component in the evolutionary legacy of a species.

Existence value: Value deriving from the knowledge that a particular biodiversity element exists and will continue to exist.

Exotic species: *see* Alien species.

Exotic invasive species: *see* Alien invasive species.

Extinct (species): A species that has lost all its individuals without producing progenies.

Extinct in the wild: A species known to survive only in cultivation well outside of its past range.

Extinction: The process by which all the individuals of a species are lost without producing progeny.

Extirpation: Process by which an individual, population, or species is totally destroyed.

Fen: Peat-producing wetlands with greater supporting vegetation composed of grasses and sedges.

Feral (Plant/microbe): Domesticated taxa that have escaped and have maintained themselves in the wild without human intervention and use.

Field gene bank: An area of land where a collection of growing plants of species needing conservation are assembled; the assemblage contains as many individuals of the target species as possible.

Fitness: The expected or relative contribution of an allele, genotype, or phenotype to the next as well as to future generations. It is a measure of relative reproductive success.

Flagship species: Popular and charismatic species of an ecosystem.

Flora: All of the plants found in a specific region.

Founder effect: The loss of genetic diversity when a new population of a species is established from only a few individuals (called founders) of an originally larger population.

Fragmentation: The breaking up of a larger habitat/ecosystem/landscape into more that one disjunct bit. A major cause of habitat/ecosystem degradation.

Gaia hypothesis: A hypothesis related to 'planetary dynamics postulating a tight interrelationship between life processes and conditions on earth that support life', i.e., a belief that biotic processes are the major regulators of physical processes of the earth. According to this hypothesis, Earth is a superorganism.

Gamma richness: Changes in species richness across larger landscape gradients.

Garrison reserve: A reserve of smaller size where intensive management is necessary to sustain it. *See* **vest pocket**.

Gene: The functiotional and fundamental unit of heredity; it is a part of the DNA molecule encoding for particular protein(s).

Gene bank: A storage facility where an organism is conserved in the form seeds, pollen, embryos, synseeds etc.; a field gene bank is one where plants are grown and conserved.

Gene flow: Unilateral or bilateral exchange of genetic material between individuals/ populations.

Gene locus: The site occupied by a specific gene on a chromosome.

Gene pool: Total genetic material of a freely interbreeding population or deme.

Gene prospecting: *see* Bioprospecting.

Gene robbing: *see* Biopiracy.

Generalist: A species that has a broad habitat range.

Genetic code: The co-ordination principle according to which the information specifying the amino acid sequences of proteins and the start and stop signals for protein synthesis is stored in the heritable nucleic acid sequences of an organism is called the 'genetic code'. The genetic code distributes 64 triplet codons to 20 amino acids, including initiation signals and three termination signals for the construction of protein molecules with specific sequences of amino acids.

Genetic diversity: Heritable genetic variation found within or between species.

Genetic drift: Random change in gene (or allele) frequencies in small populations due to chance alone.

Genetic erosion: Loss of genetic diversity over time due to various causes.

Genetic resource: Genetic materials or genetically controlled traits of organisms (domesticated, semidomesticated or wild) that are of value as a resource.

Genetic resources profile: The primary, secondary, tertiary and quaternary gene pools of crop species. The genetic resource profile is ranked as any one of the aforesaid gene pools in terms of accessibility, or the germplasms of a plant species, through conventional or biotechnological hybridisation methods.

Genetic uncertainty: Refers to random changes in the genome, mutations etc. It also includes Founder effects, Genetic drift and Inbreeding depression.

Genome: The genetic make-up of an organism; it is the sum of all genes of that organism.

Genotype: The genetic constitution of an organism.

Geographic Information System (GIS): Any computerised system of organising and analysing any spatial array of data and information.

Germplasm treasures: *see* Centres of origin.

Germplasms: Genotypes or populations of economically important plants; often maintained/stored as research materials.

Goods: The components of diversity that provide the basis for food, forage, fuel, medicine, useful chemicals, fibres, dyes, ornamentals etc.

Habitat: The ecospace selected by an organism in which it co-exists with other organisms.

Habitat degradation: Decline in habitat quality.

Habitat restoration: *see* Restoration.

Heterozygosity: A measure of genetic diversity in a population made on the basis of the number of heterozygous loci across individuals.

Heterozygous: A condition wherein an individual organism possesses two different alleles at the same gene locus.

Home garden: A small-scale conservation effort made by many indigenous communities; these contain plants of vegetable/fruit/ ornamental/medicinal value grown in the premises of a home.

Homozygous: A condition in which an individual has two of the same alleles at a given gene locus.

Hot spot: A geographic region of the world characterised by unusually high endemic species richness.

Hybrid: The product obtained after crossing two different taxa.

Ilano: Grasslands of S. America.

Inbreeding: Mating of individuals related by common ancestry.

Inbreeding depression: A decrease in the mean of a character upon inbreeding; often causes a reduction in fitness and vigour as a result of increased homozygosity.

Indeterminate (species): Species considered definitely endangered, vulnerable or rare, but information is insufficient to categorically assign it to any of these three categories.

Indicator species: A species used to gauge a particular condition in a chosen habitat/ecosystem. Example: a species that indicates very high soil salinity.

Indigenous knowledge system: All the information, knowledge, wisdom, practices, beliefs and philosophies of traditional indigenous societies and accumulated by these societies for several hundred years when in close contact with nature.

Indirect use value: 'Value derived from the role of resources and systems in supporting or protecting activities whose outputs have direct value in production or consumption' (UNEP 1995).

Information value: The genome and information contents that a biodiversity element contains reflect information value.

Inherent value: *see* Instrumental value.

Instrumental value: Value of an entity (of biodiversity) as judged by its use to humans. Also called inherent value.

Insufficiently known (species): Species that are suspected but not definitely known to belong to threatened categories.

Intraspecific diversity: *see* Genetic diversity.

Intrinsic value: Value of an entity (of biodiversity) independent of external circumstances or its value to human beings, i.e., inherent value of an entity. (= **Inherent value**).

Inventory: A formal surveying, sorting, cataloguing, quantifying and mapping of the occurrence of defined elements of biodiversity (such as genes, species etc.) at a particular point of time in a defined geographic unit.

Island biogeography theory: A theory developed by R.H.MacArthur and E.O. Wilson in 1967. As per this theory, the number of species inhabiting an island is a function of the size (area) of the island and distance from the mainland and is determined by the relationship between the rates of species immigration into the island and extinction.

Karyotype: Refers to the chromosome complement of an individual or of a related group of individuals and is characterised by the number, size and shape of its chromosomes.

Keystone species: A species whose presence or absence influences very greatly and disproportionately the other species of a habitat as well as the ecosystem process.

Land race: Crop variety of peasant farming initially derived due to domestication process and forming the base for further diversification into modern cultivars.

Landscape: Ecological mosaic of more than one ecosystem repeated in a similar form.

Landscape diversity: Diversity in a landscape; since a landscape is a mosaic of a number of heterogeneous land forms, vegetation types and land uses, it shows patterns in diversity. Also called **Pattern diversity**.

Ligation: A process by which a chosen DNA fragment is inserted into a plasmid or viral genome. This is a vital step in recombinant cloning.

Mangroves: Intertidal forested coastal wetlands of tropical and subtropical areas.

Market price: Price fixed through a market mechanism.

Marsh: Herbaceous 'swamps' or 'mires'.

Mass extinction: An exceptional loss in biodiversity of substantial size and global in extent, affecting a broad range of taxonomic groups over very short periods of geologic time.

Megadiversity centres: Countries with very great species diversity. There are 12 such countries and India is one of them.

Mesozoic: An era extending from 245 to 66 million years ago.

Metapopulation: Partially isolated small subpopulations of a larger population of a species.

Minimum Viable Population (MVP): Smallest population that has a chance of surviving for a given number of years despite the foreseeable stochastic and deterministic factors.

Monitoring: Periodic surveillance to ascertain 'the extent of compliance with a predetermined standard or degree of deviation from an expected norm' (UNEP 1995).

mt DNA: The DNA present in the mitochondria; the mt DNA is highly variable.

Mutation: Change in the genotype of an organism.

Mutualism: Interspecific relationship from which both species benefit.

Mycobiont: The fungal component of the mutualistic composite lichen thallus wherein a fungus and a photobiont live a symbiotic life.

Nandavanas: Sacred groves of medieval India, which are associated with temples and harbouring pristine biodiversity; a socioreligious mode of conservation of biodiversity.

National park: One of the types of protected areas established at the national level for the *in-situ* protection of biodiversity.

Natural reserve: *see* Protected Area.

Natural selection: The process by which the genotype in a population better adapted to the environment is selected by nature in preference to less adapted genotypes over a number of generations.

Neoendemics: Relatively recently evolved endemics.

Niche: The place in a habitat specifically occupied by a species.

Non-consumptive use value: Value which the components of biodiversity possess in terms of functions or services offered, i.e., the value of a resource that is not determined by its use.

Non-use value: 'Value relating to safeguarding the existence of assets, even though not related to their actual use in the foreseeable future' (MacArthur 1997).

Noosphere: The sphere of minds uniting every one on the planet through different types of communication networks.

Nucleotides: Building block units of DNA; consist of a sugar and phosphate backbone with a base attached.

Off-site conservation: *see* Conservation, *ex situ*.

On-site conservation: *see* Conservation, *in situ*.

Option value: The potential value of a biodiversity resource for future direct or indirect use.

Organismal diversity: *see* Species diversity.

Orthodox seed: Seed that can be stored in a seed bank for a long time without substantial loss of vitality and without genetic change. Also called **Conventional** or **Desiccation-Tolerant** seed.

Overdominance: The condition in which a heterozygote at a particular gene locus has greater fitness than homozygotes.

Palaeoendemics: Phylogenetically very early evolved endemics; also called **relict endemics**.

Pampa: *see* llano.

Participatory forest management: A movement initiated in India whereby the government and the local community, whose livelihood depends on the forest, jointly protect and conserve the forest in a sustainable manner so as to benefit both.

Passive use value: *see* Non-use value.

Passport data: This helps in establishing the origin of a germplasm. Passport data may be either **accession data** if the accession has been received from a breeder, another institution etc., or **collection data** if the accession is from a field collection. Usually the site of origin is recorded.

Pattern diversity: *see* Landscape diversity.

Periurban diversity: The biodiversity around towns and cities, i.e., at the interface between towns/ cities and the countryside.

Permian: Last period of the Palaeozoic era, 290 to 245 million years ago.

Phenetic: A term pertaining to phenotypic similarities.

Phenotype: The externally detectable traits of an organism; the result of genotype, environment or interaction of the genotype with environment.

Photobiont: The photosynthesising partner of a composite lichen thallus that lives along with a mycobiont in a symbiotic manner; the photobiont may be an alga or a cyanobacterium.

Phylogenetic method: One of the three methods of biodiversity prospecting wherein a bioprospector selects for analysis and exploration a taxon related to taxa that have already been known to be good genetic resources.

Plasmid: A genetic element that has an independent existence, apart from the main DNA, in the cell.

Plasticity: A genetically determined but environmentally induced variation in any characteristic of a an organism.

Pleistocene: A recent epoch in geological history dating back some 2.5 million years.

Ploidy: The number of chromosome sets contained by a cell/nucleus.

Plus plant: *see* Advanced cultivar.

Point richness: *see* Alpha richness.

Pollen bank: A place where pollen are stored under appropriate conditions to enhance their viability period, for purposes of use by posterity in breeding and crossing.

Polyploid: An organism possessing two or more sets of chromosomes.

Polyploidy: Possessing more than two complete sets of chromosomes.

Polytype: Any species that comprises two or more scientifically recognised entities such as subspecies or varieties.

Population: A group of individuals of a species sharing a common descent.

Population Viability Analysis (PVA): An analysis of all factors that affect directly or indirectly the survival of a small population. Also called Population Vulnerability Analysis.

Population Vulnerability Analysis (PVA): *see* Population Viability Analysis

Prairie: Grasslands of North America.

Precautionary principle: The principle that human society should take all steps to protect all elements of biodiversity irrespective of whether they are useful now or not.

Preservation: Protection of biodiversity from any kind of human activity or interference.

Primary gene pool (GP1): Represents true biological species including all cultivated, wild and weedy forms; hybrids among these are fertile and gene transfer is easy, simple and direct, and poses no problem

Primary value: The value of the system characteristics upon which all ecosystem functions depend.

Productive use value: Value given to a component of biodiversity that is commercially harvested or as a source of commercially harvestable products. Such items pass through a market.

Prokaryote: An organism whose cells do not have a distinct nucleus, e.g. Bacteria.

Protected area: A legally established area under private or public ownership that is managed to conserve specific or all elements of biodiversity.

Pseudoextinction: A process whereby a species disappears when its lineage is transformed over evolutionary time or divides into two or more separate lineages.

Psychospiritual value: Very difficult to define; it encompasses aesthetic beauty, religious and cultural value etc. of biodiversity.

Pusztas: Grasslands of Hungary.

Quaternary gene pool (GP4): The constituents are incompatible but related species. Gene transfer requires biotechnological interventions.

Quaternary period: The last period of the Coenozoic era extending from 2.5 million years ago to the present.

Quasi-option value: 'The value of the future information made available through the preservation of a resource' (UNEP 1995).

Ramsar convention: An international convention established for the purpose of identification, conservation and management of threatened wetlands of the world.

Ramsar site: A wetland site recognised by the Ramsar convention that needs to be conserved and protected.

Rangeland: Grassland of Australia.

Rapid Assessment Programme (RAP): Quick collection, analysis and dissemination of information on poorly known areas that are potentially important from the biodiversity point of view

Rapid biodiversity assessment: *see* Rapid Assessment Programme.

Rare (species): Species taxa with small populations that are not endangered or vulnerable at present, but are at risk.

Rarity: Refers to a taxon, seldom occurring either in number of individuals or in space.

Recalcitrant seed: A seed that cannot be stored under seed bank conditions; these seeds usually have a very short period of viability.

Recombinant cloning: A process by which a chosen DNA fragment is inserted into a plasmid or viral genome (ligation) and the subsequent insertion of this plasmid or viral genome into actively reproducing bacterial cells (transformation), allowing for the reproduction of the former also thousands of times.

Recreational species: Species popular for collection, growing or observation; they provide a source of recreation for people.

Red data book: A book that details information about threatened taxa at the global, regional, national or local level.

Reintroduction: Bringing an individual, population or species back to its former habitat after it had been extirpated from there.

Relaxation: Loss of species due to habitat fragmentation.

Remote sensing: Any technique employed for analysing landscape/seascape patterns and trends using aerial photography or satellite imagery.

Replication: The duplication of genomic or plasmid DNA as part of the reproductive cycle of a cell or virus.

Reserve: A multipurpose protected area (*see* Biosphere reserve).

Reserve collection: Refers to germplasm accessions other than core collections that are conserved/preserved.

Resource: A substance, material or place required by an organism for its normal well-being.

Restoration: To bring back a degraded ecological system to its original state; to reinstate a species in its original habitat.

Restoration ecology: Branch of Ecology dealing with aspects and principles of restoration of a disturbed or degraded habitat/ecosystem/landscape.

Restriction enzymes: Specific enzymes produced by bacteria to protect themselves from infecting viruses by cutting the viral DNA. Several restriction enzymes are highly specific in the DNA sequence they recognise and cut. These enzymes are advantageously used to assess genetic diversity of organisms through RFLP, RAPD or other more modern modifications of RAPD techniques.

Richness (of genetic diversity): Refers to the total number of different genotypes present in a population or a sample from it.

Rio summit: A United Nations Conference on Environment and Development held at Rio de Janeiro (Brazil) in 1992, where most important decisions (including CBD) on biodiversity were taken. Also called **Earth summit**.

Sacred grove: *see* Nandavanas.

Safe (species): An IUCN category of species; it indicates the safe ecological status of a species.

Savanna (sometimes spelled Savannah or Savana): Grasslands of Africa.

Seascape: The total diversity of different but overlapping ecosystems/habitats in the marine system and equivalent in connotation to 'Landscape' of the terrestrial system.

Secondary gene pool (GP2): Represents the group of species that can be artificially hybridised with crop species but gene transfer may be difficult. The hybrids, if produced, are weak or partially sterile.

Seed bank: A facility where seeds are stored for *ex-situ* conservation.

Services: The components of diversity that render services to human society, such as pollination, dispersal, nutrient cycling, nitrogen fixation, biogeochemical cycles etc.

Silviculture: The science of cultivating forest crops (usually timber trees).

Speciation: The evolutionary act of origin of a species.

Species abundance: The number of individuals of a species in a defined region.

Species concepts: The nature and definition of species are fraught with difficulty and a number of concepts have been proposed to date to sort out these problems. All these ideas are known as species concepts.

Species diversity: The number and diversity of species present in a defined area. Usually a synonym of species richness.

Species evenness: Equitability of different species, as given by their relative abundance in a defined region.

Species richness: The number of species in a defined region.

Species turnover: Changes in species richness between areas.

Species-area relationship: The number of species per area varies with the size of the area; the greater the area the larger the number of species. This is given by the Arrhenius equation.

Steppe: Grasslands of Eurasia.

***Sthalavriksha*:** A Sanskrit word meaning 'temple tree'. In southern India, it has been the practice to grow a particular tree in each temple invariably near the place where the presiding deity is located, so that the tree is accorded social and religious sanctity. This is one of the ways of conserving biodiversity in India.

Stochastic processes: Processes resulting from random factors.

***Sui generis* system:** An alternative to the IPR system to cover and protect exclusively indigenous knowledge systems which cannot be protected by IPRs for technical reasons.

Sustainable development: Development not only to meet the needs and aspirations of the current but also of future generations.

Swamp: Forested wetlands on waterlogged soils; woody plants are dominant here.

Synseed: An artificial or synthetic seed containing a meristem, bud, embryo or embryoid enveloped by alginate or neutral gums. It can be conserved and stored for a long time.

Systematics: The 'scientific study of the kinds and diversity of organisms and of any and all relationships among them'. This branch is involved in the recognition, comparison, classification and naming of organisms, extant or extinct.

Taxic diversity: This is an approach to estimate species diversity of an area. In this approach, greater value is attached to the presence of taxonomically isolated species or of taxonomically isolated genera. Thus an area containing taxonomically diverse taxa is considered to have greater diversity than an area with closely related species in equal numbers.

Taxon: A classification unit of any hierarchical level (such as species, genus, family etc.).

Taxonomic diversity: *see* Species diversity.

Taxonomic structure: The different levels of taxa produced as a series of hierarchical categories with species as the basic unit.

Taxonomy: The theory and practice of classifying organisms. It has four components: Classification, Nomenclature, Circumscription and Identification aids such as keys.

Technosphere: The sphere of technologies.

Tertiary gene pool (GP3): Includes all species that can be crossed with difficulty to crop species. The hybrid zygote/embryo needs embryo rescue. Gene transfer is almost impossible and requires very special techniques.

Tertiary period: The first period of the Coenozoic era, covering 66 million years to 2.5 million years ago.

Test-tube gene bank: Place where germplasms of vegetative propagules of vegetatively propagated taxa are conserved.

Theory of island biogeography: A theory proposed by MacArthur and Wilson (1963); it states that the species diversity of an island depends on its size (area), its distance from the mainland and on the rate and intensity of migration of species from the mainland to the island.

Threat index: A measure that indicates the degree of threat to a species/ ecosystem; several parameters are considered in arriving at the threat index.

Threatened species: A species believed to be at significant risk of extinction in the foreseeable future because of stochastic, deterministic or both factors or by virtue of inherent rarity.

Total economic value: The sum of all use and non-use values with due considerations of any trade-off or mutually exclusive uses. Also called **Total value** or simply **Value**.

Total environmental value: A function of primary value and total economic value of environment and its various components.

Total value: *see* Total Economic Value.

Traditional Knowledge System (TKS): *see* Indigenous knowledge system (IKS).

Traditional resource rights: These rights respect the requirements of the indigenous people whose knowledge and wisdom cannot be protected by IPRs. Here there is an internalisation of the existing or expected benefits derived from IKS.

Transformation: The insertion of plasmid or viral genome, in which a chosen DNA fragment has already been inserted, into an actively reproducing bacterial cell, allowing for the reproduction of the former also thousands of times.

Transformative value: This exists when the object of biodiversity provides 'the occasion for examining or altering a felt preference rather than simply satisfying it'.

Transgenic (plant/microbe): An organism that has received through biotechnological methods a very specific gene/groups of genes from another, usually distant organism, with a view to improving its performance/productivity.

Translocation: *see* Reintroduction.

True extinction: A phenomenon wherein all individuals of a species are lost without producing progeny.

Umbrella species: Species whose area of occupancy is very large and whose habitat requirements are very wide; if such species are conserved, many other species will receive incidental protection.

Urban diversity: The biodiversity of towns and cities.

Use value: Value obtained through the actual use of a biodiversity element.

Vavilov centres: *see* Centres of origin.

Vest pocket: A small reserve that needs intensive management care. *See* also **Garrison Reserve**; there is not much difference between Vest Pockets and Garrison Reserves.

Vulnerable (species): A taxon which is not critically endangered or endangered but is facing a high risk of extinction in the wild in the medium-term future.

Wild relative: An organism taxonomically related to a domesticated organism but living wild and serving as a potential source of useful genes to improve the domesticated organism.

Willingness to accept (WTA): The amount of compensation an individual is willing to take in exchange for giving up some good or service from Biodiversity.

Willingness to pay (WTP): The amount an individual is willing to pay to acquire some good or service from Biodiversity.

Within-species diversity: *see* Genetic diversity.

World heritage site: A type of protected area recognised by UNESCO. Such sites are historically very important and significant.

ACRONYMS AND ABBREVIATIONS

A

ABIF	Australian Biodiversity Information Facility
AEK	Areas Silvestres de Kuna Yala (Panama)
AMP – PCR	Anchored Microsatellite Primed – Polymerase Chain Reaction
AP-PCR	Arbitrarily Primed – Polymerase Chain Reaction
ARPA	Advanced Research Projects Agency
ASEANET	South East Asia Network
ASFA	Aquatic Sciences and Fisheries Abstracts
ATBI	All–Taxa Biodiversity Inventory
AV	Aesthetic value

B

BCD	Biological and Conservation Data
BDM	Biodiversity Data Management
BDS	Biodiversity Data Bank
BGCI	Botanic Gardens Conservation International, UK
BGCS	Botanic Gardens Conservation Secretariat, UK
BIMS	Biodiversity Information Management System
BIN	Biodiversity Information Network
BIN-BR	Biodiversity Information Network – Brazil
BIOCON	Biodiversity Conservation (Network)
BioNET	Biological Network of CAB
BIOS	Bacteriology Insight Orienting System
bp	Base Pairs (in DNA molecules)
BR	Biodiversity Reserve
BRAHMS	Botanical Research and Herbarium Management System
Bt	*Bacillus thuringiensis*
BV	Bequest value

C

C	Candidate (species)
C	*Circa* (Approximately)
CAB	Commonwealth Agricultural Bureau
CARINET	Caribbean Network
CASAFA	Committee on Application of Science to Agriculture, Forestry and Aquaculture
CBA	Cordillera – peoples' Bedong Alliance, Philippines
CBD	Convention on Biological diversity

CBIN	Canadian Biodiversity Information Network
CBR	Community Biodiversity Register
CCR	Community Controlled Research
cDNA	Complementary DNA
CGIAR	Consultative Group on International Agricultural Research
CHM	Clearing House Mechanism
CI	Conservation International, USA
CIDA	Canadian International Development Agency
CITES	Convention on International Trade in Endangered species of Wild Fauna and Flora
CODATA	Committee on Data for Science and Technology
COGENE	Committee on Genetic Experimentation
CONABIO	Comisión Nacional para el Conocimiento y Uso de la Biodiversidada
CPC	Centre for Plant Conservation, USA
CPD	Centre of Plant Diversity
cpDNA	Chloroplast DNA
CPGR	Commission on Plant Genetic Resources (FAO)
CR	Critically Endangered (species)
CSIRO	Council of Scientific and Industrial Research Organizations
CUV	Consumptive Use Value

D

DeV	Demand Value
DGSM	Dasholi Gram Swarajya Mandal
DHS	Distinctiveness, Homogeneity and Stability (in reference to Cultivars)
DLI	Digital Library Initiative
DNA	Deoxyribo Nucleic Acid
DSM	Deutsche Sammlung Von Mikroorganisms und Zellkulturen
DUS	Distinctiveness, Uniformity and Stability (used alternatively for DHS)
DUV	Direct Use Value
DV	Direct Value

E

EAFRINET	East African Network
EE	Environmental Education
email	Electronic Mail
EMBL	European Molecular Biology Laboratory
EN	Endangered (species)
ENHSIN	European Natural History Specimen Information System
ERIN	Environmental Resources Information Network
ERMS	European Register of Marine Species
ESA	Endangered Species Act, USA
ESU	Ecologically (and/or Evolutionarily) Significant Unit (Refers to Population)
ETE	Evolution of Terrestrial Ecosystems (Database)
EU	European Union
EuroLOOP	European Locally Organised and Operated Partnerships
EuroMAB	European Man and Biosphere Program
EV	Existence Value
EW	Extinct in the wild (refers to species)
EX	Extinct (species)
EX?	Recently become Extinct (species)

F

F	Function of
FAO	Food and Agriculture Organization of the United Nations, Rome, Italy
FINBIN	Finnish Biodiversity Information Network

FRIS	Farmers' Rights Information Service
FTP	File Transfer Protocol

G

GATT	General Agreement on Trade and Tariffs
GBA	Global Biodiversity Assessment
GBIF	Global Biodiversity Information Facility
GEF	Global Environmental Facility
GENIE	Global Environment Network Information Exchange
GLIS	Global Land Information System
GP	Gene pool
GIS	Geographical Information System
GMO	Genetically Modified Organism
GNP	Gross National Productivity
GRCs	Genetic Resources Centres
GRID	Global Resource Information Database
GRIN	Genetic Resources Information Network, USDA
GSDs	Global Species Databases

H

ha	Hectare
HMSO	Her Majesty's Stationery Office

I

I	Indeterminate (species)
I_1V	Instrumental Value
I_2V	Intrinsic Value
IABG	International Association of Botanic Gardens
IAK	Indigenous Agricultural Knowledge
IARC	International Agricultural Research Centre
IBP	International Biological Program
IBPGR	International Bureau of Plant Genetic Resources
ICSU	International Council of Scientific Unions
ICUC	International Centre for Underutilized Crops
IDA	International Depositary Authority (for microbes)
IEEP	International Environmental Education Program
IGBP	International Geosphere – Biosphere Program
IKS	Indigenous Knowledge System
ILDIS	International Legume Database and Information Service
IMF	International Monetary Fund
IMI	International Mycological Institute
INBio	Instituto Nacional de Biodiversidad, Costa Rica
INFOTERRA	Global Environmental Information Exchange Network
INTER-SSR-PCR	Inter-Simple Sequence Repeat-PCR
IOPI	International Organization for Plant Information
IPBN	Indigenous Peoples' Biodiversity Network
IPGRI	International Plant Genetic Resources Institute
IPNI	International Plant Names Index
IPRs	Intellectual Property Rights
IRRI	International Rice Research Institute, Manila, Philippines
ISBI	International Sustainable Biosphere Initiative
ITF	International Transfer Format (for Botanic Garden Plant Records)
ITIS	Integrated Taxonomic Information System
ITK	Indigenous Technological Knowledge
ITTA	International Tropical Timber Agreement
ITTO	International Timber Trade Organization

IUBMB	International Union for Biochemistry and Molecular Biology
IUBS	International Union of Biological Sciences
IUCN	International Union for the Conservation of Nature and Natural Resources
IUFRO	International Union of Forestry Research Organizations
IUIS	International Union of Immunology Societies
IUMS	International Union of Microbiological Societies
IUNS	International Union for Nutritional Sciences
IUPAB	International Union for Pure and Applied Biophysics
IUPGR	International Undertaking on Plant Genetic Resources
IUPHAR	International Union for Pharmacology
IUPS	International Union for Physiological Sciences
IUV	Indirect Use Value
IV	Indirect Value

K

K	Insufficiently Known (species)
Kb	Kilobase
KM	Kilometer

L

LOOPs	Locally Organised and Operated Partnerships
LTER	Long-term Ecological Research Network

M

M	Meter
MAB	Man and Biosphere Program
MARBID	Marine Biodiversity Database
MARS	Marine Research Stations Network
mbp	million base pairs (in DNA)
mg	Milligram
MINE	Microbial Information Network Europe
MIRCENs	Microbial Resource Centres
mm	millimeter
MP-PCR	Microsatellite Primed-Polymerase Chain Reaction
MSSRF	MS Swaminathan Research Foundation
mt DNA	Mitochondrial DNA
MTCC	Microbial Type Culture Collection
MVP	Minimum Viable Population

N

NAPRALERT	Natural Products Alert, USA
NASA	National Aeronautics and Space Administration
NBPGR	National Bureau of Plant Genetic Resources, India
NBS/AP	National Biodiversity Strategies/Action Plans
NCBI	National Centre for Biotechnology Information, US
NCSs	National Conservation Strategies
NE	Not Evaluated (species)
NEAPs	National Environmental Action Plans
ng	nanogram
NGOs	Non-Government Organizations
NIR	Network Information Retrieval
NP	National Park
NRC	National Research Council, USA
NSDI	National Spatial Data Infrastructure
NSF	National Science Foundation
NSSL	National Seed Storage Laboratory, USA
nt	Not threatened (species)
NUV	Non-consumptive Use Value
NV	Non-use Value

O, P, Q

OBIS	Ocean Biogeographic Information System
OECD	Organization for Economic Cooperation and Development, Paris
OTA	Office of Technology Assessment, USA
OV	Optional Value
PACINET	South Pacific Network
PCBs	PolyChlorinated Biphenyls
PCR	Polymerase Chain Reaction
PEET	Partnerships for Enhancing Expertise in Taxonomy
PFM	Participatory Forest Management
pg	picogram
PGRC	Plant Genetic Resources Centre
PGRFA	Plant Genetic Resources for Food and Agriculture
PROSPECT	Programmed Retrieval of species by the Property and End-use Classification of their Timbers
Ps	Psychospiritual (value)
PUV	Productive Use Value
PV	Primary Value
PVA	Population Viability Analysis or Population Vulnerability Analysis
QOV	Quasi-Option Value
QTL	Quantitative Trait Loci

R

$	US Dollar
R	Rare
RAFI	Rural Advancement Foundation International
RAMP	Random Amplification of Microsatellite Primers
RAMPO	Random Amplification of Microsatellite Primed Oligonucleotides
RAP	Rapid Assessment Program of Conservation International
RAPD	Random Amplification of Polymorphic DNA
RBA	Rapid Biodiversity Assessment
RDB	Red Data Book
RFLP	Restriction Fragment Length Polymorphism
RIKEN	The Japanese Equivalent for the Institute for Physical and chemical Research
RNA	Ribonucleic Acid
RPK	Rural Peoples' Knowledge
rRNA	Ribosomal RNA

S

S	Sedimentation Coefficient (RNA)
SA 2000	Systematics Agenda 2000
SAFRINET	South African Network
SBI	Sustainable Biosphere Initiative
SCOPE	Scientific Committee on Problems of the Environment
SCOR	Scientific Committee on Oceanic Research
SEPASAL	Survey of Economic Plants for Arid and Semi-Arid Lands
SER	Society of Ecological Restoration
SIGN	Special Interest Groups and Networks
SIN	Special Interest Network
SINGER	System-wide Information Network for Genetic Resources
SPAR	Single Primer Amplification Reaction
SRISTI	Society for Research and Initiatives for Sustainable Technologies and Institutions
SSRs	Simple Sequence Repeats (in RADP technique)
STAP	Scientific and Technical Advisory Panel (of GEF)
STMS	Sequence Tagged Microsatellite Sites
SSC	Species Survival Commission of IUCN
STRs	Simple Tandem Repeats (in RADP technique)

T	
TBK	Traditional Botanical Knowledge
TDWG	Taxonomic Databases Working Group
TEK	Traditional Ecological Knowledge
TEnV	Total Environmental Value
TEV	Total Economic Value
TFAP	Tropical Forestry Action Plan
Ti	Tumour inducing (Plasmid)
TKS	Traditional Knowledge Systems
TNC	The Nature Conservancy
TNT	Trinitrotoluene
TPC	Threatened Plants Committee
TRAFFIC	Trade Record Analysis of Flora and Fauna in Commerce
TRIPs	Trade Related Intellectual Property Rights
TRR	Traditional Resource Rights
TrV	Transformation Value
TV	Total Value
U	
UK	United Kingdom
UN	United Nations
UNCED	United Nations Conference on Environment and Development
UNDP	United Nations Development Program
UNEP	United Nations Environmental Program
UNESCO	United Nations Educational Scientific and Cultural Organization
UNO	United Nations Organization
UPOV	Union pour la Protection des Obtentions Végétales (Union for the Protection of New Varieties)
URMO	UNESCO-IOC Register of Marine Organisms
USA	United States of America
USMAB	United States Man and Biosphere Program
USPTO	United States Patents Office
USSR	(Former) United Soviet Socialistic Republic
UV	Ultra Violet (radiation)
UV	Use Value
V	
V	Value
VL	Virtual Libraries
VU	Vulnerable (species)
W	
WAIS	Wide Area Information Server
WCED	World Commission on Environment and Development
WCMC	World Conservation Monitoring Centre, UK
WDCM	World Data Center for Microorganisms
WFCC	World Federation for Culture Collections
WHC	World Heritage Convention
WHF	World Heritage Fund
WHO	World Health Organization
WHS	World Heritage Site
WIEWS	World Information and Early Warning System
WIPO	World Intellectual Property Organization
WRI	World Resources Institute
WTA	Willing to Accept
WTO	World Trade Organization
WTP	Willing to Pay
WWF	World Wide Fund for Nature
www	World Wide Web

SUBJECT INDEX

AUTHOR INDEX